Beavers: Boreal Ecosystem Engineers

Carol A. Johnston

Beavers: Boreal Ecosystem Engineers

Springer

Carol A. Johnston
Department of Natural Resource Management
South Dakota State University
Brookings, SD, USA

ISBN 978-3-319-61532-5 ISBN 978-3-319-61533-2 (eBook)
DOI 10.1007/978-3-319-61533-2

Library of Congress Control Number: 2017944360

Printed on acid-free paper

This Springer imprint is published by Springer Nature
The registered company is Springer International Publishing AG
The registered company address is: Gewerbestrasse 11, 6330 Cham, Switzerland

This book is dedicated to my children,
Andrew and Esther.

Preface

There are many books about beavers, some of which have beautiful photographs that make them suitable for display on a coffee table. This is a book about a place, the Kabetogama Peninsula of northern Minnesota, and the science of how beavers have shaped it. Although this book contains some attractive figures, it is not a coffee-table book.

Beavers have been interacting with the Kabetogama landscape since the last glaciation some 11,000 years ago, and over the long term they have probably been more influential than humans in altering this ecosystem. We have proof of beaver activity on the Kabetogama Peninsula for the past 90 years, when humans first started taking aerial photographs, but we know that beavers the size of bears co-existed with the mastodons along the margins of retreating glaciers in what is now Minnesota.

Photo by Daniel Mallwitz

We call beavers "ecosystem engineers," organisms that create or modify habitats by causing physical state changes in biotic and abiotic materials. We also call beavers "keystone species," a species whose presence and role within an ecosystem has a disproportionate effect on other organisms within the system. Although beavers are the focus of this saga, many other organisms (wolves, moose, voles, fish, plants, microbes) are players. Their interactions with each other are superimposed on a landscape template that is also a player, contributing water and soil resources that support beavers and their allies. This book is about the interactions among all these players.

My first encounters with beavers occurred when I was working as an aerial photo interpreter for the New York State Wetlands Inventory in 1974. Beavers create wetlands, so we wanted to see if we could identify wetlands that had been flooded by beaver dams. Gary Parsons, a wildlife biologist with the New York State Department of Environmental Conservation, showed me how to drive a snowmobile and guided me into the wilderness of the Adirondack Mountains to inspect beaver dams in winter. I was hooked.

I started studying Kabetogama beavers in 1986. This time, I had to learn how to drive a motor boat to cross the large lakes that surround the peninsula. My colleagues and I studied beaver works: beaver dams, beaver ponds, beaver meadows, and beaver tree cutting. We studied them in the field and from the air using contemporary and historical aerial photography. My experience as an aerial photo interpreter of wetlands was essential for mapping beaver ponds and their changes as beaver colonies spread out over the Kabetogama Peninsula.

The Kabetogama Peninsula is in Voyageurs National Park, named after the French-Canadian boatmen who began journeying through these interconnected waterways of the fur trade route over 250 years ago. The establishment of Voyageurs National Park in 1975 protected the Kabetogama beavers from trapping, providing a unique opportunity to view a landscape controlled by beaver activity as it might have been before Euro-American trappers and voyageurs arrived. The founders of Voyageurs National Park left behind a rich natural heritage that has benefited scientists, tourists, and beavers alike.

Brookings, SD, USA Carol A. Johnston

Data Availability

Datasets for many of the analyses presented here are in the Dryad data repository: Brown TN, Johnston CA (2017) Data from: Beavers: boreal ecosystem engineers. Dryad Digital Repository. http://dx.doi.org/10.5061/dryad.b6c45 GIS datasets showing beaver impoundment locations are available at: Johnston CA, Windels SK (2015) Using beaver works to estimate colony activity in boreal landscapes (online spatial data). CUAHSI Spatial Data Service, Consortium of Universities for the Advance of Hydrologic Science Inc., Medford. http://spatial.cuahsi.org/johnstonC01/

Acknowledgements

Most of the research that is reported in this book was supported by the U.S. National Science Foundation (NSF). I am especially grateful to have received an Opportunities for Promoting Understanding through Synthesis (OPUS) grant from NSF (DEB-1349240) that was the impetus for this book. Previous NSF grants received include DEB-9973357 (Isaac Schlosser, principal investigator), DEB-9615326, DEB-9119614, DEB-8817665, and DEB-8516284 (Robert Naiman, principal investigator). This effort was also supported by the National Aeronautics and Space Administration (NASA/NAGW-1828), the U.S. Department of Agriculture (NIFA Hatch Project SD00H505-15), and the National Park Service.

In addition to the authors of the chapters in this book, many people have contributed to the science behind this work, including co-investigators, technicians, and students: Michael Aho, Brian Allen, Dan Allison, John Ameel, Christopher Arens, Jason Aronson, Anastasia Bamford, Phyllis Barnidge, John Bonde, Michael Bourdaghs, Scott Bridgham, Michael Broschart, Adam C. Cahow, Brad Dewey, Amy Downing, Heather Erickson, Greg Grunwald, Cal Harth, James Kelley, Ann Lima, Tom Manning, Paul Meysembourg, Howard Mooers, Robert Naiman, Tatiana Nawrocki, John Pastor, Ryan Peterson, Gilles Pinay, Deborah Pomroy, Aaron Reichow, James Salés, Tracy Scharp, Mark Springan, Karen Updegraff, and John Wisniewski.

Contents

List of Figures

List of Tables

Contributors

Carol A. Johnston Department of Natural Resource Management, South Dakota State University, Brookings, SD, USA

Terry N. Brown Natural Resources Research Institute, University of Minnesota, Duluth, MN, USA

Kirsten Cahow-Scholtes Wisconsin Department of Natural Resources, LaCrosse, WI, USA

Larry W. Kallemeyn USGS Columbia Environmental Research Center, International Falls, MN, USA

Isaac J. Schlosser Department of Biology, University of North Dakota, Grand Forks, ND, USA

Steve K. Windels National Park Service, Voyageurs National Park, International Falls, MN, USA

Chapter 1
Legacy of Beaver–Human Interaction

Abstract Beavers have occupied the Kabetogama Peninsula for millennia, but the beaver population there has had its ups and downs since humans and beavers started to co-exist. Over-trapping of beavers during the 1600s–1800s decimated beaver populations in northern Minnesota. Logging and forest fires also changed the beaver's habitat. Beaver numbers began to recover after the fire of 1936, and trapping was discontinued with the establishment of Voyageurs National Park in 1975. Beaver density peaked in 1981 at 1.4 colonies per square kilometer, among the highest in North America.

1.1 Decimation of the Beaver

The decimation of the beaver started with hats. Beaver fur was used in the most fashionable hats of the day in the 1700s. These hats weren't like a coon-skin cap—the beaver fur was unrecognizable by the time the hats reached their wearers' heads. The shiny guard hairs were removed, exposing the soft wool undercoat beneath. The undercoat was used in hat making due to its superior felting properties. The wool was shaved from the leather pelt, felted, died, and shaped into various styles (Hawkins 2014). Later styles economized by applying a surface nap of beaver fur to a muslin base (Fig. 1.1).

The popularity of beaver hats in Europe created a demand that reached across the Atlantic. Beaver pelts were currency, the "black gold" sought after by early North American explorers. Trappers worked first in eastern North America, depleting beaver stocks in New England and Quebec. As beaver populations dwindled, trappers expanded ever westward. The lakes and rivers were their highways: along the St. Lawrence River, into the Great Lakes, and across the Grand Portage to the Pigeon River and beyond. French-Canadian "coureurs des bois" (runners of the woods) each carried two 41-kg packs of beaver pelts on their backs across the 14 km Grand Portage trail, loading 65 packs into freight canoes for transport by "voyageurs" back to Montreal, where the packs were loaded onto European-bound ships. By 1800 about 200,000 beaver pelts a year were sent from America to Europe (Longley and Moyle 1963).

C.A. Johnston, *Beavers: Boreal Ecosystem Engineers*,
DOI 10.1007/978-3-319-61533-2_1

Fig. 1.1 A beaver fur felt top hat made circa 1900 by manufacturer Drickson and Co., London. The felting is worn off the brim, but is still evident on the top and sides, imparting a sheen to the hat

Fashion is fleeting, and beaver fur was replaced by silk as a hat making material in the 1800s. By then, beaver populations were decimated throughout most of North America, even in places that seem remote today. The Hudson's Bay Company reported a beaver population crisis in the Rainy Lake region by the early 1800s: "Beaver…has been diminishing for these several years past, especially on the south side of Rainy Lake…Rainy Lake River and Lake of the Woods" (Rysgaard 1941).

The over-trapping of beaver populations during the 1700s and 1800s not only removed the animals from the landscape, it also removed their works as well. Judging by evidence from places where beaver populations have recovered, beaver dams and ponds were a common feature of North American stream systems. Without the beaver's vigilant dam maintenance, the dams would have collapsed and washed away, leaving free-flowing streams that were unlike those that existed before the demise of the beaver.

1.2 The Landscape that Beavers Inhabited

The geographic focus of this book is the Kabetogama Peninsula in what is now Voyageurs National Park (Fig. 1.2), named after the voyageurs who transported the packs of beaver pelts over the water routes to trading posts farther east. The region is renowned for its cold winter weather. With 109.4 days per year with a high temperature below freezing (0 °C), nearby International Falls has the most such days of any incorporated city in the conterminous United States. Annual precipitation averages 627 mm, annual snowfall averages 171 cm, and mean seasonal snowfall accumulation averages 56 cm (NOAA 2016).

The Kabetogama Peninsula lies near the southern edge of the Laurentian Shield, the Earth's largest mass of exposed Precambrian igneous and metamorphic rock, which extends from the Great Lakes to Hudson Bay and the Canadian Arctic.

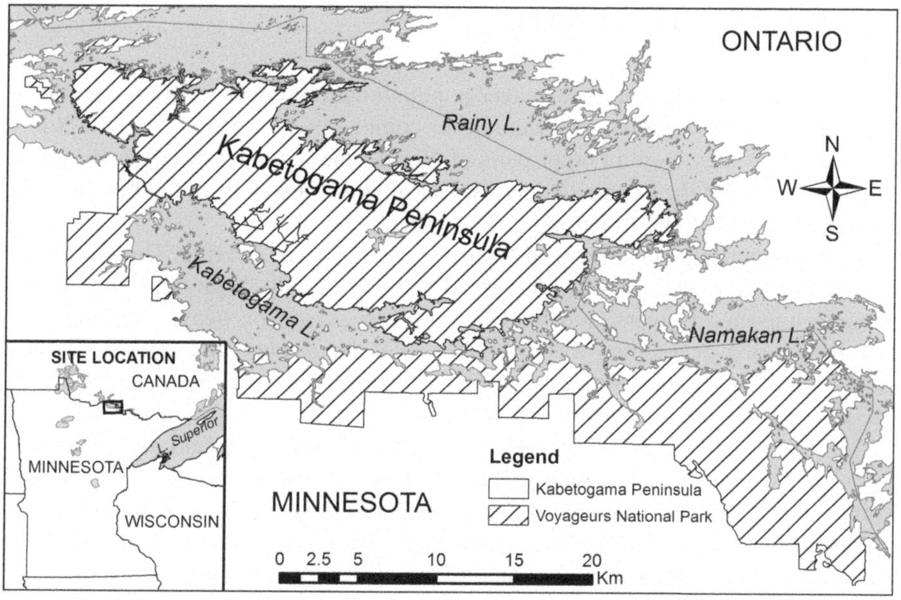

Fig. 1.2 Location of the Kabetogama Peninsula and Voyageurs National Park

It is one of the oldest continental shields on earth, with regions dating from 2.5 to 4.2 billion years (Iizuka et al. 2007). Erosion and glaciers have worn down its surface over time, producing rocky, ice-smoothed hills with an average relief of 30 m (Brittanica 2016), covered by a thin veneer of deposits left behind when continental glaciers melted at the end of the Pleistocene Ice Age 10,000 years ago (USGS 2016). Glacial deposits filled the valleys, interrupting stream drainage networks. Areas of organic soil called peatlands formed in the post-glacial landscape from the accumulation of partially decomposed vegetation in wet depressions.

The vegetation of the southern Laurentian Shield is boreal forest, a mix of conifers (pine, spruce, balsam fir) with aspen and birch trees. Prior to European settlement, the forests of the Kabetogama Peninsula were dominated by mixed red and white pine, aspen-birch, and jack pine, a species that proliferates after fire (Marschner 1930). Forest fires were common, with major fires years during prolonged summer droughts of subcontinental extent, such as those of 1864, 1910, and 1936 (Swain 1973; Heinselman 1973).

Before the Kabetogama Peninsula was declared a park, logging had altered its forests. Early loggers were selective, taking the red and white pines for sawtimber but leaving behind other species. Later logging operations sought pulpwood for paper making, and their clear-cutting promoted aspen regeneration by root suckers that sprout after a disturbance opens the forest canopy. Aspen is the preferred food of beavers, who fell the trees and carry their branches to their ponds to store in an underwater cache for overwinter food supply, so the region's logging legacy may actually have promoted the recovery of beaver populations.

1.3 Recovery of the Beaver

It took more than a century for beaver populations to recover from the assault of over-trapping. Although beavers were never completely eliminated from the Rainy Lake region as they were in other parts of North America, their numbers were few at the beginning of the twentieth century. Aerial photos taken in 1927 show only 42 beaver ponds within the Kabetogama Peninsula.

Beaver populations elsewhere in Minnesota were also recovering. The Minnesota Division of Game & Fish was authorized to issue permits beginning in 1919 for trapping nuisance beavers, and by 1939 an open season on beaver trapping was established. Beaver trapping was discontinued on the Kabetogama Peninsula when the National Park was established in 1975, and beaver colony density peaked at 1.4/km^2 during the 1980s (Johnston and Windels 2015).

As beaver populations recovered, their influence on ecosystem and landscape processes became more apparent. Although these influences had occurred for millennia, their evidence was muted during the period of beaver decimation in the eighteenth and nineteenth centuries, and only after beaver populations began to recover did people recognize the importance of beaver as an ecosystem engineer (Morgan 1868; Hood 2011; Müller-Schwarze 2011).

This book summarizes research that was done by myself and others on beaver ecosystems of Voyageurs National Park. The research is based upon aerial photo evidence starting in 1927, aerial reconnaissance flights starting in 1958, and field work starting in 1985. The book is not about beaver biology per se, but instead is about the profound ecological changes caused by beaver ponds, beaver meadows, and beaver herbivory.

1.4 The Kabetogama Peninsula: A Beaver Timeline

11,600 Years Before Present (yrs BP)
Glaciers covered the land, scraping the soil off the migmatite and biotite schist bedrock (NPS 2007). The Kabetogama Peninsula marked the limit of glacial advance for the Vermillion phase of the Rainy Lake lobe (Sado et al. 1995).

11,100 yrs BP
The glaciers had receded, and the land was inundated by Lake Agassiz, an immense ice-dammed lake that covered portions of Minnesota, North Dakota, Manitoba, and northwestern Ontario (Thorleifson 1996).

1688
With the help of Assiniboine guides, Jacques de Noyon became the first European to traverse the chain of lakes, including Rainy Lake, which lies along Minnesota's northern boundary. This soon became an important fur-trading route (Longley and Moyle 1963).

1783

The Northwest Company was organized, and established a fort and fur trading post at Grand Portage. They were joined at Grand Portage in 1797 by the rival X.Y. Company, which operated until the two companies merged in 1804 (Buck 1923).

1812

After the War of 1812, Congress passed an act excluding foreigners from the fur trade in American territory, and Astor's American Fur Company purchased the posts of the Northwest Company. The Grand Portage itself was little used by this time (Buck 1923).

1842

Fur trading in northern Minnesota came to a halt with the demise of the American Fur Company (Longley and Moyle 1963).

1927

Aerial photography taken to resolve U.S.-Canadian disputes over the level of Rainy Lake (Bullard and Scovil 1930) showed 42 beaver ponds and meadows on the Kabetogama Peninsula. The only humans were the lumberjacks cutting down the pine trees and dragging them out to the lakes by horse-drawn sleighs over ice roads. The Virginia and Rainy Lake Company established "sleigh haul" camp #131 between Wier Lake and Mica Bay (Fig. 1.3). During its 2 years of operation (1927–1929), 123 million board feet of lumber were cut there (Wyatt 1999). A huge raft of floating pine logs was assembled in Mica Bay to be towed to Hoist Bay, where the

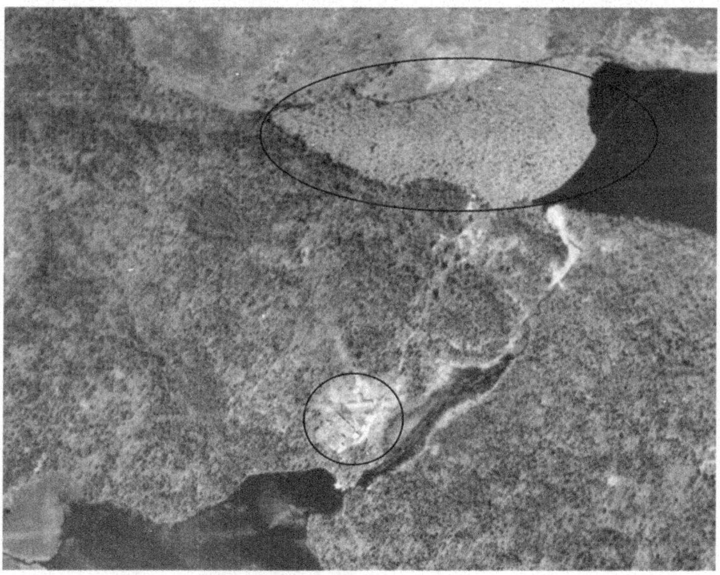

Fig. 1.3 1927 aerial photo showing Virginia and Rainy Lake Company logging camp located north of Weir Lake (*circle*). The west end of Mica Bay is filled with floating pine logs anchored by boom lines (*ellipse*). The slope north of Mica Bay has been clear-cut

logs would be lifted unto rail cars and transported south to largest white pine sawmill in the world, in Virginia, Minnesota (Wyatt 1999). The loss of the pine trees was no loss to the beavers, who avoided them anyway.

1936

The weather that summer was hot and extremely dry (Dieterich and Brown 1964). In other parts of the country, the dry wind blowing over the parched soil created the Dust Bowl. At Kabetogama, a great fire burned across the land in early August, burning 178 km^2 of the peninsula (Allen et al. 1993). News of the Kabetogama fire and other Minnesota fires spread across the country, making headlines in eastern U.S. cities (Fig. 1.4). The water of the beaver ponds didn't stop the fire; only Shoepack Lake partially blocked the fire's movement across the middle of the peninsula (Fig. 1.5). The raw bedrock that lay beneath the burned forest was exposed to the sky, no longer covered by trees and leaf litter.

1940

The beaver ponds helped heal the burnt landscape. Two-thirds of the 117 beaver ponds on the Kabetogama Peninsula were in burned areas. On the upland, seeds that had been tightly held in jack pine cones were released by the heat of the fire and germinated in the newly exposed soil (Allen et al. 1993). The Minnesota Division of Game & Fish initiated a beaver trapping season in 1938–1939, and sold 2697

TODAY'S INDEX

BUFFALO Courier EXPRESS

Buffalo's Best Newspaper

WEATHER
Partly cloudy today and tomorrow, with a slight possibility of scattered showers. Not much change in temperature.
Detailed weather report, Page 22.

VOL. CII—No. 25 BUFFALO, N. Y., FRIDAY MORNING, AUGUST 14, 1936 24 PAGES—TWO CENTS

MINNESOTA FIRES SPREAD; SETTLERS FLEEING

Flames Ravage Minnesota—Wirephoto: Top, rangers in Pine County, Minnesota, yesterday watched roaring forest flames work out of control. Here is a burst of flames breaking through the undergrowth. Below, only these stumps remain near Aurora, Minn., as mute evidence of destruction wrought by fire. Hundreds of acres of stately pines burned despite efforts of scores of men to check the flames. Rains brought a temporary halt to the devastation.

SMALL ROADS IN NINE COUNTIES TO BE CLOSED

Canadian forest fire advancing across border, driving residents to other areas

SITUATION NOW ACUTE

Thousands of small fires growing, all getting out of control of volunteer workers

St. Paul, Minn., Aug. 13 (AP)—Evacuation of several Northern Minnesota border settlers before an advancing Canadian forest fire came simultaneously tonight with a state command ordering sheriffs of nine counties to close all secondary roads to regular traffic.

Ranger Dick Willem of Warroad, Minn., notified the state forestry department that settlers north of the village, in the path of a major Canadian blaze, fled their homes down a narrow road that provides the sole access to the region.

Smoke Obscures Vision

Grover Conzet, state forester, telegraphed sheriffs of nine northern counties advising them to "post guards closing all side or secondary roads to traffic where these roads lead into large or threatening forest fires."

Simultaneously rangers announced roads in the Beltrami Island state forest, between Warroad and Baudette, were closed to regular travel. Smoke in the area has obscured vision, and sightseers are interfering with fire control operations.

A radiogram informed the state department of forestry tonight that the Kabetogama Lake blaze 40 miles west of International Falls had "torn away loose" from the 400 men stationed there.

A. F. Oppel, deputy state director who received the wire over short wave, said a smaller blaze backturned on Kabetogama Peninsula, coalesced with the principal fire and burned 3,000 feet of state hose and other equipment.

A story of firefighters playing a game of hide-and-seek with the flames came in the message. Southwest winds fanned a fire in the Lost Bay Kabetogama sector toward Mica Bay where the blazes joined as the fleeing CCC's and other laborers set up a new camp.

The original base was located at Mica Bay. From there the firefighters moved to Lost Bay, but the wind shifted from the southwest to northwest and they moved again—this time to an island in the lake. The rampaging fire, with an estimated 10,000 acres levelled, had covered from three to four miles, Oppel estimated.

Fig. 1.4 Minnesota forest fires make headlines in a Buffalo, New York newspaper, August 14, 1936

Fig. 1.5 Winds from the east blew the 1936 fire to the shoreline of Shoepack Lake, where it was halted by the large expanse of water. White patches on burned lands southeast of the lake are exposed bedrock. The forest remains intact north and west of the lake

licenses statewide. Beaver pelts sold in 1944–1945 for a staggering $48 each, a price never equaled since.

1948

Beaver colonies had expanded throughout the Kabetogama Peninsula. Clyde Creek, Sucker Creek, and Cranberry Creek were filled with beaver dams (Fig. 1.6).

1951–1953

Tularemia, an infectious bacterial disease, killed many beavers in northern Michigan and Minnesota (Stenlund 1953; Lawrence et al. 1956). Reports of dead beavers came from Ely, Hibbing, and International Falls.

1961

The Minnesota and Ontario Paper Company (later bought out by Boise Cascade) built winter logging roads across the Kabetogama Peninsula to access timber for paper manufacture in International Falls (Wyatt 1999). Spruce and fir were initially

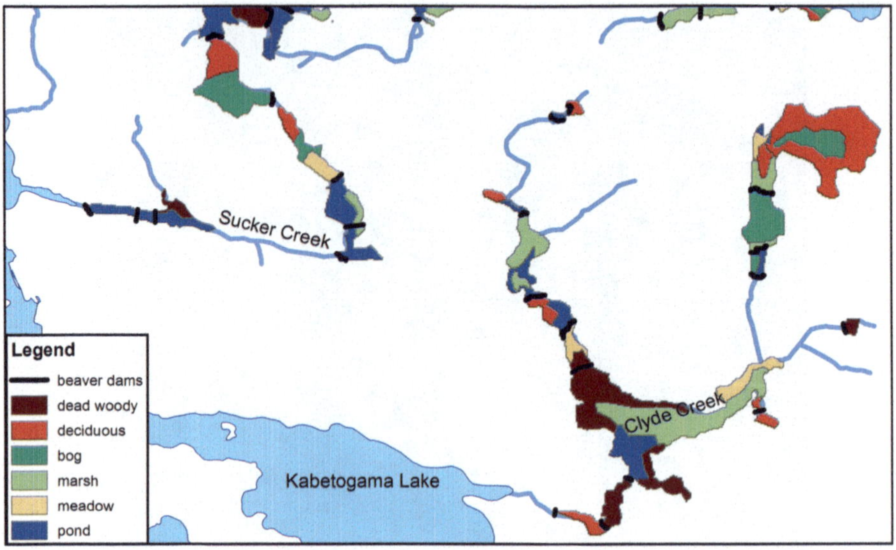

Fig. 1.6 Beaver dams and altered vegetation in Sucker and Clyde Creeks, 1948

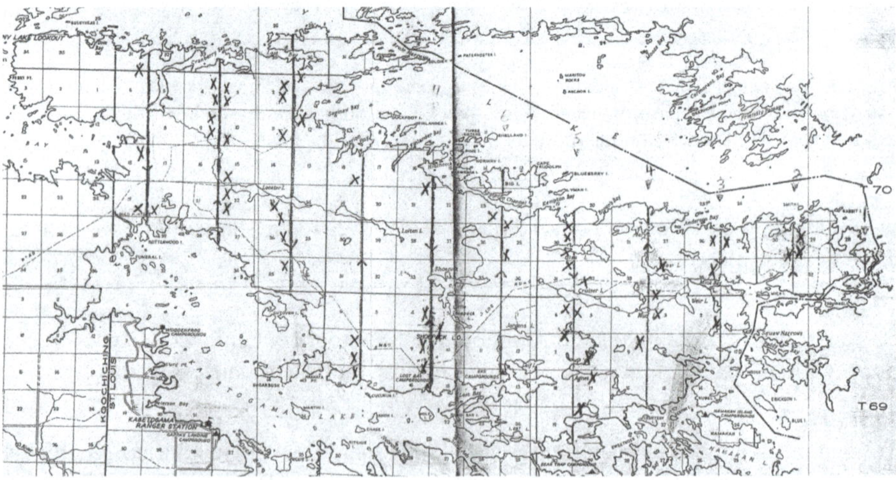

Fig. 1.7 Portion of 1969 overflight map used by Minnesota Department of Natural Resources personnel to count beaver colonies on the Kabetogama Peninsula. Each "X" denotes an observed beaver food cache

Fig. 1.8 Post-1960 logging in sections 9, 10, 15, and 16 of T69N, R20W, north of Lost Bay. *Left*: 1961 aerial photo shows parallel north–south oriented slash piles. *Right*: By the time legislation establishing Voyageurs National Park had been signed in 1971, most of this area had been clear-cut

harvested, but later advances in pulping techniques allowed aspen to be used for paper (Keays 1967). As logging became more mechanized, it became more efficient and economically practical to clear entire timber stands that would have been ignored in earlier periods. Beaver colony density was 0.8/km^2 by this time (Broschart et al. 1989), and their ponds and meadows covered about 10% of the landscape (Johnston 1994).

1969
Observers from the Minnesota Department of Natural Resources (MDNR) counted 57 active beaver colonies during aerial overflights of the Kabetogama Peninsula (Fig. 1.7).

1971
Legislation authorizing the secretary of the Department of the Interior to establish Voyageurs National Park in northern Minnesota was signed into law by President Richard M. Nixon on January 8, 1971, but it would be four more years before the park was actually established (Witzig 2004). Beaver trapping was no longer allowed on the Kabetogama Peninsula. Intensive pulp logging continued up until park establishment, clear-cutting 29 km^2 of the Kabetogama Peninsula (Fig. 1.8).

1981
Beaver colonies filled every valley of the Kabetogama Peninsula. Observers from the Minnesota Department of Natural Resources (MDNR) counted 1.4 beaver colonies per square kilometer, more than anywhere else in the state (Johnston and Windels 2015).

1986
Minnesota trappers harvested a record number of beaver pelts: 172,000 (Fig. 1.9).

Present
The beaver has regained its predominance as an ecological engineer at Voyageurs National Park.

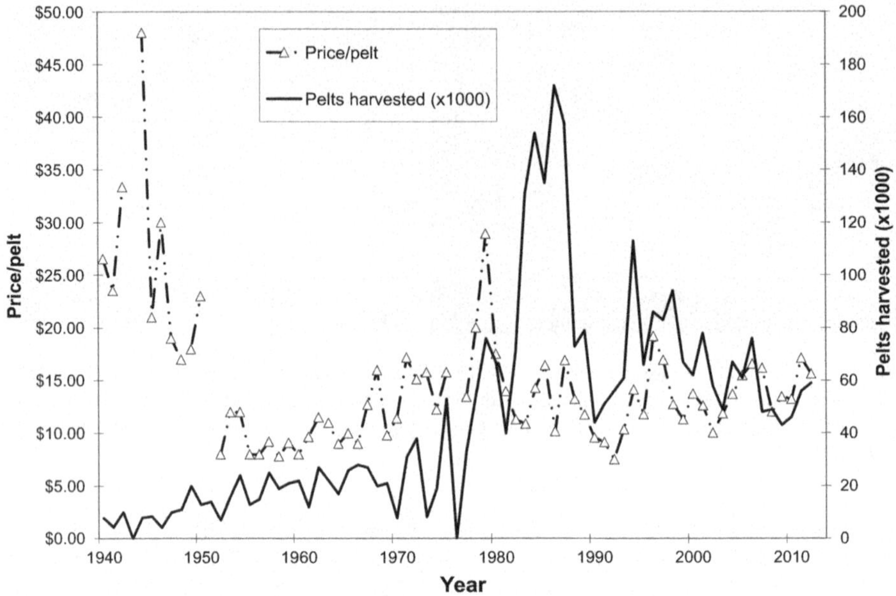

Fig. 1.9 Minnesota beaver pelt harvest and pelt prices, 1940–2013. Trapping was closed in 1943–1944 and 1976–1977. Source: unpublished data collected by the Minnesota Department of Natural Resources, Division of Fish and Wildlife—Wildlife Policy and Research Group, data provided by Margaret H. Dexter

References

Allen B, Salés J, Bonde J, Meysembourg P, Johnston CA (1993) Effects of past disturbance on present-day forests at Voyageurs National Park. In: ARC/INFO maps 1992. Environmental Systems Research Institute, Redlands, p 63

Brittanica (2016) Canadian Shield. Encyclopædia Britannica. http://www.britannica.com/place/Canadian-Shield. Accessed Jan 2016

Broschart MR, Johnston CA, Naiman RJ (1989) Prediction of beaver colony density using impounded habitat variables. J Wildl Manag 53:929–934

Buck SJ (1923) The story of the Grand Portage. Minn Hist 5(1):14–27

Bullard PC, Scovil SS (1930) Preliminary report to International Joint Commission relating to official reference re: levels of Rainy Lake and other upper waters. Ottawa

Dieterich JH, Brown JK (1964) Comparing severity of fire seasons. U.S. Forest Service, Lake States Forest Experiment Station, St. Paul

Hawkins N (2014) From fur to felt hats: the Hudson's Bay Company and the consumer revolution in Britain, 1670–1730. MA thesis, University of Ottawa, Ottawa

Heinselman ML (1973) Fire in the virgin forests of the Boundary Waters Canoe Area, Minnesota. Quat Res 3(3):329–382

Hood GA (2011) The beaver manifesto. Rocky Mountain Books, Victoria

Iizuka T, Komiya T, Ueno Y, Katayama I, Uehara Y, Maruyama S, Hirata T, Johnson SP, Dunkley DJ (2007) Geology and zircon geochronology of the Acasta Gneiss Complex, Northwestern Canada: new constraints on its tectonothermal history. Precambrian Res 153:179–208

Johnston CA (1994) Ecological engineering of wetlands by beavers. In: Mitsch WJ (ed) Global wetlands: old world and new. Elsevier Science, Amsterdam, pp 379–384
Johnston CA, Windels SK (2015) Using beaver works to estimate colony activity in boreal landscapes. J Wildl Manag 79(7):1072–1080
Keays JL (1967) Use of trembling aspen in pulp and paper manufacture. Forest Products Laboratory, Department of Forestry and Rural Development, Vancouver, British Columbia
Lawrence WH, Fay LD, Graham SA (1956) A report on the beaver die-off in Michigan. J Wildl Manag 20(2):184–187
Longley WH, Moyle JB (1963) The beaver in Minnesota. Minnesota Department of Conservation, Division of Fish and Game, St. Paul
Marschner FJ (1930) The original vegetation of Minnesota. U.S. Forest Service, North Central Forest Experiment Station, St. Paul
Morgan LH (1868) The American beaver and his works. J. B. Lippencott & Co., Philadelphia
Müller-Schwarze D (2011) The beaver: its life and impact, 2nd edn. Cornell University Press, Ithaca
NOAA (2016) NOWData - NOAA Online Weather Data. National Oceanic and Atmospheric Administration. http://w2.weather.gov/climate/xmacis.php?wfo=dlh. Accessed Jan 2016
NPS (2007) Geologic map of Voyageurs National Park. U.S. Department of the Interior, National Park Service, Geologic Resources Division
Rysgaard GN (1941) Historical notes on the beaver. Conserv Volunt 1(5):64
Sado EV, Fullerton DS, Goebel JE, Ringrose SM (1995) Quaternary geologic map of the Lake of the Woods 4° × 6° quadrangle, United States and Canada, U.S. Geological Survey Miscellaneous Investigations Series Map I-1420 (NM-15). U.S. Geological Survey, Denver
Stenlund MH (1953) Report of Minnesota beaver die-off, 1951-1952. J Wildl Manag 17(3):376–377
Swain AM (1973) A history of fire and vegetation in northeastern Minnesota as recorded in lake sediments. Quat Res 3(3):383–396
Thorleifson LH (1996) Review of Lake Agassiz history. In: Teller JT, Thorleifson LH, Matile G, Brisbin WC (eds) Sedimentology, geomorphology, and history of the central Lake Agassiz Basin, field trip B2, Guidebook. Geological Association of Canada, St. Johns, pp 55–84
USGS (2016) USGS Geology in the Parks. U.S. Geological Survey. http://geomaps.wr.usgs.gov/parks/province/laurent.html. Accessed Jan 2016
Witzig FT (2004) Voyageurs National Park: the battle to create Minnesota's national park. University of Minnesota Press, Minneapolis
Wyatt B (1999) The logging era at Voyageurs National Park: historical contexts and property types. National Park Service, Omaha

Chapter 2
Ecosystem Engineers: Beaver Ponds

Abstract The beaver population of the Kabetogama Peninsula is very dense in comparison to other areas of North America. Both biotic (forage availability) and abiotic (topography, geology, hydrology) factors contribute to this dense population. Historical (1927–2003) and contemporary aerial photos were used to map beaver ponds, beaver meadows, and other features altered by beaver dam construction. The objectives of the study were to relate the extent and type of beaver works to the beaver population present. A total of 1009 pond sites were identified within the 302 km^2 Kabetogama Peninsula where flooding by beaver dams had discernibly altered the vegetation. Cover types ranged from open water to wetland forests, depending on the depth and duration of flooding. The sites were not continuously occupied by beavers over time, but were frequently recolonized. On average, each 100 beaver colonies increased the proportion of the landscape covered by open water by 2.15%. In addition, 16 of 21 permanent lakes on the Kabetogama Peninsula had lake outlet beaver dams. Collapse of the Shoepack Lake outlet dam in 2001 released an estimated 2.16 million kL of water and drained approximately 2/3 of the lake's pre-collapse area. Beaver dams built in the first several decades of peninsula recolonization (1940–1948) created the largest ponds with the greatest potential for expansion, implying that beavers are able to optimize dam location.

2.1 Kabetogama Peninsula: A Beaver Paradise

As of 2006, the 302 km^2 Kabetogama Peninsula supported 226 active beaver colonies, and beavers had built more than a thousand dams there over time. Such density of beaver colonies is among the highest in the published literature (Table 2.1). Studies reporting higher values encompass much smaller areas or focus sampling within stream corridors, ignoring mountains or other landscape features that are unlikely to support beavers. Clearly, the Kabetogama Peninsula and surrounding area provide ideal beaver habitat.

C.A. Johnston, *Beavers: Boreal Ecosystem Engineers*,
DOI 10.1007/978-3-319-61533-2_2

Table 2.1 Beaver colony densities reported in published literature, expressed per unit area and per length of survey route

Beaver density, colonies/km^2	Location	Source
0.014	Yellowstone N.P.	Smith and Tyers (2008)
0.15–0.32	Maine	McCall et al. (1996)
0.38	Northwest Territories	Aleksiuk (1970)
0.39–0.77	Central Ontario	Voigt et al. (1976)
0.46–0.62	Newfoundland	Bergerud and Miller (1977)
3.27	Illinois	Bloomquist and Nielsen (2010)
0.33–1.38	Kabetogama Peninsula	Johnston and Windels (2015a)
Beaver density, colonies/km		
0.48	Alaska	Boyce (1981)
0.52	New York State	Müller-Schwarze (2011)
0.08–1.4	Kansas	Robel and Fox (1993)
0.25–1.69	South Dakota	Dieter (1992)
0.7–4.0	California	Busher (1987)
0.15–1.91	Tierra del Fuego	Skewes et al. (2006)
1.1–1.2	Oregon	Leidholtbruner et al. (1992)
0.01–1.36	Quebec	Jarema et al. (2009)
2.59	British Columbia	Slough and Sadleir (1977)
0.83–2.23	Kabetogama Peninsula	Broschart et al. (1989)

2.1.1 Why Do Beavers Build Dams?

Beavers are semi-aquatic mammals that are uniquely adapted to living in and around water bodies. Adult beavers are nearly invincible to natural predators within the safety of their ponds, but are susceptible to predation on land by wolves and black bears (Jenkins and Busher 1979; Smith et al. 1994). Ponds provide beavers with safe access to riparian forests, allowing beavers to browse in the upland yet retreat quickly to the safety of the water when predators approach (Johnston and Naiman 1987).

Dam-building behavior by beavers is stimulated by the sound of running water. This auditory stimulus was first reported after laboratory observations of *Castor canadensis* (Hartman and Rice 1963). Wilsson later conducted experiments using captive beavers and speakers playing recordings of flowing water; the beavers consistently built dams at the speaker locations (Wilsson 1971). Even young beavers who hadn't previously built dams were stimulated by these auditory cues. The influence of sound on beaver behavior may explain how beavers are able to optimize dam location and why they rebuild dams at previous sites.

2.1.2 *What Makes the Kabetogama Peninsula Such a Suitable Landscape for Beaver Ponds?*

Beaver ponds require three ingredients: beavers, water, and land suitable for flooding. The Kabetogama Peninsula contains many beavers, but several landscape characteristics also promote the creation and sustainability of their ponds.

2.1.2.1 It Is Completely Surrounded by Water

Migrating beavers use the large lakes that surround the Kabetogama Peninsula as aquatic highways to access its perimeter (Smith and Peterson 1988). These three lakes, Rainy, Kabetogama, and Namakan, are not only large, but they also have convoluted shorelines (Fig. 1.2). Rainy Lake is particularly large, and its many bays extend far into Canada. The cumulative shoreline of these three lakes is more than 4000 km long (Table 2.2), providing ample access to migrating beavers.

At its western end, the Kabetogama Peninsula is separated from the mainland by a stream that drains from Kabetogama Lake (normal pool elevation = 341 m) into Rainy Lake (normal pool elevation = 338 m). The "Gold Portage" along this stream bypasses its rapids (Fig. 2.1). A dam at Kettle Falls on the eastern end of the Kabetogama Peninsula separates Namakan Lake from Rainy Lake (Fig. 1.2).

2.1.2.2 It Has Complex Topography

Tilted and eroded bedrock layers of the Kabetogama Peninsula create a complex land surface with many valleys that accumulate water. The elevation transect of the west-to-east drainage divide that bisects the Kabetogama Peninsula illustrates the peninsula's complex topography (Fig. 2.1). Even though no streams cross this drainage divide, its topography is convoluted. The highest elevations, 418–419 m, occur at three separate locations along the catchment divide: Loiten Lake peak, Cruiser Lake peak, and Brown Lake peak. The fourth highest elevation, "Shoepack lookout" (410 m), was the site of a fire lookout tower until 1999 (Anonymous 2004).

Table 2.2 Area and shoreline perimeter of the three large lakes surrounding the Kabetogama Peninsula

	Area (km^2)	Perimeter (km)
Kabetogama	90	288
Namakan	98	443
Rainy	855	3300

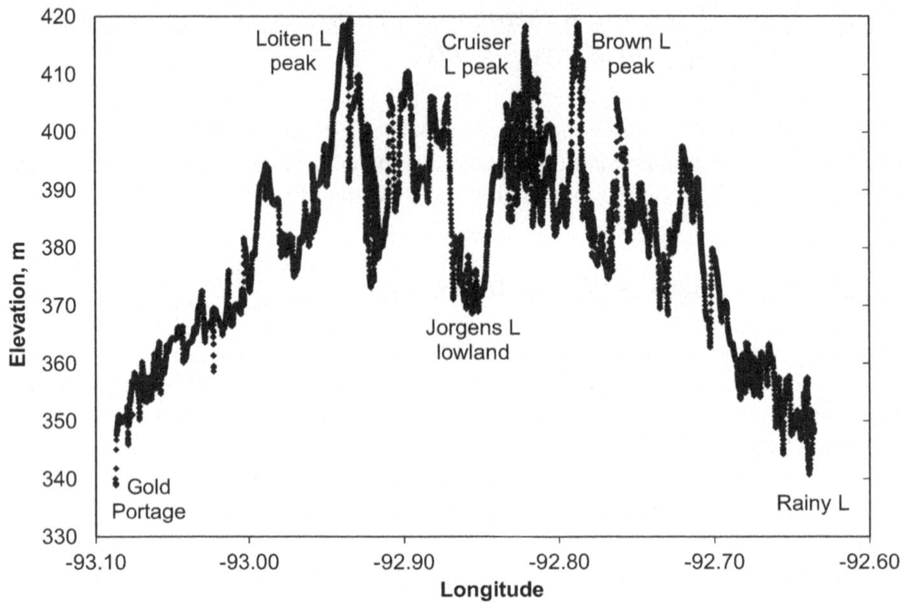

Fig. 2.1 Elevation transect of the watershed divide separating waters draining north into Rainy Lake from those draining south into Kabetogama/Namakan Lakes. I generated this by extracting the divide from the U.S. Watershed Boundary Dataset and intersecting it with one-meter digital elevation data

2.1.2.3 Its Waters Flow in Many Directions

Precipitation falling onto the Kabetogama Peninsula flows northward into Rainy Lake or southward into Kabetogama and Namakan Lakes via numerous catchments that subdivide the two halves of the peninsula (Fig. 2.2). The largest of these (28.1 km^2) drains Little Shoepack and Shoepack Lakes (catchment #8, Fig. 2.2). The longest tributary in this catchment originates in a beaver pond at 383.7 m, dropping 47 m in elevation as it flows down the 13.2 km stream path to its outlet on Rainy Lake. Only the final 0.2 km segment of the stream path is free-flowing: the upper stream flows through contiguous beaver ponds, beaver meadows, and Shoepack Lake before reaching its outlet.

The Cranberry Creek catchment (catchment #3, Fig. 2.2) is the second largest catchment on the Kabetogama Peninsula (27.4 km^2). Its waters drain northwest through Loiten, Quill, War Club, and Locator Lakes along a geologic fault line, emptying into Cranberry Bay and thence to Rainy Lake. Although beavers occupy all four of these lakes (Johnston and Windels 2015a), most of the beaver dams in the catchment are located downstream of the lakes on Cranberry Creek and its tributaries.

Three other named creeks flow off the peninsula into Kabetogama Peninsula: Deer Creek, Clyde Creek, and Sucker Creek (catchments 22–24, Fig. 2.2). As with the Shoepack Lake catchment, they are almost completely impounded by contiguous

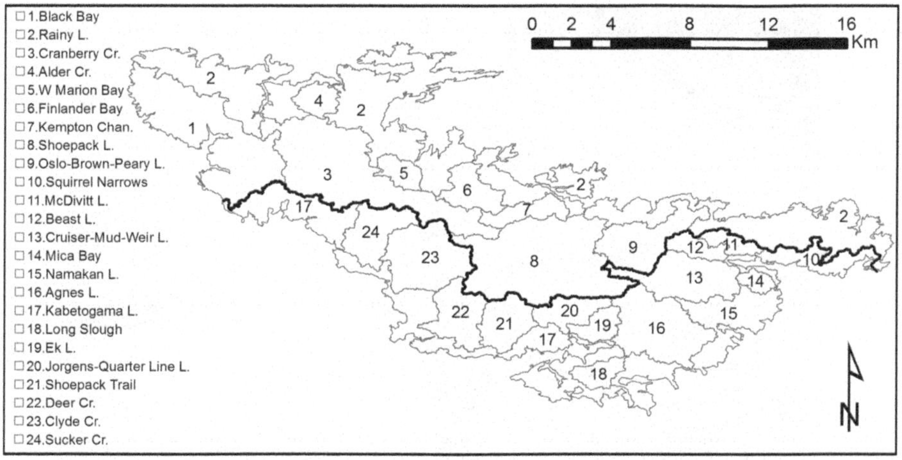

Fig. 2.2 Major catchments of the Kabetogama Peninsula, based on government databases and my unpublished map of beaver pond catchments. *Bold black line* is the watershed divide separating the Rainy Lake and Kabetogama/Namakan Lakes drainage basins that I used to prepare Fig. 2.1

beaver ponds and beaver meadows (Fig. 1.6). Other catchments of the Kabetogama Peninsula are drained by unnamed streams; I have informally named them based on their internal lakes, outlet bays, or other features (Fig. 2.2).

2.1.2.4 It Has Impervious Bedrock

Voyageurs National Park is at the southern end of the Laurentian Shield, a large area of some of the oldest Precambrian rocks in North America that form the ancient core of the continent (Fig. 2.3). The igneous and metamorphic bedrock consists of granite, biotite schist, and migmatite (interlayered granite and biotite schist) (Hemstad et al. 2002). This bedrock is not water-bearing except where faulted (Kanivetsky 1979), and forms an impervious base layer that keeps wetlands and water bodies perched on the bedrock surface.

2.1.2.5 It Has Impervious Soils

The Kabetogama Peninsula was submerged under the waters of early Glacial Lake Agassiz about 11,000 years before present (Teller 1985). Lake Agassiz formed during retreat of the last continental glacier, when meltwater backed up because the north-flowing rivers that drained into Hudson Bay were blocked by ice. Clayey glaciolacustrine sediments accumulated in low-lying areas of the Kabetogama Peninsula, especially on its western end (Fig. 2.4). These fine-textured soils transmit water very slowly.

Fig. 2.3 (**a**) Granite exposed by fresh rock cut along road to Ash River Visitor Center. (**b**) The root wad of a mature tree blown over in a windstorm illustrates shallow upland soils over bedrock

2.1.2.6 There Is Abundant Woody Browse

Beavers do not hibernate, instead relying on a submerged pile of twigs and branches ("cache") that they access underwater during the long winter. Quaking aspen (*Populus tremuloides*) is greatly preferred as a source of this woody browse (Slough and Sadleir 1977; Allen 1983), and Quaking Aspen-Paper Birch Forest is the most abundant forest type in Voyageurs National Park (Faber-Langendoen et al. 2007b).

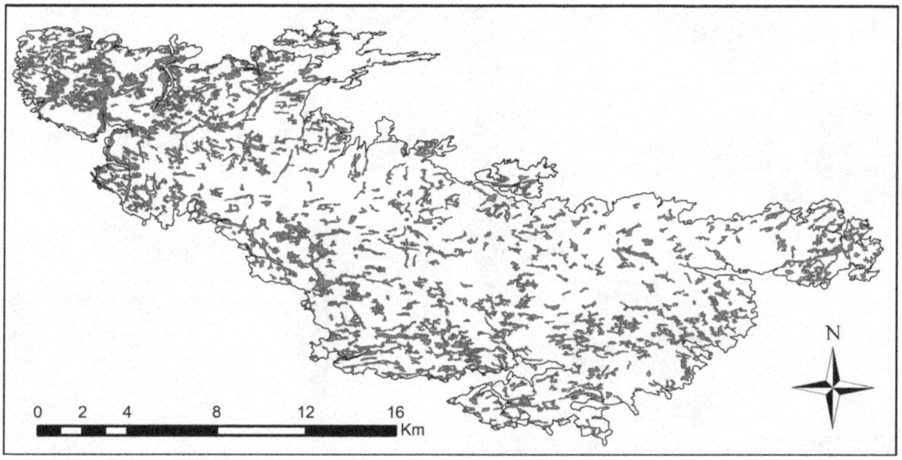

Fig. 2.4 Soils of the Kabetogama Peninsula mapped as having glaciolacustrine parent materials, derived from Soil Survey Geographic Database (SSURGO) for Voyageurs National Park

Other forest alliances at Voyageurs National Park that contain *P. tremuloides* are Spruce Fir-Aspen Forest, White Pine-Red Pine-Quaking Aspen-Paper Birch Forest, and Jack Pine-Aspen Forest.

2.2 Beaver Dam Characteristics

2.2.1 Dam Construction Materials

Beaver dams are typically constructed of mud-reinforced sticks, but dams constructed entirely of rocks and even cornstalks have been reported (Rue 2002; Jung and Staniforth 2010). In his classic book about the American beaver, Lewis H. Morgan described construction of a stick-dam, consisting of "…interlaced stick and pole work upon the lower face, with an embankment of earth intermixed with the same materials on the upper, or water face of the dam" (Morgan 1868). Morgan also described a solid-bank dam type, in which "the large amount of earth and mud, used to strengthen the work, buries and conceals the greater part of the brush and poles used to bind the embankment together."

Tree saplings and branches are a primary construction material used in beaver dams of the Kabetogama Peninsula. Branches are laid across the downstream face of the dam, parallel to the water flow (Fig. 2.5, left photo). Beavers also mound mud and grasses to make beaver dams, particularly in the early stages of construction (Fig. 2.5, right photo). In central Ontario, Doucet and colleagues (Doucet et al. 1994) reported that beavers preferentially used coniferous species and speckled alder in dam construction, reserving trembling aspen and other deciduous species

Fig. 2.5 Examples of beaver dam construction materials. *Left*: Abandoned wood and mud dam in the Kabetogama Lake drainage basin. *Right*: Dam constructed of mud and grass at a pond in the Clyde Creek drainage basin (catchment #23, Fig. 2.2)

for their food caches. I observed no such species differentiation on the Kabetogama Peninsula, but aspen branches were usually debarked prior to their use in dams; aspen bark is an important beaver foodstuff.

2.2.2 Beaver Dams on Streams

The iconic beaver dam blocks a stream, causing the flowing waters to slow and spread out, thereby creating a pond out of an area that was previously terrestrial (Fig. 2.6). Trees that are flooded die because their roots are deprived of oxygen, and trees around the pond may be cut by beavers for food or construction materials, further opening up the canopy. Stream-blocking beaver dams are the most common type on the Kabetogama Peninsula.

Solitary beaver ponds are uncommon on the Kabetogama Peninsula, however. Beaver ponds typically fill entire drainageways, so that the wetland vegetation at the upper end of one beaver pond abuts the next upstream dam. In the oblique aerial

Fig. 2.6 Pond creation by stream damming

Fig. 2.7 The longest beaver dam on the Kabetogama Peninsula, 309 m (catchment #16, Fig. 2.2)

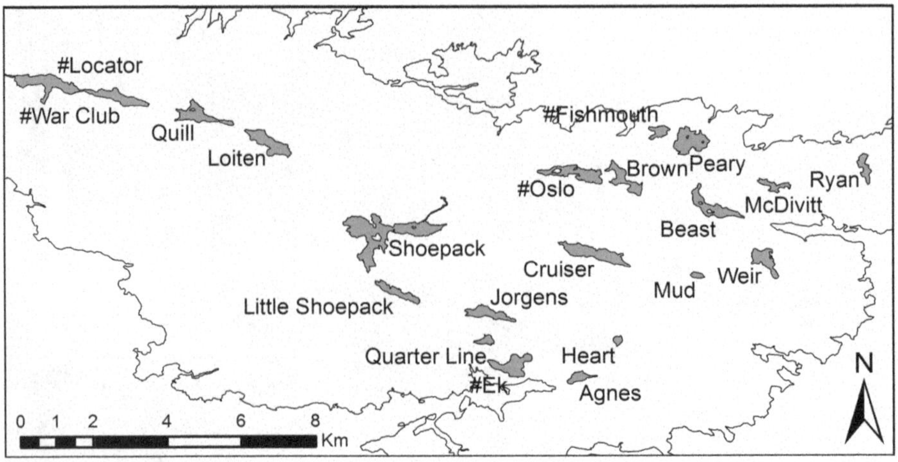

Fig. 2.8 Permanent lakes of the Kabetogama Peninsula. All lakes except those preceded by # have beaver dams at their outlets

view shown in Fig. 2.7, three beaver dams within a 0.5 km stretch of stream, each one meter higher than the next, create a massive pond/wetland complex (catchment #16, Fig. 2.2).

2.2.3 *Lake Outlet Beaver Dams*

Outlet beaver dams are a common feature of boreal lakes (Morgan 1868; Bertolo et al. 2008). Approximately half of 1085 "drainage lakes" sampled in the Adirondack Mountains of New York had beaver dams at their outlets (Kretser et al. 1989). In southern Finland, lake outlets were dammed more frequently than streams (Vehkaoja et al. 2015). Despite these geographically widespread examples, the prevalence and ecological influence of lake-outlet beaver dams is relatively unknown.

Of the 21 named permanent lakes and ponds on the Kabetogama Peninsula, 16 have beaver dams at their outlets (Fig. 2.8). In many cases, the purpose of such dams is not evident, because the dam building does little more than raise the lake level. For example, the beaver dam at the outlet of Ryan Lake has hardly affected its size and shape because of that lake's steep side slopes. However, beaver dams at the outlets of lakes with adjacent wetlands can flood extensive shoreline areas, greatly increasing lake area or the accessibility of upland food resources.

A downstream beaver dam raised the water level of Mud Lake, lifting the floating peat mat along its northern shore and creating a water moat that separated the peat mat from the upland (Fig. 2.9). This moat allowed beavers to swim directly to the upland and access trees for forage, rather than crossing through the dense ericaceous shrubs of the peat mat. The water level was so high that a chunk of peat mat broke

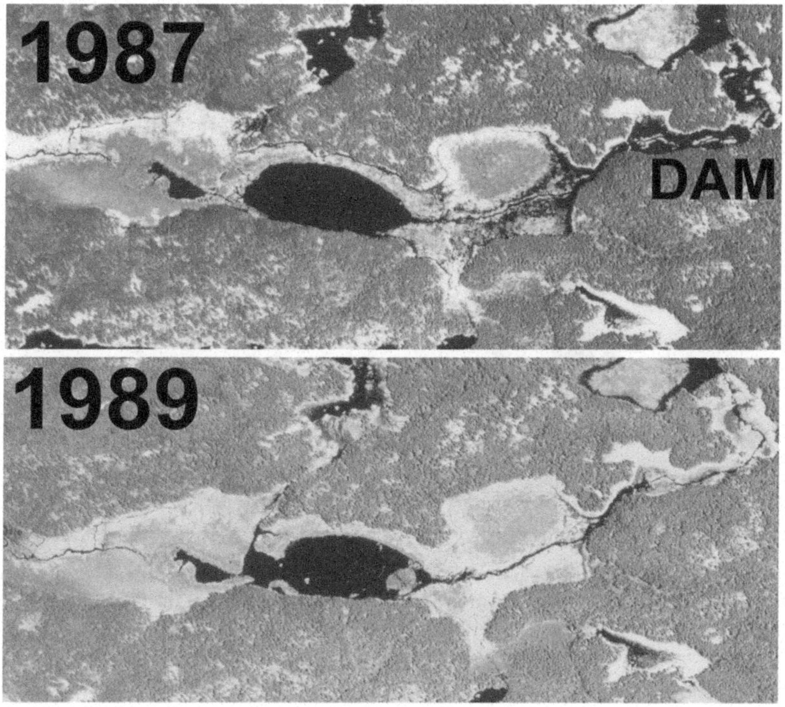

Fig. 2.9 High water levels caused by a downstream beaver dam in 1987 (*top photo*), caused a peat mat to break off and float to the opposite side of Mud Lake (*bottom photo*), changing its shoreline configuration as of 1989

off and floated to the opposite end of the lake, where it became grounded when water levels receded, thereby changing Mud Lake's shoreline configuration.

An outlet beaver dam has profoundly affected Shoepack Lake, a lake that is surrounded by extensive wetlands that flood when its outlet dam is raised. A map generated using a 1×1-m digital elevation model derived from LiDAR data (Minnesota Department of Natural Resources 2015) shows areas with an elevation of 363.89 m (lake level as of the DEM acquisition date in Spring 2011) to 366.89 m, the area that would be flooded by a 3-m dam (Fig. 2.10). The flooding depicted from these elevation data matches the limits of flooding visible on a 1992 aerial photograph and altered vegetation that persisted in 2003 after the dam breach (Fig. 2.11). Based on this evidence, I calculated that the outlet beaver dam approximately tripled the area flooded, from 126 ha to 366 ha.

Nick Frohnauer was studying the muskellunge fishery of Shoepack Lake in 2001, and described its outlet dam as having a 2.16-m head above a bedrock sill; this sill limited how low the lake could drain (Frohnauer et al. 2007). The dam burst during his study due to a 12.5-cm rain event on July 23, 2001, "engendering the loss of roughly 2.16×10^6 kL of water." The cycle of dam construction and collapse had probably happened before, because Frohnauer noted remnant beaver dams near the

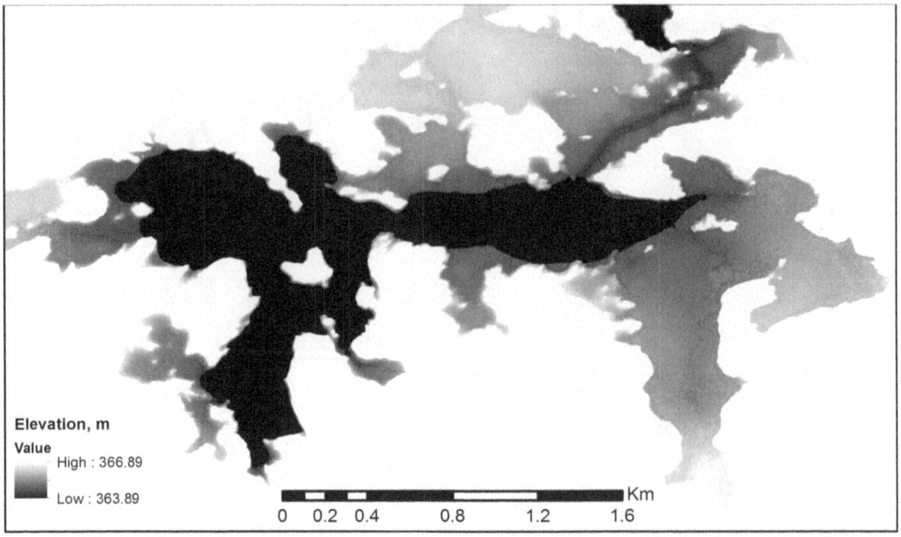

Fig. 2.10 Areas flooded by Shoepack Lake outlet beaver dam, generated from a 1-m digital elevation model, showing areas with an elevation of 363.89 m (lake level as of the DEM acquisition date in Spring 2011) to 367 m

outlet. The location of the Shoepack Lake outlet beaver dam has been very stable over time: a dam was present there as of the earliest aerial photo date in 1927. This location represents a key topographic breakpoint at a narrows along the outlet stream at which a short dam can back up a substantial amount of water.

In a landscape with developed human infrastructure, such a dam collapse could have catastrophic downstream results (Butler 1989). The Shoepack Lake dam collapse did appear to break a few downstream beaver dams, but a major dam that was 3.8 km downstream remained intact, presumably because of the many intervening beaver meadows that reduced the kinetic energy of the flood wave.

2.2.4 Beaver Dams in Peatlands

When beavers impound wetlands where vegetation is adapted to water level fluctuations or is capable of floating up and down with changes in the water level, vegetation alteration may be minimal (Johnston and Naiman 1987). Extensive peatlands (i.e., wetlands with organic soil) to the north and east of Shoepack Lake became wetter as a result of the outlet beaver dam, but retained their vegetation and were not converted to open water (1992 photo, Fig. 2.11). The vegetation in these areas consists of "Leatherleaf-Sweet Gale Shore Fen" and bog shrub species that form floating mats that lift with rising water levels (Faber-Langendoen et al. 2007a). Although the wetter conditions may alter the vegetation of floating mats (Reddoch and Reddoch 2005), these peatlands remain quite resilient to the disturbance caused by

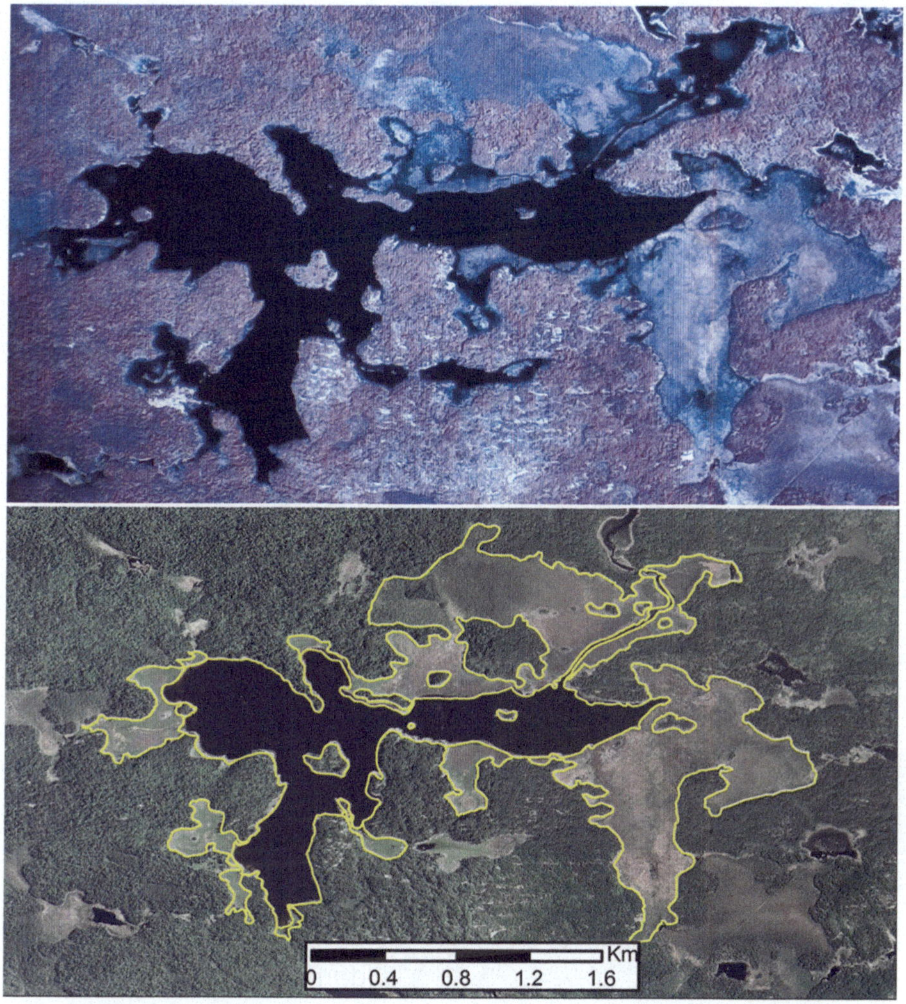

Fig. 2.11 The 2001 collapse of an outlet beaver dam reduced the impounded area of Shoepack Lake and adjacent wetlands from 366 ha in 1992 (*top*) to only 126 ha in 2003 (*bottom*). The *yellow outline* is the maximum extent of flooding

flooding. Experimental flooding of a treed bog complex in northeastern Ontario showed that shallow flooding of bog vegetation led to quick re-establishment of open bog vegetation in the floating peat mats, such that there was little loss of total biomass (Asada et al. 2005).

Beaver dams also impound wetlands that are not associated with lakes. Between 1940 and 1988, beaver dams on the Kabetogama Peninsula flooded 2183 ha of pre-existing wetlands as opposed to only 1504 ha of uplands (Johnston 1994). One-third of the wetlands flooded were converted into ponds, whereas two-thirds of the impounded wetlands were merely made wetter. Beaver dams at peatland outlets

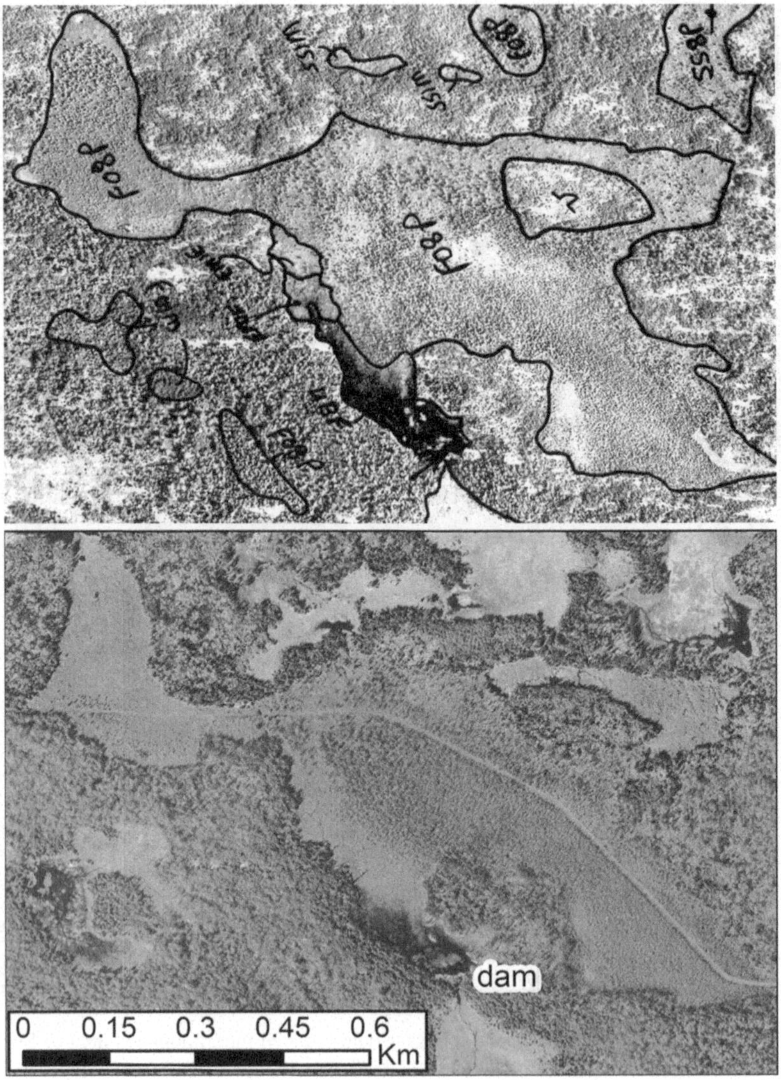

Fig. 2.12 *Top*: A beaver dam (marked by *arrow*) and pond (*dark area* labeled UBF) flood the edge of a black spruce peatland (labeled FO8P) on this 1940 aerial photo. *Bottom*: This 2008 aerial photo shows little change in the beaver dam and pond area over the 68 intervening years, although a winter logging road now traverses the peatland

push water back into the peatland, causing localized vegetation alteration and sometimes tree death, but the boundaries of beaver flooding impact are difficult to delineate because the changes are gradual (Fig. 2.12).

Large peatlands can also act as barriers to beaver activity because their surface waters are not channelized, making them less conducive to dam construction. A 363 ha area west of Shoepack Lake contains only a single beaver dam because of the large peatlands present.

2.2.5 *Beaver Dam Persistence*

In contrast with beaver dams in the western U.S. that are frequently washed out by high water levels or channel migration, the beaver dams of the Kabetogama Peninsula are quite persistent; the same locations are used over and over again. In some cases, many generations of beaver colonies maintain the dam over time. For example, a persistent beaver dam occurs south of Oslo Lake in catchment #9 (Fig. 2.2). A smaller beaver dam existed at this spot in 1927 and 1940, and the present dam has held water continuously since 1949 (Fig. 2.13). Aerial censusing of beaver lodges in 1984, 2000, and 2006 showed this site to have an active beaver colony in all 3 years.

Some abandoned beaver dams continue to hold water despite lack of beaver maintenance, as long as the dam is not breached. Over the period of aerial photo record used to study the Kabetogama beaver ponds, the only beaver dams that have completely washed out without a trace are along the downstream reaches of Cranberry Creek (catchment #3, Fig. 2.2). Some beaver dams disappear from view because they are overtopped by water from a taller dam downstream. River otters (*Lontra canadensis*) dig passages through beaver dams for under-ice access between adjacent water bodies, but normally this doesn't drain the entire pond (Reid et al. 1988).

I analyzed the historical beaver dam map that was prepared by Lewis H. Morgan in 1868 (Morgan 1868) to address the question, "Can the artifacts of beaver engineering be detected after a century or more?" I compared Morgan's map with contemporary digital imagery for the Ishpeming, Michigan region (Johnston 2015). Of the 64 beaver dam and pond sites mapped in the 1860s, 72% were still discernible in 2014. Land use changes that altered the terrain (mining, residential development) or stream paths (channelization) were the main sources of beaver pond loss. This remarkable consistency of beaver pond placement over the last 150 years is evidence of the beaver's resilience.

The persistence of beaver presence over geologic time has been demonstrated by relict beaver dams found buried in peat or sediment deposits, indicating a legacy of beaver activity. Sites described occur in Finland (Aalto et al. 1989), southeastern New England (Kaye 1962), western U.S. mountains (James and Lanman 2012; Persico and Meyer 2009), the Canadian Yukon (Lewkowicz and Coultish 2004), and a former bog buried under urban debris in Columbus, Ohio (Garrison 1967).

2.2.6 *Large Beaver Dams*

Lewis Morgan was particularly impressed with the length of a 79-m beaver dam in northern Michigan (Morgan 1868), which still exists today (Johnston 2015). However, Morgan's dam is only one-tenth the size of the world's longest beaver dam, 850 m, which was found in 2007 using DigitalGlobe satellite imagery for Northern Alberta, Canada (Thie 2016). That beaver dam backs up surface water in a large alluvial fan wetland, which lacks a discernable inlet or outlet stream.

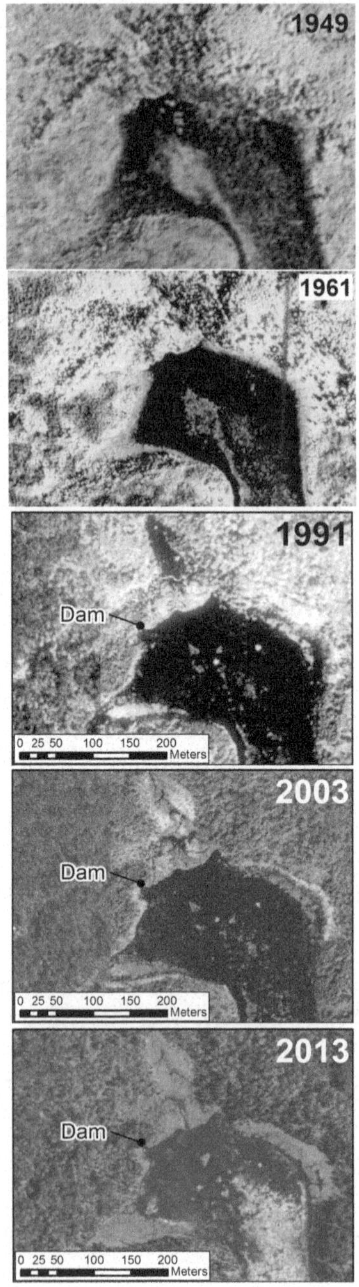

Fig. 2.13 Persistent beaver dam and pond near Oslo Lake, 1949–2013

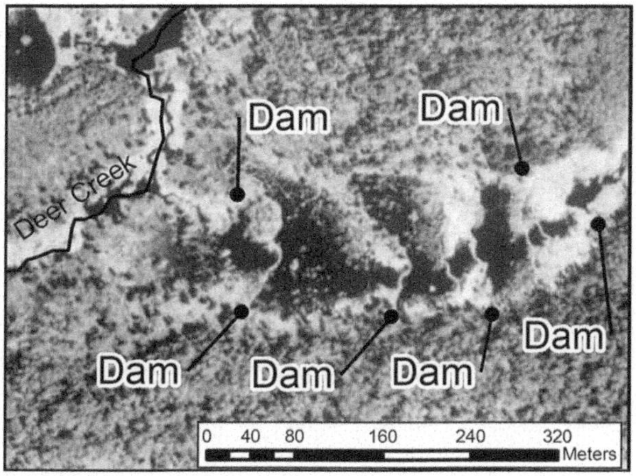

Fig. 2.14 Multiple beaver dams spanning a 140-m wide lowland near Deer Creek, 1992 (catchment #22, Fig. 2.2)

The longest beaver dam on the Kabetogama Peninsula measured 309 m (Fig. 2.7), and was located at the confluence of two tributaries in catchment #16 (Fig. 2.2). Aerial photos showed that the dam was breached sometime between 1997 and 2003, and had not been repaired as of 2015.

The tallest beaver dam I have personally observed, estimated to be 4.5 m tall, was in the Adirondack Mountains of New York. A 2.16 m beaver dam was documented on the Kabetogama Peninsula (Frohnauer et al. 2007), and the tallest dam mentioned by Morgan's 1868 beaver treatise was 3.7 m (Morgan 1868).

2.2.7 *Unusual Beaver Dams*

In contrast to the Shoepack Lake outlet dam, the location of some beaver dams seems illogical given the effort required to construct them versus the habitat gains achieved. An unusual series of beaver dams was observed on a minor tributary of Deer Creek (Fig. 2.14). They are very close together: there are six primary dams and two minor dams within a 350 length of stream. The dams span a 140-m wide valley, and the longest is about 160 m. The dams were initiated in 1961 and were full of water in 1991, but had drained and reverted to wet meadow by 2013.

I have observed several Kabetogama beaver ponds with multiple outlets on drainage divides, an arrangement which seems to defy the laws of hydrology (Johnston 2000). This phenomenon can be clearly seen in Fig. 2.15, where dams occur at the opposite ends of an impounded black ash swamp in level terrain that drains to Black Bay (catchment #1, Fig. 2.2). This pond receives water from an upslope wetland ("inlet"), but is not on a mapped stream. Such ponds are usually shallow and in suboptimal locations: this pond was initiated in 1987, and was abandoned after 2008.

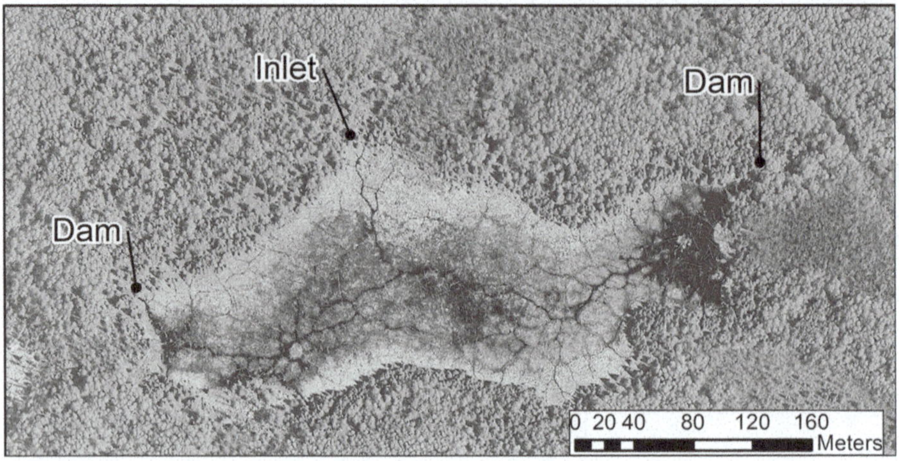

Fig. 2.15 Beaver pond with outlet dams on opposite ends. This dam occurs in very level terrain that drains to Black Bay (catchment #1, Fig. 2.2). Fallen dead tree boles appear throughout, but are easiest to see in the water ponded at the eastern end. Aerial photo taken in 2009

Fig. 2.16 Section of the Agnes Lake trail that was flooded by a beaver dam

Beaver flooding of roads and other infrastructure can be a problem in developed areas (Boyles and Savitzky 2008; Jensen et al. 2001; Johnston 2012). Although there is little built infrastructure on the Kabetogama Peninsula, beaver flooding has impacted trails (Fig. 2.16). When beaver pond expansion started to flood the access road to the

Fig. 2.17 Curved beaver dam constructed by beavers between sandbags placed along the old Ash River Visitor Center access road, 1992

Ash River Visitor Center on the mainland, resourceful park staff used beaver psychology to trick the beavers into constructing a remedy to the problem. They placed piles of sandbags at regular intervals along the side of the road, and the beavers built a dam between them, using the sandbag piles as anchors. The result was a dam that curved along the side of the road (Fig. 2.17). The water level in the pond was at eye level to drivers passing the dam! When the access road was relocated in the 1990s, the road was wisely moved to higher ground, but the curved dam still remains.

2.3 Beaver Pond Characteristics

2.3.1 Beaver Impoundments, Pond Sites, and Clusters

Most of the analyses that follow are based on data extracted from maps that my colleagues and I prepared for the Kabetogama Peninsula by examining aerial photos from multiple years (Table 2.3).

We delineated areas in which elevated water levels from beaver dams had flooded, killed, or otherwise altered the vegetation present, and classified them according to National Wetlands Inventory conventions (Cowardin et al. 1979). These patches were delineated manually or electronically on aerial photo prints or digital images, and digitized using an ArcMap Geographic Information System (GIS) (Johnston and Naiman 1990b). We called the final product "beaver impoundment" maps, because they show more than just ponds. There were 3521 individual polygons on the 2005 beaver impoundment map, the most recent one prepared. GIS shapefile versions of these impoundment maps (1940–2005) can be viewed online and downloaded (Johnston and Windels 2015b).

Table 2.3 Aerial imagery used to generate beaver impoundment maps

Flown	Scale	Type	Commissioning agency
May 1927	1:9082	BW film	Department of the Interior, Canada
Jul 1940	1:20,000	BW film	U.S. Agric. Stabilization and Conservation Service
Sep 1948, Aug 1949	1:15,840	BWIR film	St. Louis County, Minnesota
Aug 1961	1:15,840	BWIR film	St. Louis County, Minnesota
Jun 1972	1:15,840	BWIR film	Voyageurs National Park
Jul 1981	1:24,000	Color film	U.S. Forest Service
May 1986	1:24,000	CIR film	Natural Resources Research Institute
May 1987	1:24,000	CIR film	Natural Resources Research Institute
May 1988	1:24,000	CIR film	Natural Resources Research Institute
May 1989	1:24,000	CIR film	Natural Resources Research Institute
May 1990	1:24,000	CIR film	Natural Resources Research Institute
Sep 1997	1:15,840	CIR film	Minnesota Department of Natural Resources
May 2003	1 m pixels	Color digital	National Agricultural Imagery Program
Sep, Oct 2005	1:15,840	CIR film	Minnesota Department of Natural Resources

BW black and white panchromatic, *BWIR* black and white infrared, *CIR* color infrared

The beaver impoundment maps did not distinguish individual dams; they delineated polygons of contiguous vegetation of the same class. To identify the complex of vegetation and open water associated with an individual dam, we created digital maps of the local catchments draining to each beaver dam and intersected the beaver impoundment maps with them to generate a "pond site" map. Pond sites are bounded at the downstream end by the beaver dam that caused the ponding, and at the upstream end by unimpounded vegetation or the next upstream beaver dam. Pond sites usually contain a complex of water and wetland vegetation, and may or may not contain an active beaver colony in any single year.

As of 1997, there were 1009 pond sites on the Kabetogama Peninsula. The average area per pond site was 3.6 ha, and the maximum pond site area was 39.3 ha. A frequency distribution by pond site area showed the smallest sizes to be most abundant, with only 42 pond sites having areas of 12 ha or greater (Fig. 2.18).

Contiguous beaver-altered areas were merged into a single "cluster" by dissolving the internal GIS boundaries separating adjacent impoundment polygons. Each cluster usually contained more than one pond site. In 1940, there were 71 pond sites but only 47 clusters, a ratio of 1.5 pond sites/cluster (Fig. 2.19a). By 1986, the 835 pond sites were contained in only 346 clusters, a ratio of 2.4 pond sites/cluster. The number of clusters grew at a slower rate than the number of pond sites because beavers were attaching new pond sites onto existing clusters in addition to creating new clusters. Although pond site density increased by an order of magnitude between 1940 and 1986, from 0.3 pond sites/km^2 to 3.0 pond sites/km^2, cluster density peninsula-wide remained at about 1.1 clusters/km^2 after 1961. Cluster density actually declined after 1972 in one region that was densely populated by beavers (Fig. 2.19b). This decrease, within a 382 ha area that encompassed catchments 21–24 (Fig. 2.2),

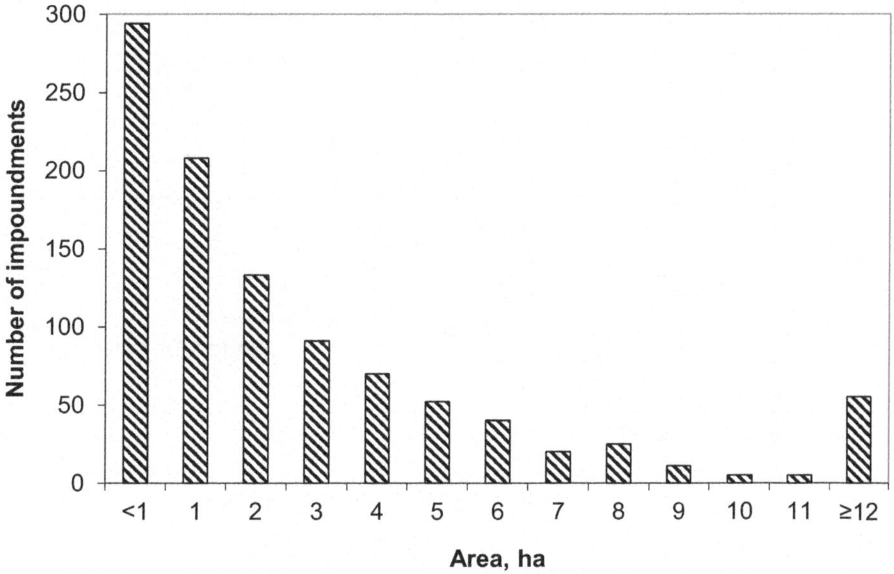

Fig. 2.18 Size frequency distribution, area per pond site as of 1997

occurred because beavers were building ponds that were connecting the clusters together, filling entire valleys with beaver impoundments. These beaver impoundment corridors are favored as travel routes by gray wolves (see Chap. 10).

2.3.2 Influence of Geomorphology on Pond Shape and Boundaries

Land suitable for flooding, one of the three ingredients required for beaver ponds, affects the size and shape of beaver ponds. The Shoepack Lake outlet dam was only about 50 m long, yet it altered water levels 4.4 km upstream due to the large size of Shoepack Lake and its surrounding wetlands (Figs. 2.10 and 2.11). In contrast, the Deer Creek tributary dams were as much as three times longer than the Shoepack Lake outlet dam, but raised water levels only 50–110 m upslope (Fig. 2.14). Stream channel geomorphology has been used to model the capacity of riverscapes to support beaver dams (Macfarlane et al. 2017; Jakes et al. 2007).

I examined the influence of geomorphology on beaver pond configuration, contrasting the shape and boundaries of ponds created by flooding uplands versus those created by flooding wetlands (Johnston and Naiman 1987). Streams flowing through uplands tend to have V-shaped valleys, so beaver ponds created by flooding uplands have steeply sloping bottoms and straight, abrupt side boundaries (Fig. 2.20). In contrast, the flat topography of beaver-flooded wetlands allows a low beaver dam to impound a relatively large area with shallow

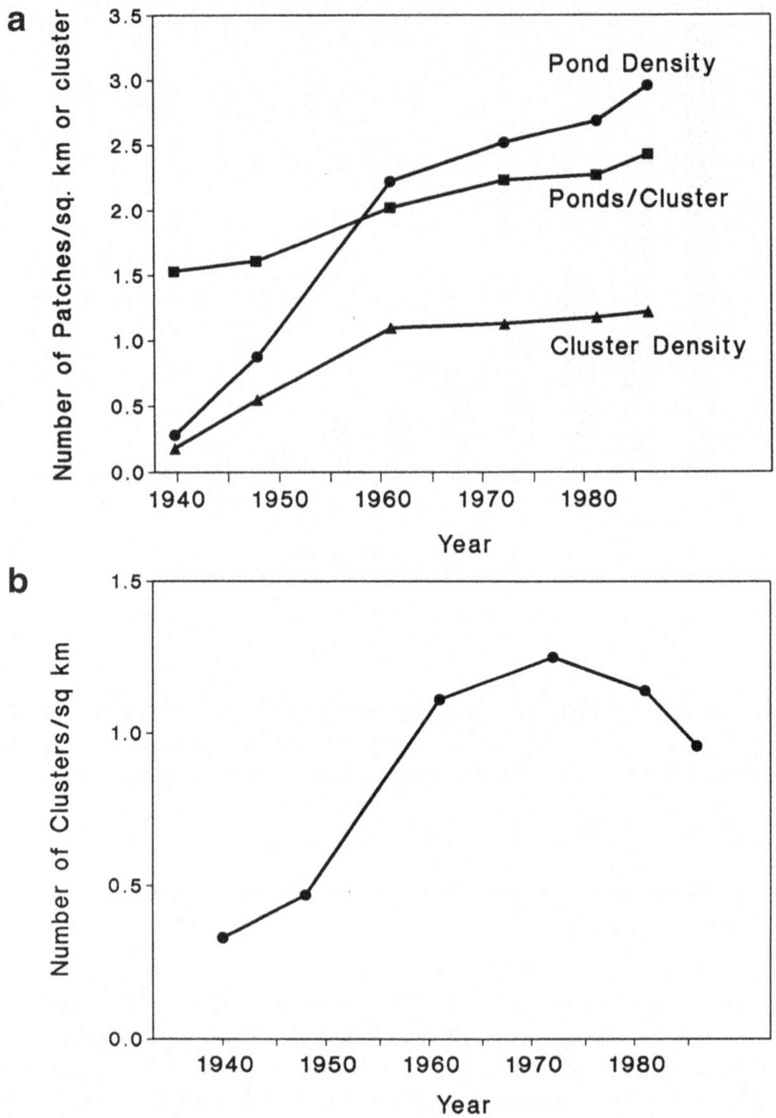

Fig. 2.19 (**a**) Density of pond sites and pond site clusters per unit land area (number of patches/km^2) and average number of pond sites per cluster within the Kabetogama Peninsula, 1940–1986. (**b**) Density of clusters (number of patches/km^2) in a 382 region that was densely populated by beavers, 1940–1986

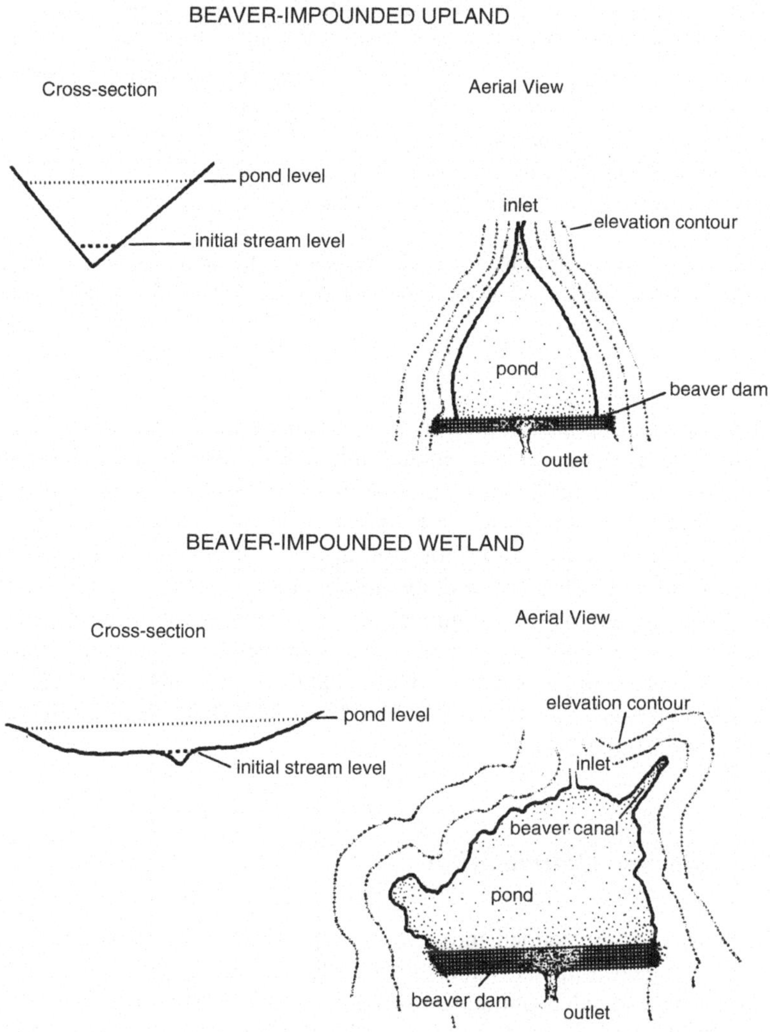

Fig. 2.20 Cross-sectional and plan views of a hypothetical beaver pond created by flooding an upland versus a wetland. From Johnston CA, Naiman RJ (1987) Boundary dynamics at the aquatic-terrestrial interface: the influence of beaver and geomorphology. Landscape Ecology 1(1):47–57, with permission of Springer

Table 2.4 Comparison of beaver foraging patterns in the riparian zones around a flooded wetland (Arnold Pond) and a flooded upland pond (Ash Pond) near Duluth, Minnesota

Characteristic	Wetland pond	Upland pond
1. Perimeter of upland riparian zone, meters	800	500
2. Mean maximum foraging radius, meters (± S.D.)	16.1 ± 10.9	25.8 ± 12.7
3. Area of riparian foraging, sq. meters (row 1 * row 2)	12,880	12,900
4. Foraging occurrence within 20 m of pond (% of plots browsed)	33%	83%

From Johnston CA, Naiman RJ (1987) Boundary dynamics at the aquatic-terrestrial interface: the influence of beaver and geomorphology. Landscape Ecology 1(1):47–57, with permission of Springer

water. Boundaries of ponds in beaver-impounded wetlands are more gradual and diffuse than ponds in beaver-impounded uplands because the land flooded has a gentle slope. This can affect the "permeability" of the pond boundary to beaver movement into the riparian zone: a beaver threatened by predators can quickly return to the deep water of an impounded upland pond, whereas the distance is longer to the safety of deep water in an impounded wetland pond. This can affect riparian foraging patterns. A comparison of woody plant foraging around a upland beaver pond versus a wetland beaver pond showed that, even though the area subject to browsing was nearly identical at the two ponds (about 12,900 m^2), riparian foraging was less evenly distributed around the flooded wetland than the flooded upland (Table 2.4).

2.3.3 *Pond Site Optimization*

Beaver ponds can be grouped into cohorts representing their creation sequence, such that all the beaver ponds visible on the earliest aerial photos are the first cohort, and beaver ponds added as of the next aerial photo date are the second cohort, etc. I did this with the beaver ponds created on the Kabetogama Peninsula using aerial photos taken at approximately decadal intervals from 1940 through 1986 (Johnston and Naiman 1990a).

The number of pond sites increased dramatically over the half-century period, from 71 sites in 1940 to 835 sites in 1986. The rate of new pond creation was greater in 1940–1961 (25 new sites/year) than it was in 1961–1986 (10 new sites/year). New ponds established in 1940, 1948, or 1961 were also significantly larger ($F = 8.85$, $df = 5$, $P = 0.0001$) than new ponds established in 1972, 1981, or 1986, so that the average area of a pond established in 1986 (1.2 ha) was only one-third the average area of a new pond in 1948 (3.9 ha). Not only were 1940-cohort ponds significantly larger to begin with, they also grew at a faster rate once established, doubling in average area

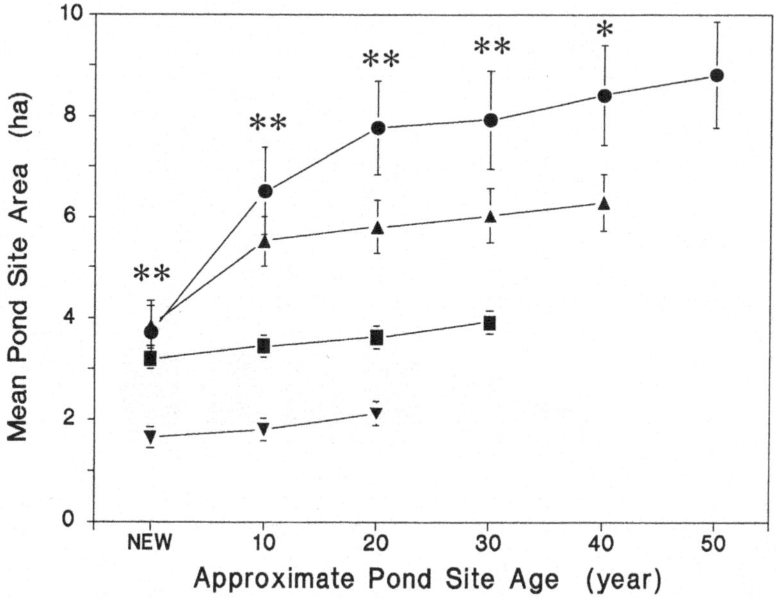

Fig. 2.21 Average pond-site area, by pond cohort. *Circle* = 1940 pond cohort ($y = 1.27 \ln(x) + 3.72$), *triangle* = 1948 pond cohort ($y = 0.63 \ln(x) + 3.85$), *square* = 1961 pond cohort ($y = 0.18 \ln(x) + 3.16$), *inverted triangle* = 1972 pond cohort ($y = 0.14 \ln(x) + 1.62$). * and ** indicate significant effects of pond-site age on average pond area (Kruskal–Wallis one-way ANOVA) at 0.05 and 0.01 significance levels. From Johnston CA, Naiman RJ (1990a) Aquatic patch creation in relation to beaver population trends. Ecology 71:1617–1621, with permission of John Wiley & Sons

after only two decades (Fig. 2.21). The 1948-cohort ponds also grew rapidly during the first decade after establishment, but the growth of the 1961 and 1972-cohort ponds was small (0.1–0.8 ha/decade) and linear over time (Johnston and Naiman 1990a).

The combination of high rates of pond creation, larger initial pond area, and rapid growth made the 1940, 1948, and 1961 pond cohorts the most spatially influential (Fig. 2.22). As of 1986, these cohorts constituted 75% of the total number of ponds and 90% of the total area impounded. The establishment of new ponds was the primary cause of increased cumulative pond area prior to 1961 (70% of the total increase), the rest being due to the enlargement of existing ponds. Therefore, ponds constructed by beaver during their first few decades of occupancy have the greatest impact on the landscape (Johnston and Naiman 1990a).

How is it possible that beavers are able to optimize their dam location so as to create the largest ponds first? It has been established that beavers build dams in response to the sound of running water (Wilsson 1971; Hartman 1975), so the stream pour points at these optimal dam sites probably convey auditory clues to beavers that prompt them to build there.

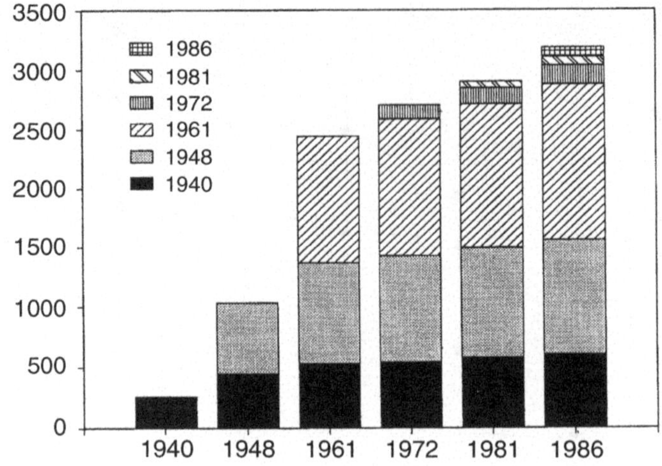

Fig. 2.22 Cumulative pond area, by age class. From Johnston CA, Naiman RJ (1990a) Aquatic patch creation in relation to beaver population trends. Ecology 71:1617–1621, with permission of John Wiley & Sons

2.4 Vegetation Alteration by Beaver Dams

2.4.1 Cover Types of Beaver Impoundments

The raised water level of beaver dams not only creates open water ponds, but also affects a variety of riparian vegetation communities. Beaver dam construction in forested regions causes a fairly predictable sequence of vegetation changes: (1) flooded forest, (2) tree death and toppling, (3) pond containing submergent, floating-leaved, and emergent wetland plants, (4) drained pond with exposed sediments, and (5) drained pond revegetated to grasses and sedges (Johnston 1994; Sturtevant 1998; Little et al. 2012). Beaver colonies move around to new locations and often reflood areas that they have previously abandoned, so that at any one time the beaver-impacted landscape is a shifting mosaic of these various vegetation stages (Naiman et al. 1988).

The impoundment maps classified the cover types of beaver-altered ponds and wetlands (Table 2.5). Open water beaver ponds were relatively easy to detect, but the boundaries of beaver-altered wetland vegetation were more subtle. For example, the beaver dam that impounded the peatland in Fig. 2.12 not only created open water (labeled UBF), but also raised the water level sufficiently to alter the vegetation in an upslope marsh (labeled EM1F) and wet meadow (labeled EM1E). These cover types were bounded along the north and northeast by peatland that was judged to be unaffected by the raised water level, labeled FO8P. The beaver impoundment encompassed the open water pond, marsh, and wet meadow, but the FO8P peatland

Table 2.5 National Wetlands Inventory (NWI) classification codes used, and grouping into six general classes

General class	NWI class and subclass	NWI description	NWI water regime(s)
Dead woody	PSS5	Scrub/shrub dead	B, E, F
	PFO5	Forested dead	B, E, F
Pond	PAB3	Aquatic bed rooted vascular	F, H
	PUB	Unconsolidated bottom (i.e. open water)	F, H
Meadow	PEM1	Emergent persistent	A, B, E
Marsh	PEM1	Emergent persistent	F
Bog	PSS3	Scrub/shrub broad-leaved evergreen	A, B, E, F
	PSS4	Scrub/shrub needle-leaved evergreen	A, B, E, F
	PFO3	Forested broad-leaved evergreen	A, B, E, F
	PFO4	Forested needle-leaved evergreen	A, B, E, F
Deciduous swamp	PSS1	Scrub/shrub broad-leaved deciduous	A, B, E, F
	PFO1	Forested broad-leaved deciduous	A, B, E, F

The "System" for all NWI classes used was Palustrine (P). Water regimes: A—temporarily flooded, B—saturated, E—seasonally flooded-saturated, F—semi-permanently flooded, H—permanently flooded. Additional information about classes at http://www.fws.gov/wetlands/Data/Wetland-Codes.html. From Johnston CA, Windels SK (2015a) Using beaver works to estimate colony activity in boreal landscapes. Journal of Wildlife Management 79 (7):1072–1080, with permission of John Wiley & Sons

was not included in it. GIS analysis was used to quantify the cover types mapped for a 250 km^2 area of the Kabetogama Peninsula (Johnston and Naiman 1990b; Johnston and Windels 2015a, b).

The cumulative area affected by beaver dams increased rapidly from 1940 to 1961, increased at a slower pace from 1961 to 1986, and was relatively stable at about 32 km^2 (12.8% of the landscape) after 1986 (Fig. 2.23). The area of open water pond increased from only 0.12 km^2 in 1940 to 9.8 km^2 in 1990, when it constituted 30% of the landscape, but decreased to half that area in 2003 and 2005. These pond changes were associated with the increase, stabilization, and decrease in the Kabetogama beaver population (Johnston and Windels 2015a).

The Kabetogama Peninsula was mostly wooded prior to the expansion of beavers, vegetated by deciduous and coniferous upland forests, conifer bogs, and swamps (Marschner 1930). Therefore, it is surprising that only about one-third of beaver-impounded vegetation is woody (i.e., deciduous, bog, or dead woody) (Fig. 2.23). Although woody vegetation can persist in floating bog and fen mats, trees rooted in mineral soil die under prolonged flooding (Figs. 2.15 and 2.16). Therefore, beaver-impounded forests are generally short-lived, converting to open water ponds or herbaceous wetlands as the trees topple (Fig. 2.24, top photo). Impounded live deciduous trees and shrubs covered about 5 km^2 during 1961 through 1981, but the area in this cover type was lower after 1981. Impounded bog area has been small but relatively stable over time.

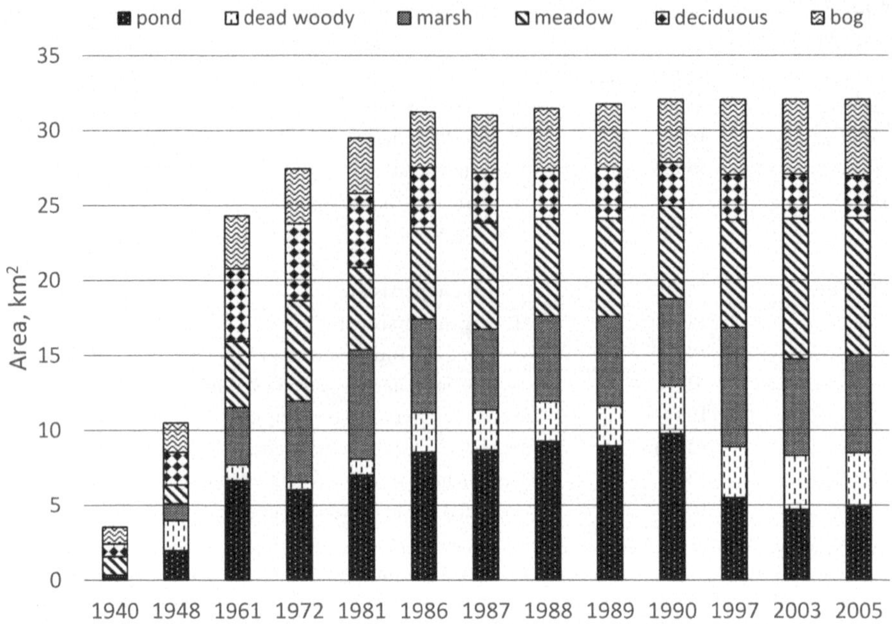

Fig. 2.23 Area of major vegetation types associated with beaver impoundments within a 250 km^2 area of the Kabetogama Peninsula, 1940–2005

2.4.2 Beaver Meadows

Beaver meadows (Fig. 2.24, bottom photo) have been described by a number of researchers around the world (Westbrook et al. 2011; Little et al. 2012; Polvi and Wohl 2012; Simonavičiūtė and Ulevičius 2007; Wright et al. 2003), and are the most abundant cover type of Kabetogama pond sites (Fig. 2.23). Beaver meadows are established when herbaceous vegetation quickly becomes rooted on the exposed bottom sediments of drained beaver ponds. Alternatively, the death of woody plants in floating peat mats can increase the dominance of sedges and other herbaceous species (Mitchell and Niering 1993; Reddoch and Reddoch 2005). The Field Guide to the Plant Community Types of Voyageurs National Park (Faber-Langendoen et al. 2007a) includes two main types of herbaceous wetland vegetation associated with beaver impoundments, Northern Sedge Wet Meadow and Canada Bluejoint Eastern Meadow. Individual plant species occurring in these herbaceous communities are further described in Chap. 6.

On the Kabetogama Peninsula, wet meadows constituted 12–29% of impoundment area between 1948 and 2005, reaching their maximum area in 2003 and 2005 as the beaver population dropped and abandoned ponds drained (Johnston and Windels 2015a). Marshes, including floating sedge mats, constituted 20–25% of

Fig. 2.24 Field photographs of beaver impoundment vegetation. *Top*: This beaver pond with standing dead trees was first created in 1992, 11 years before this photo was taken. *Bottom*: White pine (*Pinus strobus*) saplings and willow (*Salix* spp.) shrubs begin to encroach on the edges of a beaver meadow that had persisted for 30 years

impoundment area between 1972 and 2005. In the absence of renewed disturbance, beaver meadows may succeed to shrubs and trees (Remillard et al. 1987), but this process is very slow on the Kabetogama Peninsula. Woody plant succession was only beginning to occur at one of the field sites that we studied, called Found Pond, after 30 years of being a beaver meadow.

2.4.3 Vegetation of Active Versus Inactive Beaver Pond Sites

Beavers are of course a key ingredient in the creation of beaver ponds, so it makes sense that pond sites that are actively being maintained by a resident beaver colony should differ from those that have been abandoned. Comprehensive beaver colony maps had been prepared for the National Park Service using autumn aerial censuses of beaver lodges in 1984, 2000, and 2006. By pairing that information with the pond site maps, we were able to analyze the vegetation of active versus inactive beaver pond sites (Johnston and Windels 2015a).

For each of the three dates, active pond sites were larger and contained greater pond and marsh area than did inactive pond sites (*t* tests, $P < 0.001$; Fig. 2.25a). Sites active two or three times (i.e., active in 1984, 2000, and 2006) were significantly larger than the two other site groups, and the area per pond site of open water pond and marsh increased significantly with increasing frequency of beaver occupancy (Fig. 2.25b). The area of wet meadow vegetation was greater in twice-active than inactive pond sites, but wet meadow as a fraction of total pond site area was greatest in once-active and inactive pond sites. On average, bog vegetation constituted 12–16% of pond site area regardless of beaver activity, and the area of bog vegetation was significantly greater in thrice-active than in inactive pond sites. Dead woody vegetation was not significantly different across the three groups. These data illustrate that the same vegetation classes occur in both active and inactive pond sites, but that the area and relative proportion of open water and herbaceous vegetation classes differ across levels of beaver activity.

Classification (CART) trees were developed to estimate the activity of individual pond sites using the three lodge maps and the beaver impoundment cover type maps, with overall accuracies of 78–82% (Johnston and Windels 2015a). The CART model derived from the 2000 lodge map paired with the 2003 beaver impoundment map retained only two variables: pond area per site and pond plus marsh area per site. The first split in the tree classified as inactive those sites with very small pond areas (<0.255 ha) and the second split classified as inactive those sites with pond plus marsh areas <2.633 ha (Fig. 2.26). The remaining sites were estimated to be active. CART models developed for the other two dates also incorporated pond and pond plus marsh area, as well as the area of woody deciduous cover (Johnston and Windels 2015a).

2.4.4 Changes in Beaver Population Alter Beaver Impoundment Cover Types

Knowing that beavers alter the relative proportion of vegetation cover, my colleagues and I modeled beaver landscape alteration over a longer time period, using aerial beaver lodge counts that had been conducted for the Kabetogama Peninsula

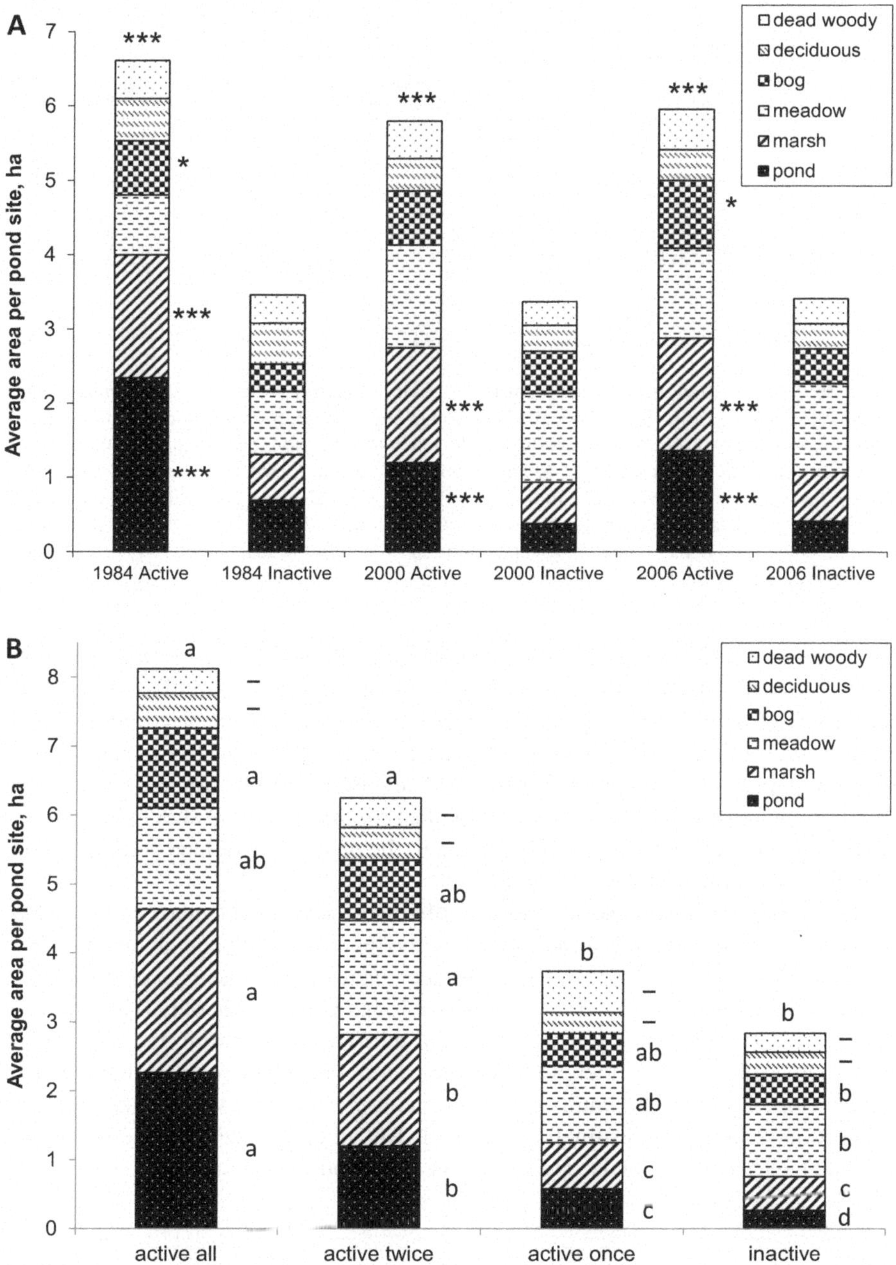

Fig. 2.25 Vegetation types associated with active versus inactive beaver pond sites. (**a**) Average area of vegetation types for active versus inactive pond sites during lodge map census years 1984, 2000, and 2006. Bars or portions of bars with * indicate significant differences (***$P < 0.001$, *$P < 0.05$). (**b**) Average area of vegetation types for beaver pond sites that had been active once, twice, thrice, or inactive according to lodge maps. Bars or portions of bars sharing the same lowercase letter are not significantly different (Tukey's multiple comparisons of means, $P < 0.05$). – = no significant difference among groups. From Johnston CA, Windels SK (2015a) Using beaver works to estimate colony activity in boreal landscapes. Journal of Wildlife Management 79 (7):1072–1080, with permission of John Wiley & Sons

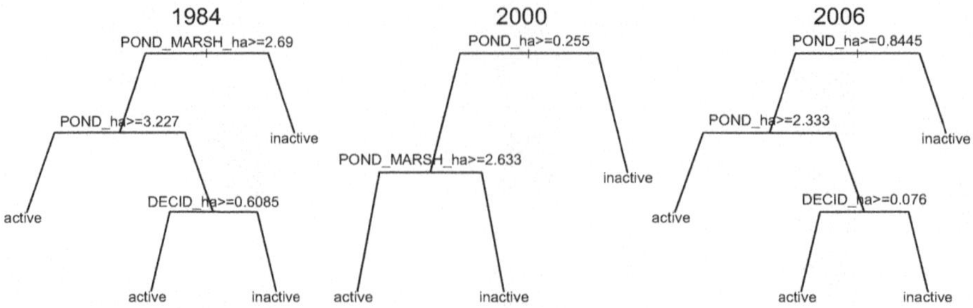

Fig. 2.26 Classification trees for distinguishing activity of beaver pond sites based on associated cover types, derived from 1984, 2000, and 2006 lodge maps and their corresponding impoundment maps. Sites meeting the decision criterion go to the left split, whereas those not meeting the criterion go to the right. DECID_ha = area of beaver-impounded deciduous woody vegetation in hectares, POND_ha = beaver pond water area in hectares, POND_MARSH_ha = area of beaver-impounded marsh and pond area in hectares. From Johnston CA, Windels SK (2015a) Using beaver works to estimate colony activity in boreal landscapes. Journal of Wildlife Management 79 (7):1072–1080, with permission of John Wiley & Sons

annually since 1958 (Berg 1987). In contrast to the three lodge maps that were used in the previous analysis, the annual lodge counts were not spatially explicit. Our initial model matched the aerial beaver colony counts with beaver impoundment maps representing approximately 10-year intervals between 1940 and 1986, a period of rapid beaver population increase. The initial regression model showed that beaver populations were best predicted by the proportion of land area in impounded shallow marsh, and were negatively related to the proportion of land area in seasonally flooded meadow (Broschart et al. 1989).

This model failed, however, when applied to more recent maps (1987–2005) because the vegetation alteration caused by the recolonization and abandonment of pond sites evolved as the beaver landscape matured and the beaver population peaked and declined (Fig. 2.27). Total active lodge density was only 0.33/km^2 during the first transect survey in 1958, increased to 1.1/km^2 during the mid-1980s, then decreased to 0.9/km^2 thereafter. The new landscape-scale regression model that we developed related the number of beaver colonies to open water pond area as a proportion of land area, estimating that each 100 additional beaver colonies would impound 2.15% of the landscape (Johnston and Windels 2015a).

2.5 Beaver Excavation

Beavers engineering includes excavation. Beavers excavate the mud behind their dams and around their lodges, applying the mud to the structures. Beavers also excavate canals so as to create deeper water access to riparian foraging areas. Beaver canals that are not obscured by overlying water or overhanging vegetation are often visible on detailed aerial imagery (Figs. 2.15 and 2.28a). Beaver canals are also detectable using vertically precise digital elevation model (DEM) data. Figure 2.28b

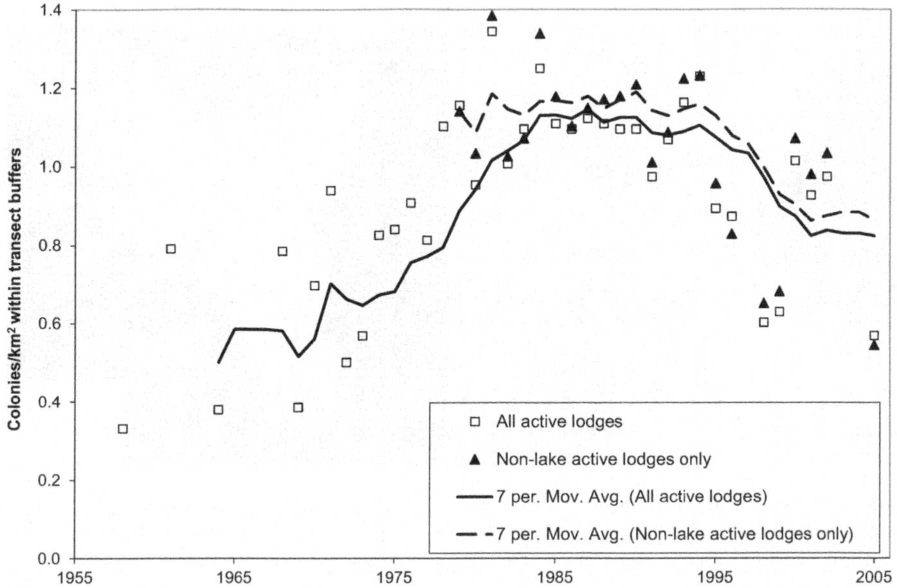

Fig. 2.27 Beaver colony density at Voyageurs National Park, Minnesota based on aerial survey transects, 1958–2005. Point symbols indicate average annual colony density per square km, including and excluding lake lodges. Linear trends are seven-period moving averages of the point data. From Johnston CA, Windels SK (2015a) Using beaver works to estimate colony activity in boreal landscapes. Journal of Wildlife Management 79 (7):1072–1080, with permission of John Wiley & Sons

is color-coded to represent elevations within a 1.1 m range, using a LIDAR-derived DEM depicting 1 cm elevation differences. The beaver canals are 20–50 cm deep (Fig. 2.28). Beaver canal excavation to access upland food reserves can increase average wetland perimeters by over 575% (Hood and Larson 2015).

2.6 Conclusions

- Beaver ponds require three ingredients: beavers, water, and land suitable for flooding.
- The Kabetogama Peninsula is an ideal area for beaver ponds because: (1) it is completely surrounded by water, (2) it has complex topography, (3) its waters flow in many directions, (4) it has impervious bedrock, (5) it has impervious soils, and (6) there is abundant woody browse.
- Beavers build dams on streams, in peatlands, and at lake outlets. When beaver dams flood peatlands, the vegetation often forms floating mats that lift with the rising water level.
- The lake outlet dam at Shoepack Lake raised water levels 4.4 km upstream and tripled the lake's surface area by flooding adjacent wetlands.

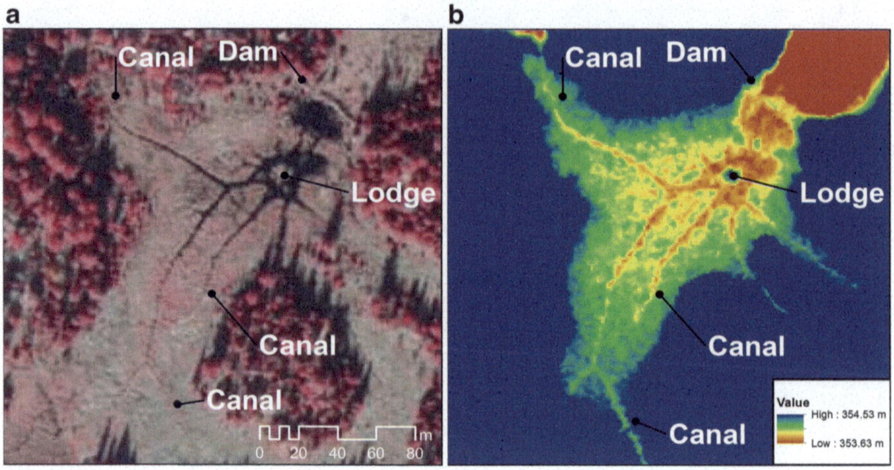

Fig. 2.28 Beaver-excavated features. (**a**) Water appears as black patches behind dam, around lodge, and in linear canals in this color infrared image. (**b**) LIDAR-derived DEM has been color-coded to show beaver-excavated canals that are 20–50 cm deep

- The longest beaver dam on the Kabetogama Peninsula is 309 m, but the longest beaver dam in the world is 850 m.
- Beaver dams on the Kabetogama Peninsula are quite persistent. Some have been continuously occupied by beaver colonies. Others go through cycles of abandonment and rebuilding. Dams may be breached by high water flows, but are usually rebuilt at the same location.
- Some beaver ponds have dams at multiple outlets draining in different directions.
- Beaver impoundments have altered 12.8% of the Kabetogama Peninsula over time.
- When colonizing a new area, beavers choose to flood first the sites that will create the largest ponds with the greatest potential for areal growth. Smaller ponds are created after the best sites are used up.

References

Aalto M, Donner J, Hirvas H, Niemelä J (1989) An interglacial beaver dam deposit at Vimpeli, Ostrobothnia, Finland. Geological Survey of Finland, Espoo

Aleksiuk M (1970) The seasonal food regime of arctic beavers. Ecology 51(2):264–270

Allen AA (1983) Habitat suitability index models: beaver. U.S. Department of the Interior, Division of Biological Services, Washington

Anonymous (2004) Wanted: a new home; lookout tower available for acquisition. Minnesota history interpreter, vol 32. Minnesota Historical Society, St. Paul

Asada T, Warner BG, Schiff SL (2005) Effects of shallow flooding on vegetation and carbon pools in boreal peatlands. Appl Veg Sci 8(2):199–208

Berg WE (1987) Counting our northern wildlife. The Minnesota Volunteer 50 (March–April):26–31
Bergerud AT, Miller DR (1977) Population dynamics of Newfoundland beaver. Can J Zool/Revue Canadienne De Zoologie 55:1480–1492
Bertolo A, Magnan P, Plante M (2008) Linking the occurrence of brook trout with isolation and extinction in small Boreal Shield lakes. Freshw Biol 53(2):304–321
Bloomquist CK, Nielsen CK (2010) Demography of unexploited beavers in southern Illinois. J Wildl Manag 74(2):228–235
Boyce MS (1981) Beaver life-history responses to exploitation. J Appl Ecol 18:749–753
Boyles SL, Savitzky BA (2008) An analysis of the efficacy and comparative costs of using flow devices to resolve conflicts with North American beavers along roadways in the coastal plain of Virginia. In: Timm RM, Madon MB (eds) Proceedings of the 23rd vertebrate pest conference. University of California, Davis, San Diego, pp 47–52
Broschart MR, Johnston CA, Naiman RJ (1989) Prediction of beaver colony density using impounded habitat variables. J Wildl Manag 53:929–934
Busher PE (1987) Population parameters and family composition of beaver in California. J Mammal 68(4):860–864
Butler DR (1989) The failure of beaver dams and resulting outburst flooding: a geomorphic hazard of the southeastern Piedmont. Geogr Bull 31(1):29–38
Cowardin LM, Carter V, Golet FC, LaRoe ET (1979) Classification of wetlands and deepwater habitats of the United States. U.S. Fish and Wildlife Service, Washington
Dieter CD (1992) Population characteristics of beavers in Eastern South-Dakota. Am Midl Nat 128(1):191–196
Doucet CM, Adams IT, Fryxell JM (1994) Beaver dam and cache composition: Are woody species used differently? Ecoscience 1(3):268–270
Faber-Langendoen D, Aaseng N, Hop K, Lew-Smith M (2007a) Field guide to the plant community types of Voyageurs National Park. U.S. Geological Survey, Reston
Faber-Langendoen D, Aaseng N, Hop K, Lew-Smith M, Drake J (2007b) Vegetation classification, mapping, and monitoring at Voyageurs National Park, Minnesota: an application of the U.S. National Vegetation Classification. Appl Veg Sci 10(3):361–374
Frohnauer NK, Pierce CL, Kallemeyn LW (2007) Population dynamics and angler exploitation of the unique muskellunge population in Shoepack Lake, Voyageurs National Park, Minnesota. N Am J Fish Manag 27(1):63–76
Garrison GC (1967) Pollen stratigraphy and age of an early postglacial beaver site near Columbus, Ohio. Ohio J Sci 67(2):96–105
Hartman AM (1975) Analysis of conditions leading to the regulation of water flow by a beaver. Psychol Rec 25(2):427–431
Hartman AM, Rice HK (1963) Observations of a beaver (*Castor canadensis* Carolinesis) care, 24-hour activity records and performance on a simple learning task. Psychol Rec 13:317–328
Hemstad CB, Southwick DL, Ojakangas RW (2002) M-125 Bedrock geologic map of Voyageurs National Park and vicinity, Minnesota. Minnesota Geological Survey, St Paul
Hood GA, Larson DG (2015) Ecological engineering and aquatic connectivity: a new perspective from beaver-modified wetlands. Freshw Biol 60(1):198–208
Jakes AF, Snodgrass JW, Burger J (2007) *Castor canadensis* (Beaver) impoundment associated with geomorphology of southeastern streams. Southeast Nat 6(2):271–282
James CD, Lanman RB (2012) Novel physical evidence that beaver historically were native to the Sierra Nevada. California Fish and Game 98(2):129–132
Jarema SI, Samson J, McGill BJ, Humphries MM (2009) Variation in abundance across a species' range predicts climate change responses in the range interior will exceed those at the edge: a case study with North American beaver. Glob Chang Biol 15(2):508–522
Jenkins SH, Busher PE (1979) Castor canadensis. Mamm Species 120:1–8
Jensen PG, Curtis PD, Lehnert ME, Hamelin DL (2001) Habitat and structural factors influencing beaver interference with highway culverts. Wildl Soc Bull 29(2):654–664
Johnston CA (1994) Ecological engineering of wetlands by beavers. In: Mitsch WJ (ed) Global wetlands: old world and new. Elsevier Science, Amsterdam, pp 379–384

Johnston CA (2000) Wetland soil and landscape alteration by beavers. In: Richardson JL, Vepraskas MJ (eds) Wetland soils: their genesis, morphology, hydrology, landscapes, and classification. CRC Press, Boca Raton, pp 391–408

Johnston CA (2012) Beaver wetlands. In: Batzer DP, Baldwin A (eds) Wetland habitats of North America: ecology and conservation concerns. University of California Press, Berkeley, pp 161–171

Johnston CA (2015) Fate of 150 year old beaver ponds in the Laurentian Great Lakes region. Wetlands 35:1013–1019

Johnston CA, Naiman RJ (1987) Boundary dynamics at the aquatic-terrestrial interface: the influence of beaver and geomorphology. Landsc Ecol 1(1):47–57

Johnston CA, Naiman RJ (1990a) Aquatic patch creation in relation to beaver population trends. Ecology 71:1617–1621

Johnston CA, Naiman RJ (1990b) The use of a geographic information system to analyze long-term landscape alteration by beaver. Landsc Ecol 4:5–19

Johnston CA, Windels SK (2015a) Using beaver works to estimate colony activity in boreal landscapes. J Wildl Manag 79(7):1072–1080

Johnston CA, Windels SK (2015b) Using beaver works to estimate colony activity in boreal landscapes (online spatial data). CUAHSI Spatial Data Service, Consortium of Universities for the Advancement of Hydrologic Science Inc., Medford. http://spatial.cuahsi.org/johnstonC01/

Jung TS, Staniforth JA (2010) Unusual beaver, *Castor canadensis*, dams in central Yukon. Can Field Nat 124(3):274–275

Kanivetsky R (1979) Hydrogeologic map of Minnesota, bedrock hydrogeology, S-5. Minnesota geological survey, University of Minnesota, St. Paul

Kaye CA (1962) Early postglacial beavers in southeastern New England. Science 138(3543): 906–907

Kretser W, Gallagher J, Nicolette J (1989) Adirondack lakes study 1984–1987, an evaluation of fish communities and water chemistry. Adirondack Lakes Survey Corporation, Ray Brook

Leidholtbruner K, Hibbs DE, McComb WC (1992) Beaver dam locations and their effects on distribution and abundance of Coho salmon fry in 2 coastal Oregon streams. Northwest Sci 66(4):218–223

Lewkowicz AG, Coultish TL (2004) Beaver damming and palsa dynamics in a subarctic mountainous environment, Wolf Creek, Yukon Territory, Canada. Arct Antarct Alp Res 36(2):208–218

Little AM, Guntenspergen GR, Allen TFH (2012) Wetland vegetation dynamics in response to beaver (*Castor canadensis*) activity at multiple scales. Ecoscience 19(3):246–257

Macfarlane WW, Wheaton JM, Bouwes N, Jensen ML, Gilbert JT, Hough-Snee N, Shivik JA (2017) Modeling the capacity of riverscapes to support beaver dams. Geomorphology 277:72–99

Marschner FJ (1930) The original vegetation of Minnesota. U.S. Forest Service, North Central Forest Experiment Station, St. Paul

McCall TC, Hodgman TP, Diefenbach DR, Owen RB (1996) Beaver populations and their relation to wetland habitat and breeding waterfowl in Maine. Wetlands 16(2):163–172

Minnesota Department of Natural Resources (2015) LiDAR Elevation, Arrowhead Region, NE Minnesota, 2011. http://www.mngeo.state.mn.us/chouse/metadata/lidar_arrowhead2011.html

Mitchell CC, Niering WA (1993) Vegetation change in a topogenic bog following beaver flooding. Bull Torrey Bot Club 120(2):136–147

Morgan LH (1868) The American beaver and his works. J. B. Lippencott & Co., Philadelphia

Müller-Schwarze D (2011) The beaver: its life and impact, 2nd edn. Cornell University Press, Ithaca

Naiman RJ, Johnston CA, Kelley JC (1988) Alteration of North American streams by beaver. Bioscience 38:753–762

Persico L, Meyer G (2009) Holocene beaver damming, fluvial geomorphology, and climate in Yellowstone National Park, Wyoming. Quat Res 71(3):340–353

Polvi LE, Wohl E (2012) The beaver meadow complex revisited – the role of beavers in post-glacial floodplain development. Earth Surf Process Landf 37(3):332–346

Reddoch JM, Reddoch AH (2005) Consequences of beaver, *Castor canadensis*, flooding on a small shore fen in southwestern Quebec. Can Field Nat 119(3):385–394

Reid DG, Herrero SM, Code TE (1988) River otters as agents of water loss from beaver ponds. J Mammal 69(1):100–107

Remillard MM, Gruendling GK, Bogucki DJ (1987) Disturbance by beaver (*Castor canadensis* Kuhl) and increased landscape heterogeneity. In: Turner MG (ed) Landscape heterogeneity and disturbance. Ecological studies, vol 64. Springer, New York, pp 103–122

Robel RJ, Fox LB (1993) Comparison of aerial and ground survey techniques to determine beaver colony densities in Kansas. Southwest Nat 38(4):357–361

Rue LL (2002) Beavers. Voyageur Press, Stillwater

Simonavičiūtė L, Ulevičius A (2007) Structure of phytocenoses in beaver meadows in Lithuania. Ekologija 53(3):34–44

Skewes O, Gonzalez F, Olave R, Ávila A, Vargas V, Paulsen P, König H (2006) Abundance and distribution of American beaver, Castor canadensis (Kuhl 1820), in Tierra del Fuego and Navarino islands, Chile. Eur J Wildl Res 52(4):292–296

Slough BG, Sadleir RMFS (1977) A land capability classification system for beaver (*Castor canadensis* Kuhl). Can J Zool/Revue Canadienne De Zoologie 55(8):1324–1335

Smith DW, Peterson RO (1988) The effects of regulated lake levels on beaver in Voyagears National Park, Minnesota. U.S. Department of the Interior, National Park Service, Midwest Regional Office, Omaha

Smith DW, Tyers DB (2008) The beavers of Yellowstone. Yellowstone. Science 16(3):4–15

Smith DW, Trauba DR, Anderson RK, Peterson RO (1994) Black bear predation on beavers on an island in lake superior. Am Midl Nat 132(2):248–255

Sturtevant BR (1998) A model of wetland vegetation dynamics in simulated beaver impoundments. Ecol Model 112(2–3):195–225

Teller JT (1985) Glacial Lake Agassiz and its influence on the Great Lakes. In: Karrow PF, Calkin PE (eds) Quaternary evolution of the Great Lakes. Geological Association of Canada, St. John's. Special Paper 30, pp 1–16

Thie J (2016) Exploring beaver habitat and distribution with Google Earth: the longest beaver dam in the world. EcoInformatics Inc. http://www.geostrategis.com/p_beavers-longestdam.htm. Accessed 26 Feb 2016

Vehkaoja M, Nummi P, Rask M, Tulonen T, Arvola L (2015) Spatiotemporal dynamics of boreal landscapes with ecosystem engineers: beavers influence the biogeochemistry of small lakes. Biogeochemistry 124(1):405–415

Voigt DR, Kolenosky GB, Pimlott DH (1976) Changes in summer foods of wolves in central Ontario. J Wildl Manag 40:663–668

Westbrook CJ, Cooper DJ, Baker BW (2011) Beaver assisted river valley formation. River Res Appl 27(2):247–256

Wilsson L (1971) Observations and experiments on the ethology of the European beaver (*Castor fiber* L.) Viltrevy 8:115–266

Wright JP, Flecker AS, Jones CG (2003) Local vs. landscape controls on plant species richness in beaver meadows. Ecology 84(12):3162–3173

Chapter 3
Altering the Water Cycle

Carol A. Johnston, Terry N. Brown, and Kirsten Cahow-Scholtes

Abstract Beaver ponds require a supply of water, which ultimately comes from precipitation. Although most of northern Minnesota's precipitation is received in the summer months, snowmelt is the main driver of peak streamflow in the spring. I analyzed NOAA snowpack records and USGS streamflow records to show that peak annual streamflow was significantly related to maximum snow water equivalent during the winters of 2005–2016. We conducted intensive field studies of water fluxes in a Kabetogama Peninsula beaver meadow, measuring evapotranspiration, pond level, groundwater, streamflow, and runoff from watershed side slopes. Evapotranspiration was a significant source of water loss from the beaver meadow during late summer, with discernable diurnal effects on streamflow, water table, and pond level. Groundwater measurements showed recharge (i.e., water flowing from the surface into the ground) occurring within most of the beaver meadow, but periodic discharge (i.e., water flowing from the ground toward the surface) in sites closer to the stream flowing through the beaver meadow. "Edge collectors" positioned around the perimeter of the beaver meadow showed that lateral fluxes of runoff into the beaver meadow were spatially variable, such that the temporal pattern of flow in a gently sloping portion of the catchment with >0.3 m of soil was different than that for three sites with steeper slopes and shallower soils over bedrock. A spatially explicit model that routed flow within the beaver meadow catchment successfully replicated this difference.

C.A. Johnston (✉)
Department of Natural Resource Management, South Dakota State University, Brookings, SD 57006, USA
e-mail: carol.johnston@sdstate.edu

T.N. Brown
Natural Resources Research Institute, University of Minnesota, Duluth, MN 55811, USA

K. Cahow-Scholtes
Wisconsin Department of Natural Resources, LaCrosse, WI 54601, USA

C.A. Johnston, *Beavers: Boreal Ecosystem Engineers*,
DOI 10.1007/978-3-319-61533-2_3

3.1 Introduction

Water is the second most important ingredient, besides beavers, needed to create beaver ponds. Beavers can't move water upstream, so water must flow into their ponds from precipitation (Pr) and inputs from the surrounding land: inflowing streams (St_{in}), lateral runoff from adjacent slopes (Lr), and groundwater inputs (Gw_{in}) (Fig. 3.1a). Outputs from beaver ponds include outflowing streams (St_{out}), groundwater outputs (Gw_{out}), and evapotranspiration (ET). The difference between inputs and outputs represents storage in the beaver pond itself (ΔP_L). A water balance calculates the interaction of these terms as:

$$\left(\text{Pr}+\text{St}_{in}+\text{Lr}+\text{Gw}_{in}\right)-\left(\text{ET}+\text{St}_{out}+\text{Gw}_{out}\right)=\Delta P_{\text{L}} \tag{3.1}$$

Overwinter, inputs decrease because precipitation is locked up in the snowpack and there is no lateral runoff, but outputs also decrease because of the lack of evapotranspiration (Fig. 3.1b).

Water exchanges between the landscape and the atmosphere are altered by the presence of beaver ponds and meadows. The creation of ponds and wetlands by beaver dams increases the water surface area that receives direct precipitation inputs (Hood and Bayley 2008), as opposed to precipitation that is intercepted by forest vegetation or that soaks into the ground before eventually reaching ground and surface waters. Ponds and wetlands also have evapotranspiration rates comparable to potential evapotranspiration (PET), whereas evapotranspiration from boreal forest vegetation is a fraction of PET (Neff 1996).

Using stable isotopes (^{18}O and deuterium) and H_4SiO_4 concentrations, researchers in the Adirondack Mountains of New York determined that a headwater beaver pond (i.e., lacking an inlet stream) altered most aspects of catchment hydrology when compared to a nearby catchment without a beaver pond, but that the effects varied seasonally (Burns and McDonnell 1998). The beaver pond provided minimal runoff retention during snowmelt (because it was already full of water), but retained nearly all precipitation event water during late spring and autumn rain events. Water previously stored within each catchment (pre-event water) was the predominant component of streamflow except during a rain-on-snow event. At base flow, mean catchment residence time was comparable between a beaver-impounded and a non-beaver-impounded catchment (~100 days).

Beaver dams are more porous than human-constructed dams, particularly when the lack of maintenance due to abandonment erodes the mud that makes the dam water-tight. Actively maintained dams are usually of the overflow or gap flow type, but may become underflow or throughflow types as the dam deteriorates (Woo and Waddington 1990). These dam types have different effects on stream hydrographs: overflow dams have steep hydrograph rises and fast recessions after the peaks with little effect on channel flows once the water level tops the dam crest, whereas underflow dams cause a longer delay to peak flow downstream because the dam restricts outflow, retaining the increasing water supply from upstream and greatly moderating the flow downstream. At the conclusion of a storm, discharge below the underflow dam continues to rise as the ponded water is released (Woo and Waddington

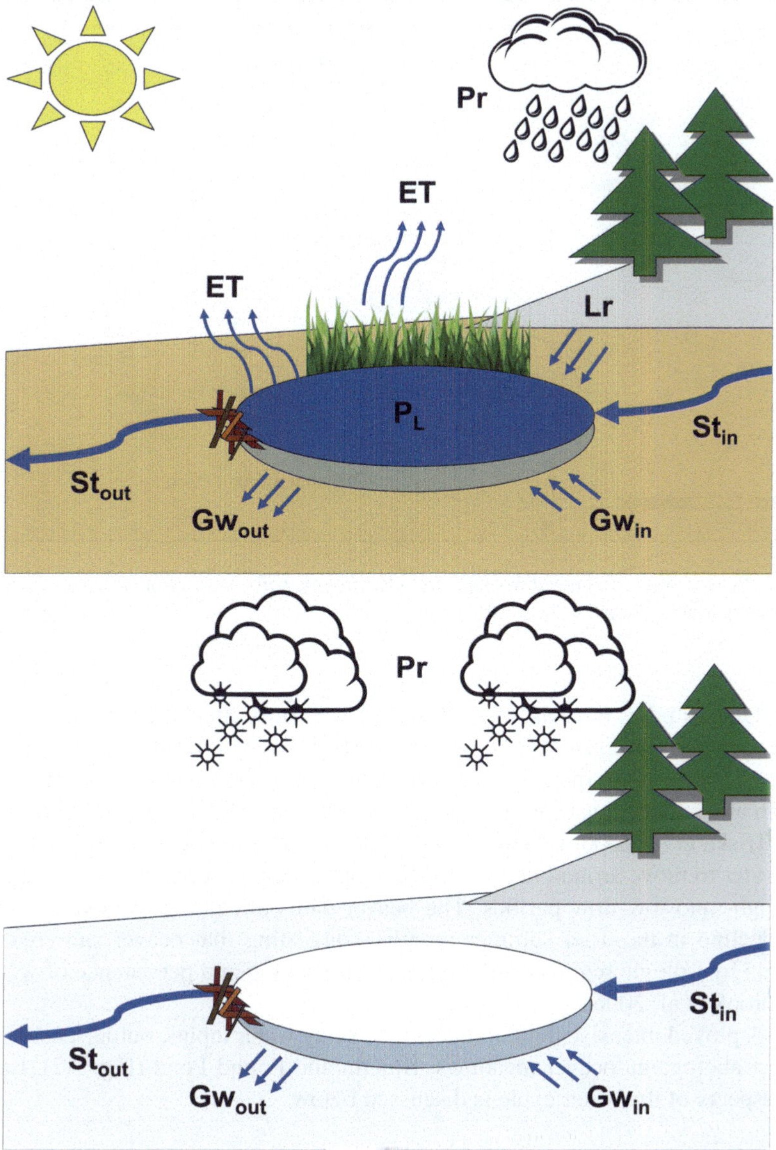

Fig. 3.1 Water budget for a beaver pond. *Top*: growing season. *Bottom*: winter. *ET* evapotranspiration, GW_{in} groundwater inputs, GW_{out} groundwater outputs, *Lr* lateral runoff, P_L pond water level, *Pr* precipitation, St_{in} stream inputs, St_{out} stream outputs

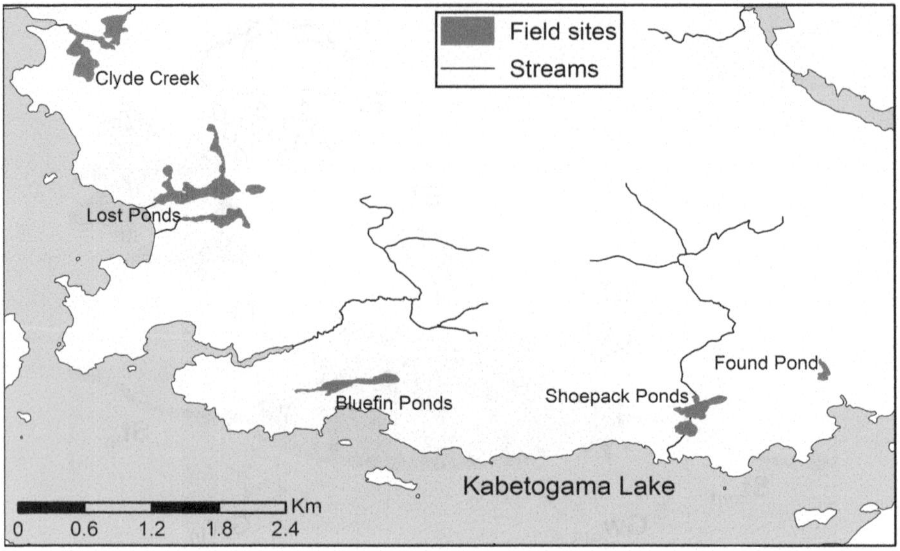

Fig. 3.2 Beaver ponds and meadows studied in hydrology, soils, vegetation, and biogeochemical field investigations (Chaps. 3, 5, 6, 7, 8)

1990). Other studies have also found that beaver dams attenuate streamflow peaks, releasing water more gradually downstream (Nyssen et al. 2011).

Beaver dams raise riparian water tables, and can increase stream water inputs to riparian areas that limit water table declines (Hill and Duval 2009; Majerova et al. 2015; Triska et al. 2000; Lowry 1993). Beaver dams on the Colorado River caused river water to move around them as surface runoff and groundwater seepage during both high- and low-flow periods. The beaver dams attenuated the expected water table decline in the drier summer months, suggesting that beaver can create and maintain hydrologic regimes suitable for the formation and persistence of wetlands (Westbrook et al. 2006).

We deployed intensive field monitoring to study water inputs, outputs, and storage at two Kabetogama beaver meadows, Bluefin and Found Pond (Fig. 3.2). Each of these aspects of the water cycle is discussed below.

3.2 Precipitation

3.2.1 Rainfall and Snowfall

The available water supply starts with precipitation. Average annual precipitation for International Falls over the period of continuous record (1940–2015) was 62.7 cm (NOAA 2016d), of which 67% falls during the five summer months of May through September (Fig. 3.3a) (Arguez et al. 2010). These are the same five months

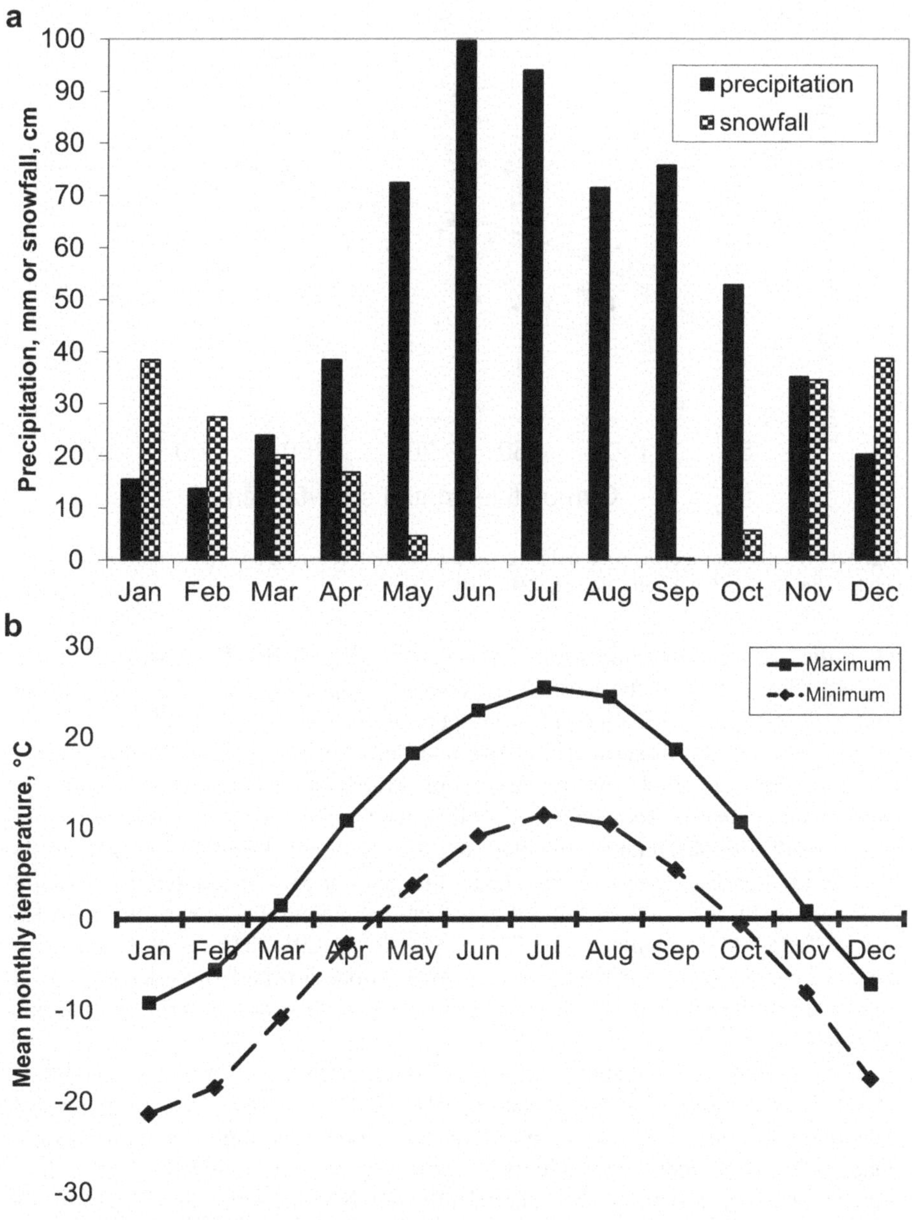

Fig. 3.3 Climate normals for the International Falls MN airport weather station, 1981–2010. (**a**) Precipitation (**b**) Temperature

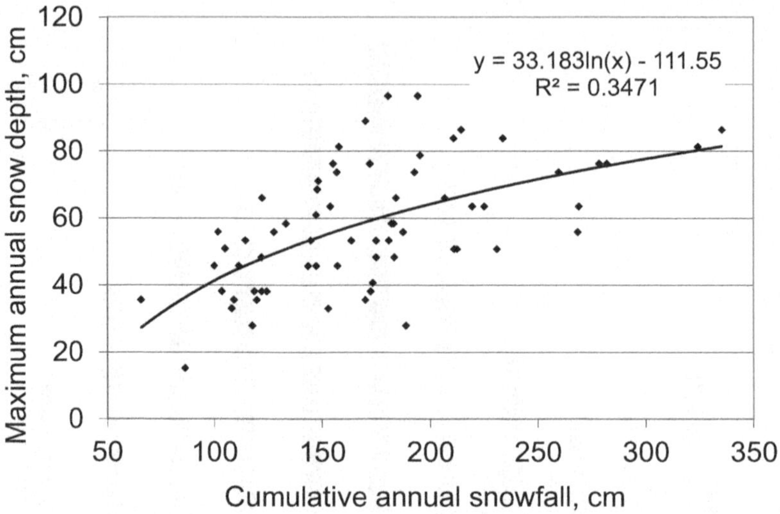

Fig. 3.4 Relationship between cumulative annual snowfall received and maximum annual snow depth at the International Falls airport, 1948–2015

when the average low temperature exceeds 0°C (Fig. 3.3b). International Falls is dubbed "the icebox of the nation" because it is reputedly the coldest city in the conterminous U.S., with a mean low temperature in January of −21.4°C.

Precipitation falls as snow from November into April (Fig. 3.3a). Winter precipitation amounts are small, but the cold temperatures cause the snow to accumulate overwinter, so that the snowpack melts all at once in the spring. Using snowfall and snow depth measurements collected by the National Weather Service at the International Falls airport, I calculated that the maximum annual snow depth increased with the natural logarithm of annual snowfall: snowpack = 33.183 ln(snowfall)–111.55 (units = cm, $R^2 = 0.347$, $P < 0.001$). The relationship is not perfect because of factors that affect snow compaction and snow loss via wind erosion or sublimation as it lies on the ground, but it illustrates that the snowpack is cumulative.

The maximum annual snow depth at the International Falls airport ranged from 15 to 97 cm within the period of record, 1948–2016 (Fig. 3.4). I used regime shift detection (Rodionov 2006) to analyze long-term snowpack trends, identifying rapid changes from one relatively stable mean value to another. I found that, despite interannual variation, maximum snowpack depth centered on a mean of 57.1 cm for the first 48 years (1948–1997). This stable period was followed by a period of thinner snowpack (mean = 37.4 cm) during 1998 through 2008. The two years with the thinnest snowpack, 1999 and 2007, were strong "La Niña" years, but I could find no general relationship with the Oceanic Niño Index. In 2009 there was a shift to deeper than normal snowpack, averaging 69.5 cm (Fig. 3.5).

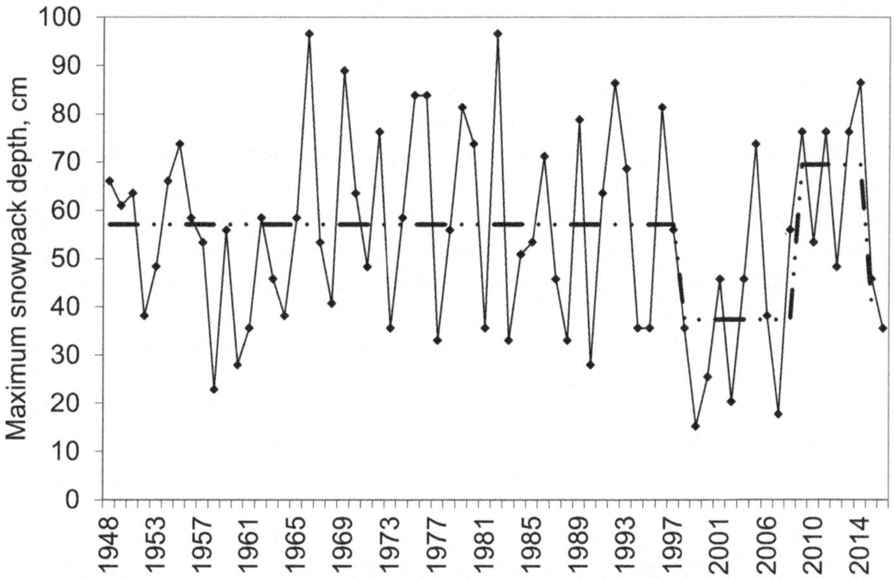

Fig. 3.5 Maximum annual snow depth, 1948–2016, as measured at the International Falls, MN airport. *Dashed line* is the weighted regime mean. Shifts in means were analyzed using the following parameters: significance level = 0.10, cut-off length = 6 years, Huber's weight parameter = 2, and red noise (serial correlation) estimation using the inverse proportionality with four corrections (IP4) method

We mapped snow depth at Found meadow, one of our field study sites, at the end of winter on March 19, 1997 (Fig. 3.6). Latitude and longitude coordinates were recorded using a fixed integer kinematic GPS with sub-decimeter repeatability at >95% probability, and snow depths were measured with a meter stick. The snowpack was at its maximum depth at this time, and measured 56 cm at the International Falls weather station (NOAA 2016a). Snow depth in the beaver meadow varied from 35 to 91 cm, and was greatest along the south edge of the meadow, which I attribute to shading by the adjacent tree cover due to the low winter sun angles at this latitude. Snow depth was least over the residual pond (dark area west of the outlet dam), because snow tends to become incorporated into the ice overlying water bodies; the water equivalent remains, but in the form of ice rather than snow. Other possible causes of snow depth variation include redistribution by wind and the presence of thatch beneath the snowpack that insulates and lifts it off the ground. The variability in snow distribution makes it difficult to accurately quantify snow water inputs from field measurements.

3.2.2 *Why Is Snowpack So Important?*

Snowfall has little water content, but snow accumulates overwinter due to the extreme cold (Fig. 3.3). The average snowpack at International Falls contains an accumulated snow water equivalent (SWE) that is typically 10–15 cm by the time the snow melts in the spring (NOAA 2016c). The soil is usually frozen during snowmelt, providing an impervious surface that promotes runoff. Thus, even though winter precipitation represents only about one-third of total annual precipitation, snowmelt drives the peak flows of rivers and streams in the region because water is released quickly and doesn't soak into the ground.

This spring runoff is very important to streamflow dynamics. Streamflow records that have been collected since 1909 by the U.S. Geological Survey for the Little Fork River, Minnesota (USGS station 05131500) (U.S. Geological Survey 2017), about 40 km west of Lake Kabetogama, illustrate that the highest flows usually occur in April (Fig. 3.7). Mean discharge rates exceeding 120 $m^3 s^{-1}$ occur on average between April 20 and April 30.

Discharge records for recent years also show that peak discharge nearly always occurred early in the spring (Fig. 3.8). In four of the years, peak flows occurred as soon as flow measurements commenced in the spring. The year 2010 was anomalous because its recorded peak flow was only 105 $m^3 s^{-1}$ and did not occur until July, but that is probably an artifact of monitoring: USGS did not begin flow measurements that year until April 1, but NOAA records showed that the snowpack was completely melted by March 17. This data point was excluded from subsequent analyses.

I used linear regression to compare these recent peak streamflow data for the Little Fork River (Fig. 3.8) with snowpack characteristics published by NOAA's National Operational Hydrologic Remote Sensing Center for the Kabetogama station for the winters of 2005–2016 (NOAA 2016c). Peak annual streamflow was significantly related to maximum snow water equivalent, the depth of water that would result if the snowpack instantaneously melted (Fig. 3.9). Peak annual streamflow was also significantly related to the date of snowmelt initiation: the later the start of snowmelt, the greater the peak streamflow. Thicker snowpacks were self-insulating, and thus tended to melt later in the spring than did thinner snowpacks. Annual snow accumulation clearly has an important influence on the flow of boreal streams.

3.3 Evapotranspiration

Just as moisture can drop out of the sky in the form of precipitation, it can also return to the sky via evapotranspiration (ET, evaporation from land and water and transpiration by plants). A nationwide analysis using climate and land cover data calculated mean annual actual evapotranspiration (AET) for northern Minnesota of

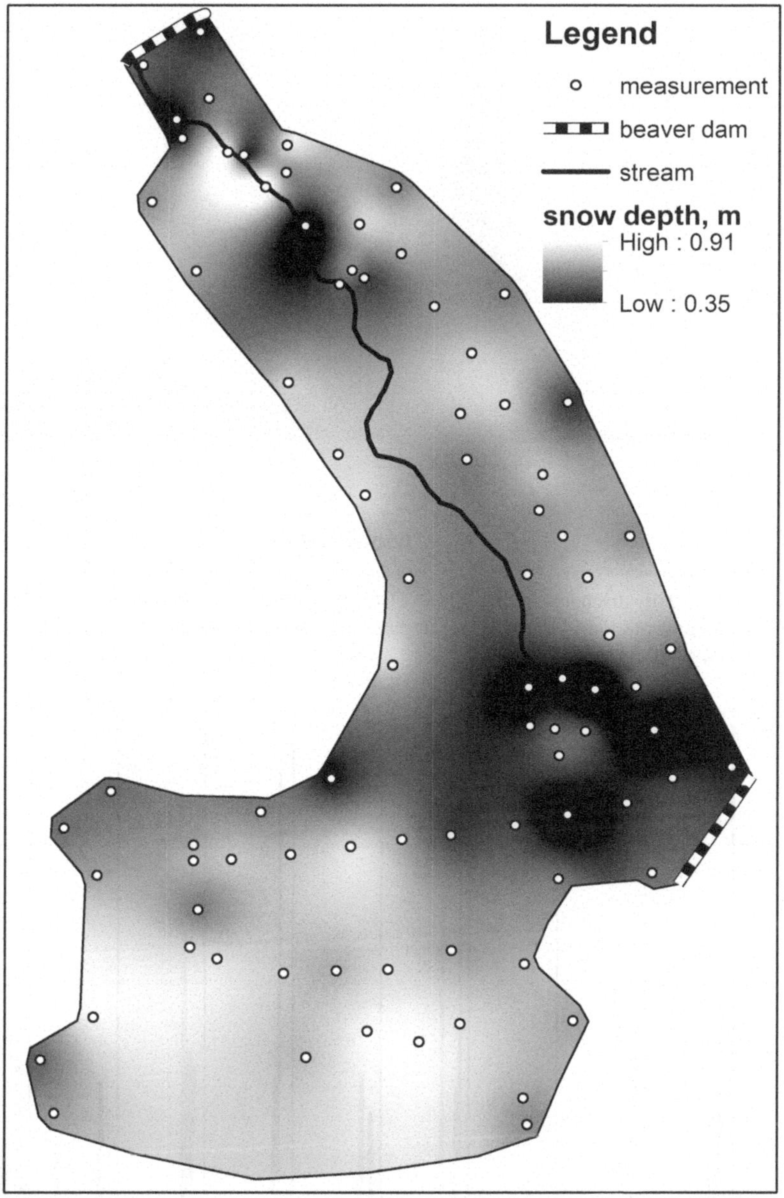

Fig. 3.6 Snow depths measured at Found Pond on 19 March 1997

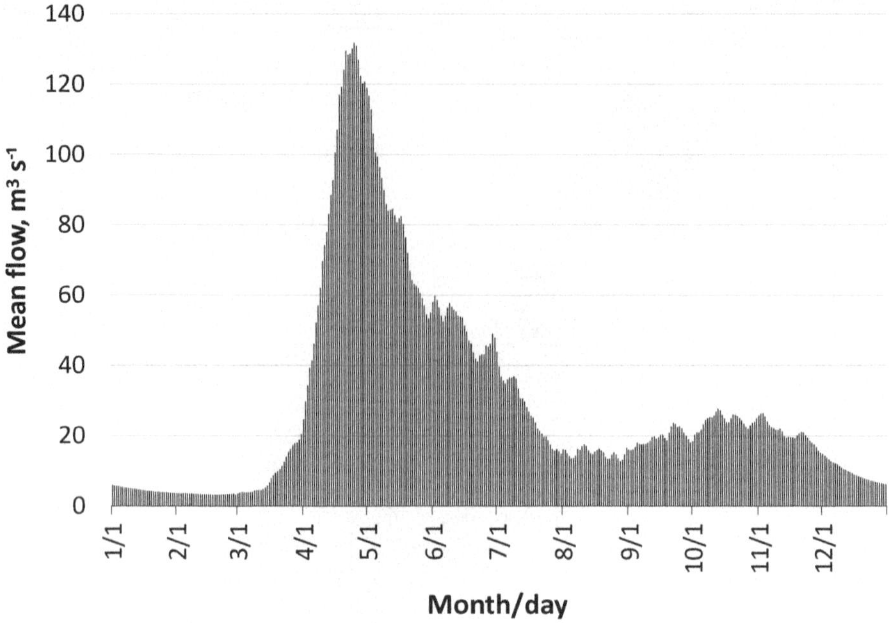

Fig. 3.7 Average daily discharge in m^3/s, 1909–2015, measured at the Little Fork River, Littlefork, MN (USGS gauging station 05131500). Mean discharge rates exceeding 120 m^3/s occur between April 20 and April 30

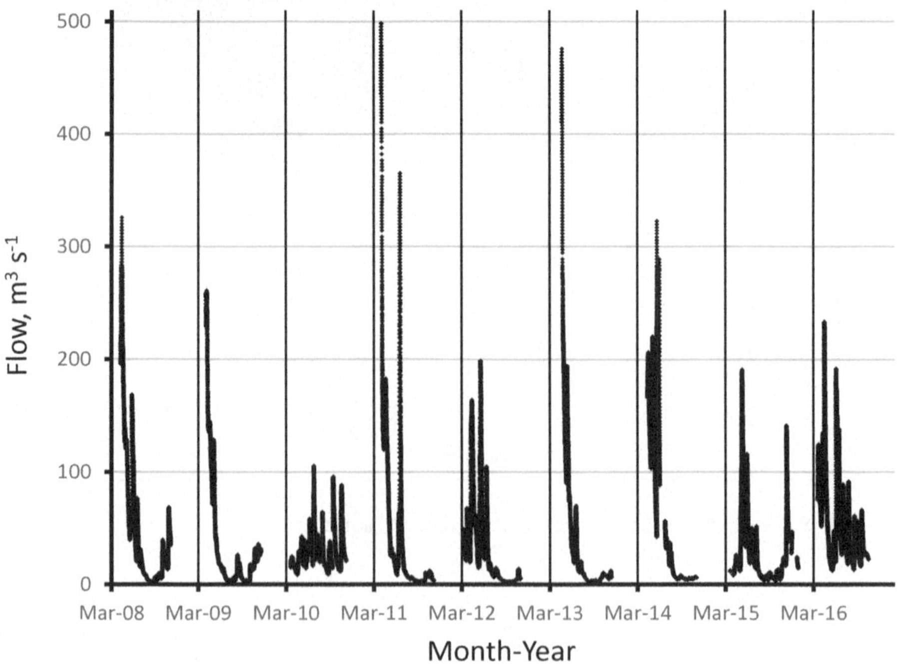

Fig. 3.8 Discharge (m^3 s^{-1}) measured every 30 min on the Little Fork River at Littlefork, Minnesota, March 2008 through Oct 2016

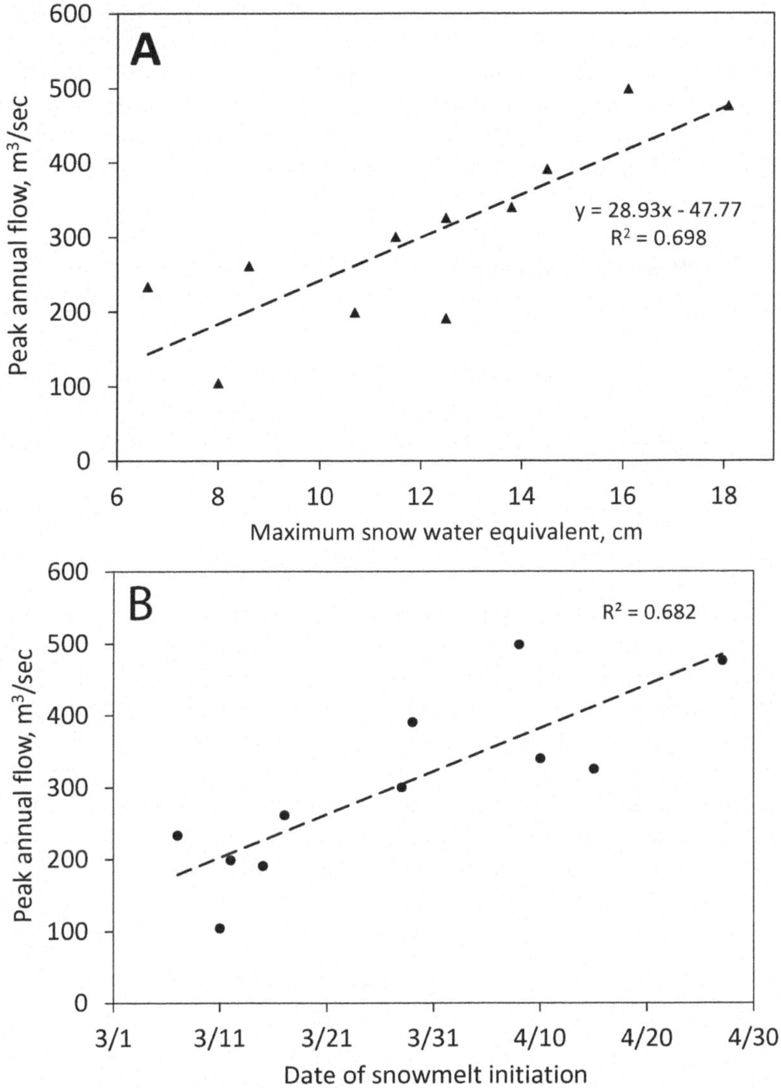

Fig. 3.9 Linear regressions between snowpack characteristics measured at the Kabetogama Visitor's Center and peak annual flow measured on the Little Fork River, 2005–2016. (**a**) Peak annual flow as a function of maximum snow water equivalent. (**b**) Peak annual flow as a function of snowmelt initiation date

about 41–50 cm, which constitutes about 60% of precipitation inputs (Sanford and Selnick 2013). Large lakes have higher ET rates than the overall landscape; calculated ET from large lakes near International Falls was 65–78 cm per year (Dadaser-Celik and Stefan 2008). Wetlands also have very high ET rates because transpiring plants are rooted in the water table, but their ET rates probably do not exceed those of open water surfaces (Abtew 2012). Nevertheless, ET is a very important process in emergent wetlands, and commonly accounts for the greatest water loss in many wetland systems (Goulden et al. 2007; Brown et al. 2010). Methods for measuring evapotranspiration have been reviewed by several papers (Drexler et al. 2004; Winter et al. 1995; Rosenberry et al. 2004; Kool et al. 2014).

3.3.1 Potential Evapotranspiration Versus Water Balance Unknown Term

We compared two different methods of quantifying ET: estimation of potential ET (PET) from temperature data (Eq. 3.2) versus ET as the unknown term in a water balance analysis (Eq. 3.1). The PET estimation was based on the Malmström method (Malmström 1969):

$$\mathrm{PET} = 0.409\left[e_{\mathrm{sat}}\left(T_{\mathrm{a}}\right)\right] \tag{3.2}$$

where $e_{\mathrm{sat}}(T_{\mathrm{a}})$ is the saturation vapor pressure at a given temperature. We calculated PET (Table 3.1) using the average weekly temperature, measured at a weather station (Campbell Scientific, Logan, UT) located within the Found meadow. The weeks with lower PET values (e.g., 1 Jun–7 Jun 1998) were those with cooler temperatures.

We compared the 1997 PET values with actual evapotranspiration (AET) calculated as the unknown term in a water balance (Cahow no date). The two values matched adequately during the first four weeks, but were very different in week five (Fig. 3.10). This difference is because PET assumes unlimited water availability, whereas AET was low by the end of August because soil water was depleted, there was very little rain, and the inlet stream had dried up.

Table 3.1 Potential evapotranspiration (PET) calculated by the Malmström method (Cahow and Mooers 1998)

Dates	PET (mm/day)
27 Jul–2 Aug 1997	3.53
3 Aug–9 Aug 1997	3.72
10 Aug–16 Aug 1997	2.61
17 Aug–23 Aug 1997	2.48
24 Aug–30 Aug 1997	3.02
25 May–31 May 1998	2.78
1 Jun–7 Jun 1998	1.81
8 Jun–14 Jun 1998	3.43
15 Jun–21 Jun 1998	3.81
22 Jun–28 Jun 1998	3.60
29 Jun–5 Jul 1998	3.85
6 Jul–12 Jul 1998	4.67
13 Jul–19 Jul 1998	4.04
20 Jul–26 Jul 1998	3.37

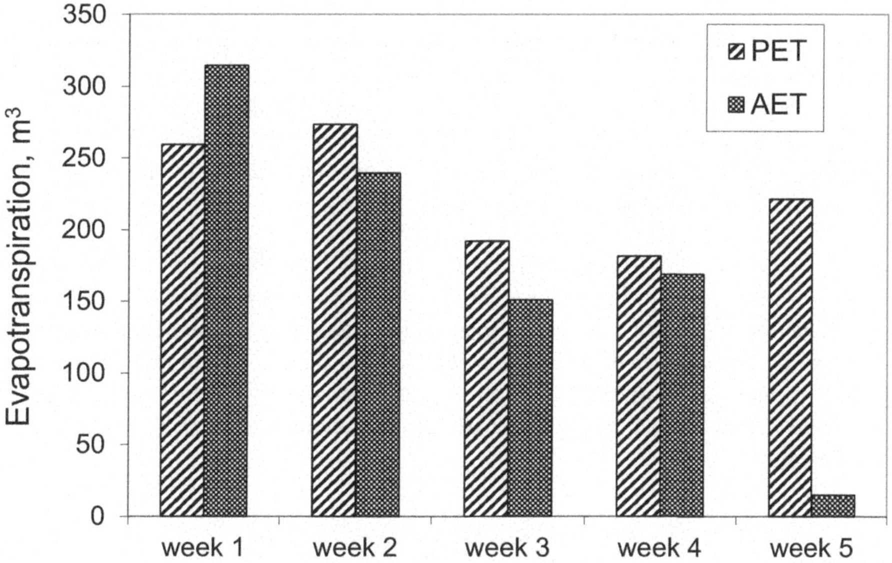

Fig. 3.10 Comparison of actual evapotranspiration (AET) calculated as the residual term in the water balance equation and potential evapotranspiration (PET) calculated by the Malmström method

3.3.2 Chamber ET

Box 3.3.2: Chamber ET Methods

The hemispheric chamber was commissioned by collaborator Howard Mooers (University of Minnesota-Duluth) after Stannard's design (Stannard 1988), and molded from clear acrylic by Cadillac Plastics (St. Paul, MN). The chamber had a 1-m diameter base, 50-cm height, 0.785 m^2 areal footprint, and 0.2618 m^3 volume. A temperature and relative humidity probe was suspended 10 cm from the top of the chamber, and wired to an external CR10X data-logger (Campbell Sci., Logan, UT). Two 12-volt fans, controlled by an external rheostat, were mounted on opposite sites of the chamber to mix the air within. A small utility box housed the rheostat, the data-logger, and a control keypad. Four stackable cylinders were used to elevate the chamber above higher vegetation to a maximum of 150 cm (Fig. 3.11).

To make an ET measurement, the chamber was positioned over a site type, and temperature and relative humidity were recorded every second for ~2 min. The apparatus was then moved, and at least two replicate measurements were made for each site type. Measurements were made throughout the day from about 10:30 am to 5:00 pm, and grouped into four time of day categories: morning, noon, early afternoon, late afternoon.

Data were downloaded to a computer, and vapor density (g m^{-3}) was calculated as the product of the relative humidity fraction times the saturation vapor density (SVD), the mass of water the air can hold at a given temperature. SVD was estimated from the empirical equation:

$$\mathrm{SVD} = 0.0214T^2 + 0.1944T + 5.0893 \tag{3.3}$$

The vapor density time series was plotted, and the instantaneous evapotranspiration rate was determined as the slope of the maximum rate of vapor density increase (Fig. 3.12). For woolgrass (*Scirpus cyperinus*) and cattail (*Typha latifolia*) site types at lower Bluefin, instantaneous ET (ET_{ins}) was calculated as:

$$\mathrm{ET_{ins}} = 3.6\frac{M \times V}{A} \tag{3.4}$$

where M = the steepest slope of the vapor density time series (g m^{-3} s^{-1}) and V = the volume of the hemispheric chamber and cylinder stack (m^3), A = the ground area covered by the chamber (0.785 m^2), and 3.6 is a conversion factor that converts g water m^{-2} s^{-1} to an hourly ET rate (mm/h).

Fig. 3.11 Evapotranspiration chamber and 30-cm cylinder in use at Found meadow

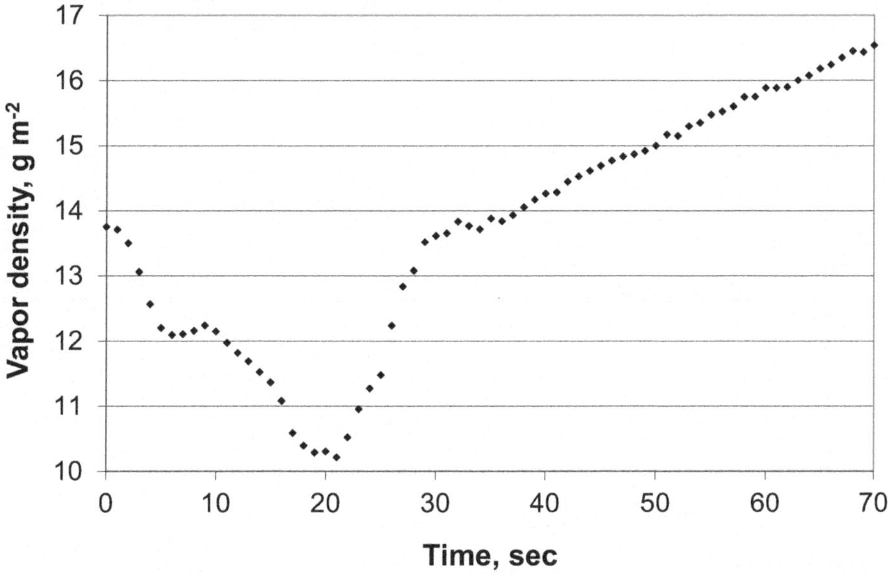

Fig. 3.12 Vapor density changes over time within a hemispheric chamber positioned over *Typha latifolia*. The initial drop (0–20 s) represents the aeration of the chamber to achieve ambient humidity, and the first rise (20–28 s) is the increase that is isolated and used for calculating ET. The increase in vapor density is more gradual after 29 s because plants react to the accumulating humidity within the chamber by reducing their transpiration rates

A chamber method (Box 3.3.2) was also used to measure ET from different combinations of vegetation cover and soil wetness, hereafter called site types, at the Bluefin and Found beaver meadows. The maximum vapor density flux rates (g m^{-3} s^{-1}) were compared for six different site types: dry, medium-dry, and wet bluejoint grass (*Calamagrostis canadensis*), goldenrod (*Solidago*), willow (*Salix*), and lake sedge (*Carex lacustris*) (Allison et al. 1999). There was a significant effect of site type (2-way ANOVA, $F = 30.86$, $P < 0.001$), but not time of day ($F = 0.270$, $P = 0.846$) or their interaction ($F = 0.698$, $P = 0.762$). The three site types with *Calamagrostis* vegetation had significantly different evapotranspiration rates (Fig. 3.13), controlled by soil wetness and leaf area. The "wet *Calamagrostis*" sites, located along the stream flowing into the residual Found Pond, had groundwater at or above the soil surface and high biomass (see Chap. 6) that promoted evapotranspiration. The "dry *Calamagrostis*" sites had much deeper water tables and sparser, shorter vegetation. Therefore, ET is not a function of plant species *per se*, but rather the combination of site wetness and biomass.

The *Carex* and *Salix* site types both had intermediate vapor density flux rates (Fig. 3.13), but for different reasons. The *Carex* sites were all located along the stream flowing through the upper Bluefin meadow, and had very high water tables.

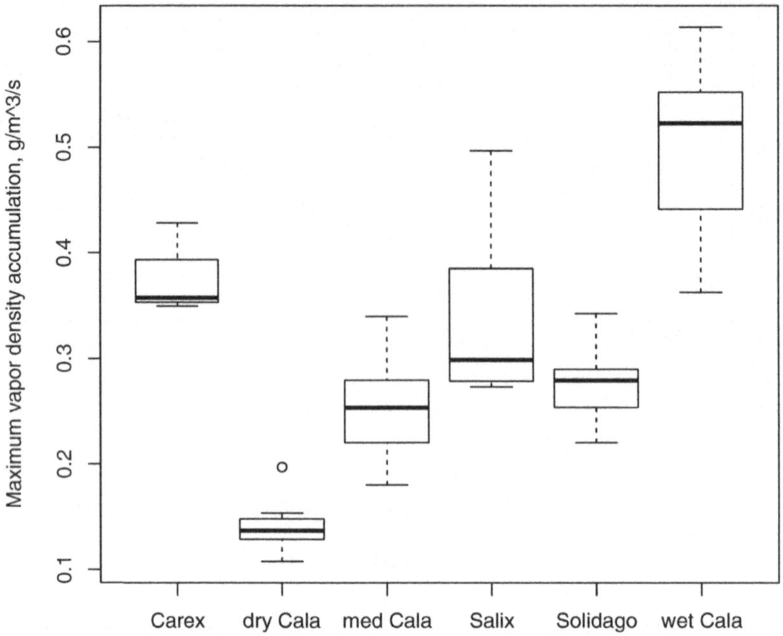

Fig. 3.13 Boxplots showing maximum vapor density accumulation rates (g m^{-3} s^{-1}) for chamber measurements made at different site types at Bluefin and Found meadows. Carex = lake sedge, dry Cala = dry *Calamagrostis*, med Cala = medium *Calamagrostis*, Salix = willow, Solidago = goldenrod, wet Cala = wet *Calamagrostis*. Each boxplot shows the median (heavy *horizontal line*), *upper* and *lower* quartiles (*top* and *bottom* of box), and minimum and maximum values (whiskers)

The *Salix* sites were on drier soils at Found meadow, but were presumably able to access deeper water reserves via their longer roots. The vapor density flux rate of *Solidago* overlapped with those of *Carex*, *Salix*, and medium *Calamagrostis*.

The mean hourly ET_{ins} rates calculated for *Typha* by the chamber method were very high, 1.67 mm/h (Fig. 3.14). The mean hourly ET_{ins} rates for *Scirpus*, which was located on higher, drier ground within the lower Bluefin meadow were 1.0 mm/h. Although these ET_{ins} rates seem high compared to rates computed by other methods, afternoon ET_{ins} rates as high as 20 mm/day were measured by Masoner and Stannard (Masoner and Stannard 2010) using this chamber method. It should be noted that these hourly rates cannot be simply multiplied by 24 to calculate a daily rate, because little evapotranspiration would occur during the evening and early morning hours due to low sunlight and cool temperatures. The beaver meadows' high relative humidity overnight caused heavy dew to condense and a fog layer to form over them in the morning (Fig. 3.15). Extrapolating chamber measured values to daily rates requires making ~8 chamber measurements per hour between 0700 and 2000 h and using a time-weighted approach to determine the evapotranspiration occurring between instantaneous chamber measurements (Masoner and Stannard 2010).

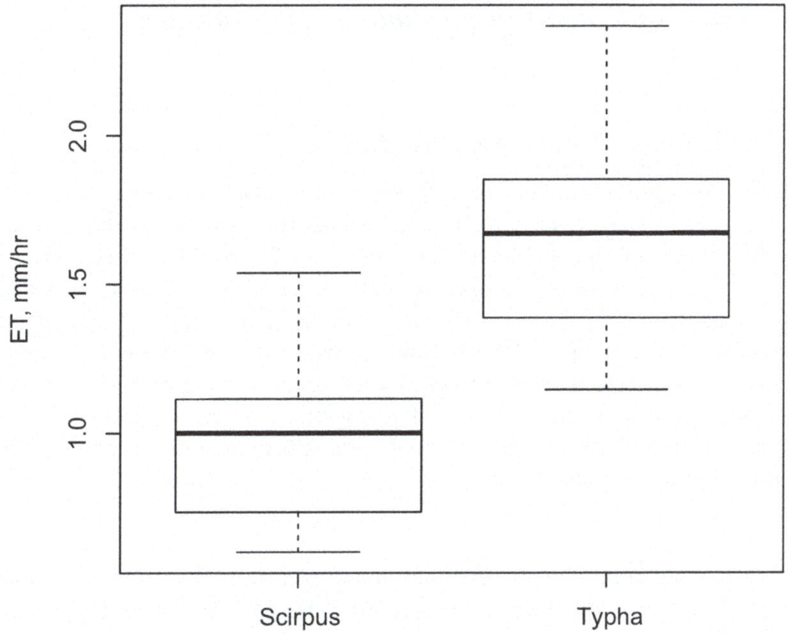

Fig. 3.14 Boxplots showing evapotranspiration (ET) values for *Scirpus cyperinus* and *Typha latifolia* at lower Bluefin. Each boxplot shows the median (heavy *horizontal line*), *upper* and *lower* quartiles (*top* and *bottom* of box), and minimum and maximum values (whiskers). *Typha* ET is significantly greater than that of *Scirpus* (ANOVA, $F = 26.27$, $P < 0.001$)

Fig. 3.15 A local fog bank blankets the beaver meadow in the morning

3.3.3 *Effect of Evapotranspiration on Groundwater*

Box 3.3.3: WL 40 Water Level Recorders

WL40 series programmable recorders (Remote Data Systems, Whiteville, NC) are self-contained units with a rigid capacitive sensor housed in a casing constructed of 7.6-cm diameter PVC pipe with 0.025 cm well screen slots (Fig. 3.16). Data were stored in a non-volatile RAM located in a waterproof data storage unit above the sensor, and downloaded via infrared connection to a handheld instrument (HP48G, Hewlett Packard, Palo Alto, California). Depth to water table was determined by reference to a calibration mark permanently marked on the exterior of the well, with an operating range of 102 cm. Well boreholes were excavated with a posthole digger and backfilled with sand after well placement.

Water levels were continuously recorded in a *Typha* stand during the first week of August, 1997 using a WL40 water level recorder (Box 3.3.3). The only rain received during this period was 2 mm on August 4. The data showed two temporal trends: (1)

Fig. 3.16 Downloading water level data from the WL40 recorder at Found Pond

diurnal fluctuations of the water table and (2) an overall water table decline of 8.9 cm over the 7-day period (Fig. 3.17). Each diurnal fluctuation consisted of a drawdown phase starting at about 1130 h, and a recovery phase starting at about 2200 h. Linear regressions were fit to each of these phases, and the rates of discharge and recharge were computed assuming a porosity value of 40%. The ET rates obtained from this analysis were 0.477–0.533 g m^{-2} s^{-1}, comparable to the chamber measurements for *Typha* of 0.465–0.550 g m^{-2} s^{-1} (Allison et al. 1999).

Studies of other wetlands have shown evapotranspiration-induced diurnal water table fluctuations, but those studies were in arid to semi-arid regions and supported phreatophytes, deep-rooted plants that draw moisture from groundwater (Butler et al. 2007; Moreo et al. 2007; Loheide et al. 2005; Stannard and Weltz 2006; Meybloom 1966). Loheide (2008) described a method to calculate sub-daily evapotranspiration from diurnal patterns of groundwater depression in wetlands and riparian zones.

3.3.4 Effect of Evapotranspiration on Stream Water Inflow

Found meadow is located on a small stream that delivers water from several upstream beaver ponds (Fig. 3.18). Stream water inputs were measured using a WL40 water level recorder coupled with a V-notched weir (Box 3.3.4). Stage data were recorded at the inlet weir for July 27–Aug 7, 1997, a dry period when only trace amounts of rain were received (0.25 mm on August 3 and 0.5 mm on August 5). These water levels exhibited a similar diurnal pattern (Fig. 3.19) as was observed in the *Typha* groundwater well (Fig. 3.17). This means that daytime evapotranspiration was sufficient to reduce streamflow. Streamflow dropped through the afternoon to a minimum daily flow at about 7:30 pm. Flow remained relatively constant overnight, then increased between 10:00 am and noon. Flow rate peaks decreased by 3.3 cm over time, indicating an overall decline in stream inputs (Fig. 3.19).

Box 3.3.4: V-Notch Weirs

Stream water inputs were measured with a wooden 90° V-notched weir positioned within the beaver meadow 30 m down-stream from the inlet dam, and stream water outputs were measured using the same setup 100 m downstream of the outlet dam (Aronson and Mooers 1998). WL40 water level recorders, programmed to record water levels every 30 min, were installed just upstream of each weir to record the stage of the stream, and also within the residual pond to measure its storage (Fig. 3.18). The standard 90° V-notched weir equation was used to convert stage into discharge:

$$Q = 1.379\,\mathrm{H}^{5/2} \qquad (3.5)$$

where Q = discharge (m^3 s^{-1}), and H = height of the water (m) above the V-notch in the weir (Fetter 1994).

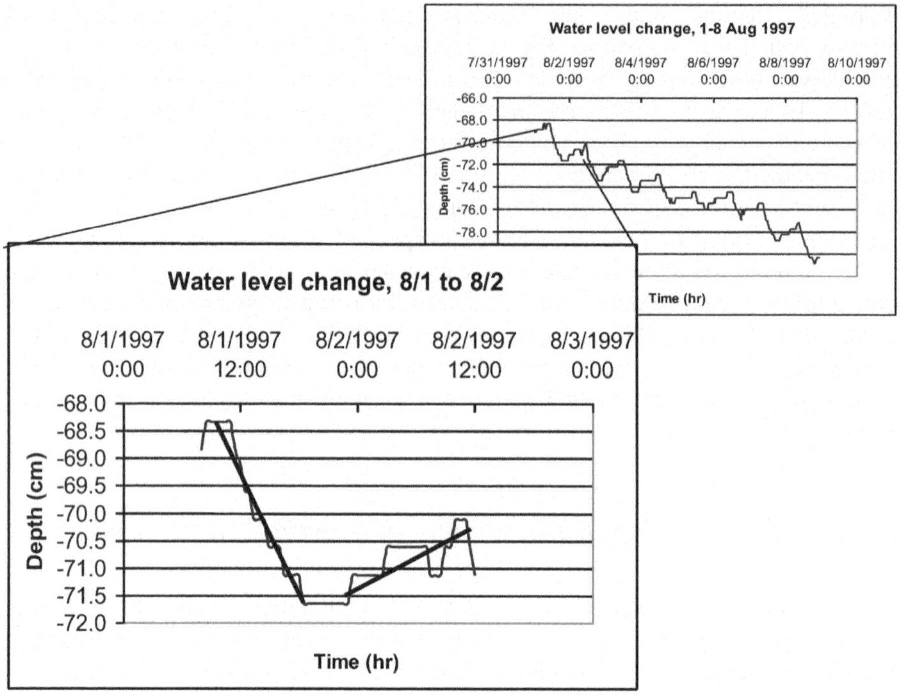

Fig. 3.17 Daytime drawdown and nighttime recharge of groundwater due to *Typha* evapotranspiration near Bluefin pond during August 1997

3.3.5 Effect of Evapotranspiration on Pond Levels

Water levels in the residual pond downstream of the inlet stream at Found meadow also exhibited diurnal fluctuations that temporally matched those of the inlet stream but were dampened in magnitude due to the greater area of the pond surface (Fig. 3.19). The amplitude of the diurnal fluctuations was usually a centimeter or less, but the overall trend was a decrease in pond level by 1.03 cm per day.

Pond levels recorded in Bluefin pond similarly exhibited diurnal fluctuations. These were particularly pronounced during late August 2000, when the pond level changed by as much as 4.5 cm daily (Fig. 3.20). Unlike Found pond, Bluefin is a headwater pond that receives no stream water inputs from upstream ponds, so this drawdown and recovery was due entirely to evapotranspiration and groundwater flow within the Bluefin meadow catchment.

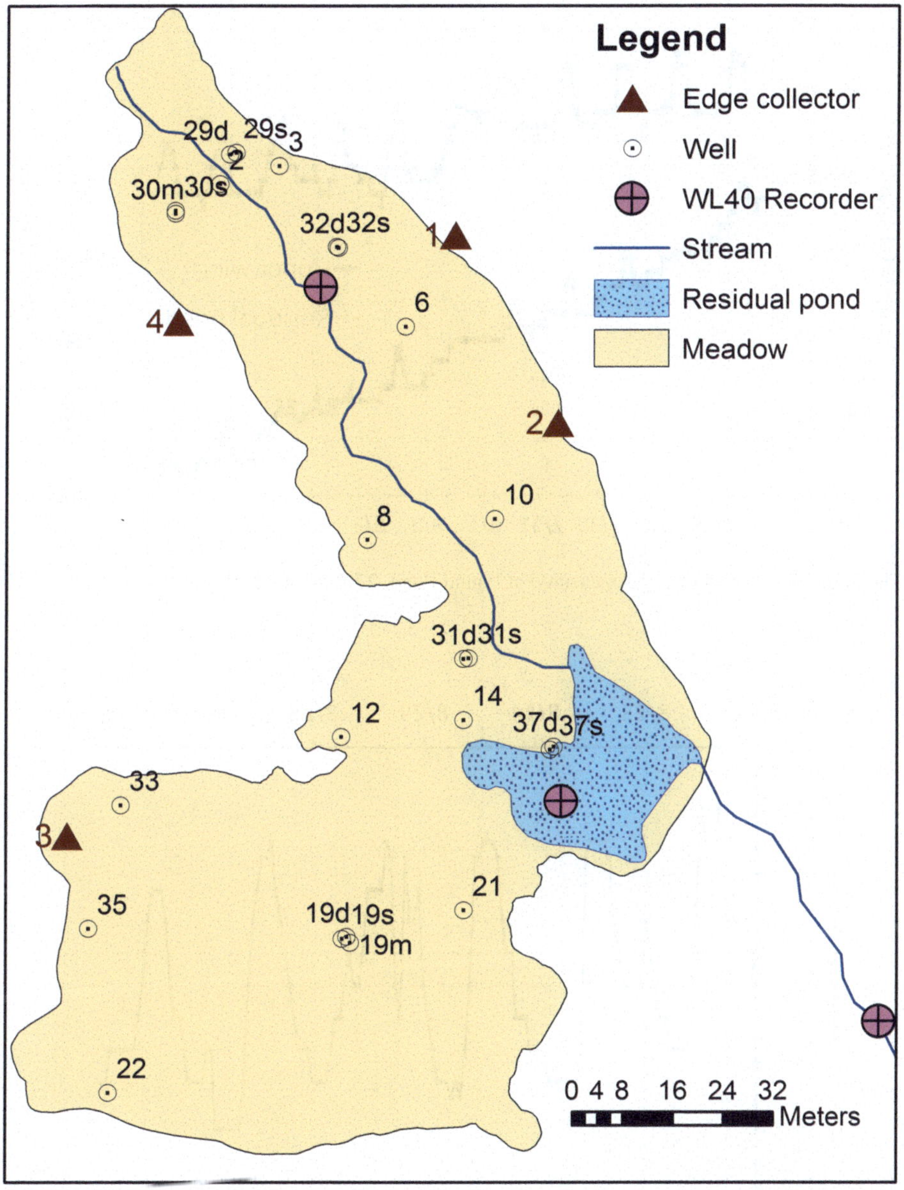

Fig. 3.18 Location of selected wells (see Table 3.3), surface runoff "edge collectors," and WL40 water level recorders in Found meadow. Well numbers followed by a letter indicate piezometer nests with shallow (s), medium (m), or deep (d) wells. WL40 recorders were associated with the residual pond and the inlet and outlet weirs

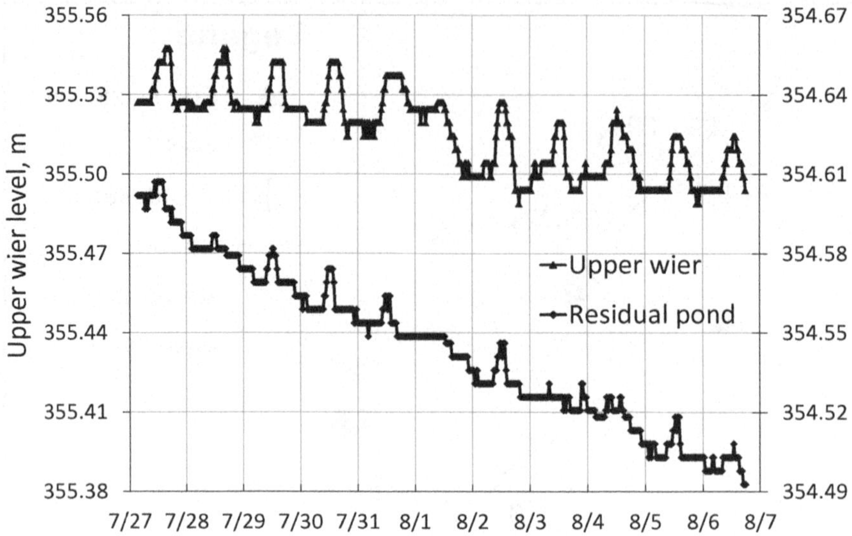

Fig. 3.19 Water level changes measured at Found Pond, 27 Jul–7 Aug 1997

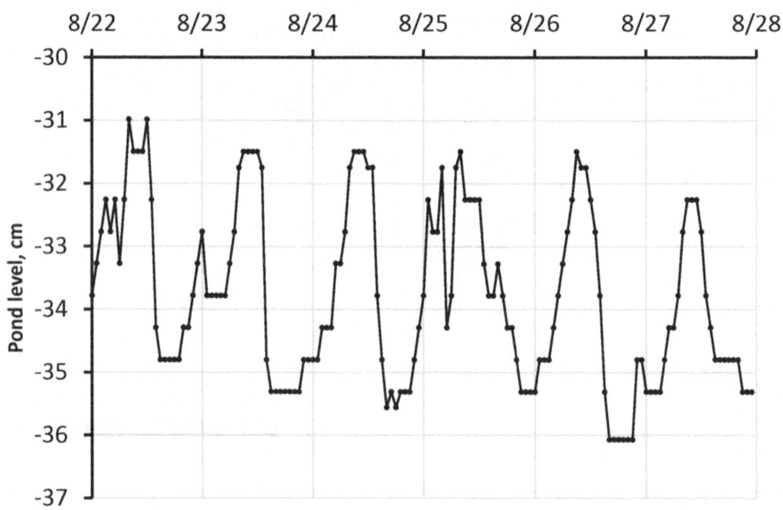

Fig. 3.20 Extreme diurnal Bluefin pond level fluctuations due to evapotranspiration, 22–28 August 2000

3.4 Water Storage in Residual Ponds

3.4.1 Estimating Pond Water Storage

Karran et al. (2017) examined water storage in beaver ponds from 40 different sites in North America and South America, including six sites on the Kabetogama Peninsula with areas ranging from 1504 to 12,985 m^2 and volumes ranging from 264 to 4740 m^3. This analysis showed that the following equation worked best to estimate beaver pond surface-water storage:

$$V_{max} = \frac{A_{max} \times h_{max}}{1 + 2/p} \tag{3.6}$$

where V_{max} is the pond's maximum volume, A_{max} is the pond's maximum surface area, h_{max} is the dam height, and p is a dimensionless morphometry coefficient that represents the shape of the bathymetric curve, calculated to be 0.91 for beaver ponds. Where dam height is unavailable, pond volume can be estimated from surface area as:

$$V_{max} = 0.0955 A_{max}^{\;1.1562} \tag{3.7}$$

3.4.2 Seasonal Water Level Change in Residual Ponds

The water levels of the residual ponds at the Bluefin and Found meadows changed in response to snowmelt and precipitation events during the ice-free season (Figs. 3.21 and 3.22). Year-round monitoring at Bluefin pond showed that water levels were low but remained relatively constant at −48 cm and −53 cm during the winters of 1998–1999 and 1999–2000, probably constrained by ice covering the pond (Fig. 3.21). Little water flow would be expected in the winter because precipitation is solidified as snow and ice (Fig. 3.1). However, the lowest water levels occurred in the late summers of 1998 and 1999. In 1998, water levels in Bluefin Pond dropped more or less continuously from a peak value (set at 0) on May 19 to −57 cm on Aug 10. Although there were several rain storms in June 1998 they had little effect on the downward trajectory of the pond level. The Historical Palmer Hydrological Drought Indices (PHDI) (NOAA 2016b) for April through July of 1998 indicated moderate drought, and for August and September indicated severe drought, consistent with the rapid drop and persistent low levels of the pond that year. With the exception of a brief increase after rains at the end of August, the pond was at its lowest level (−57 cm) for more than two months in 1998 (August 10–October 14).

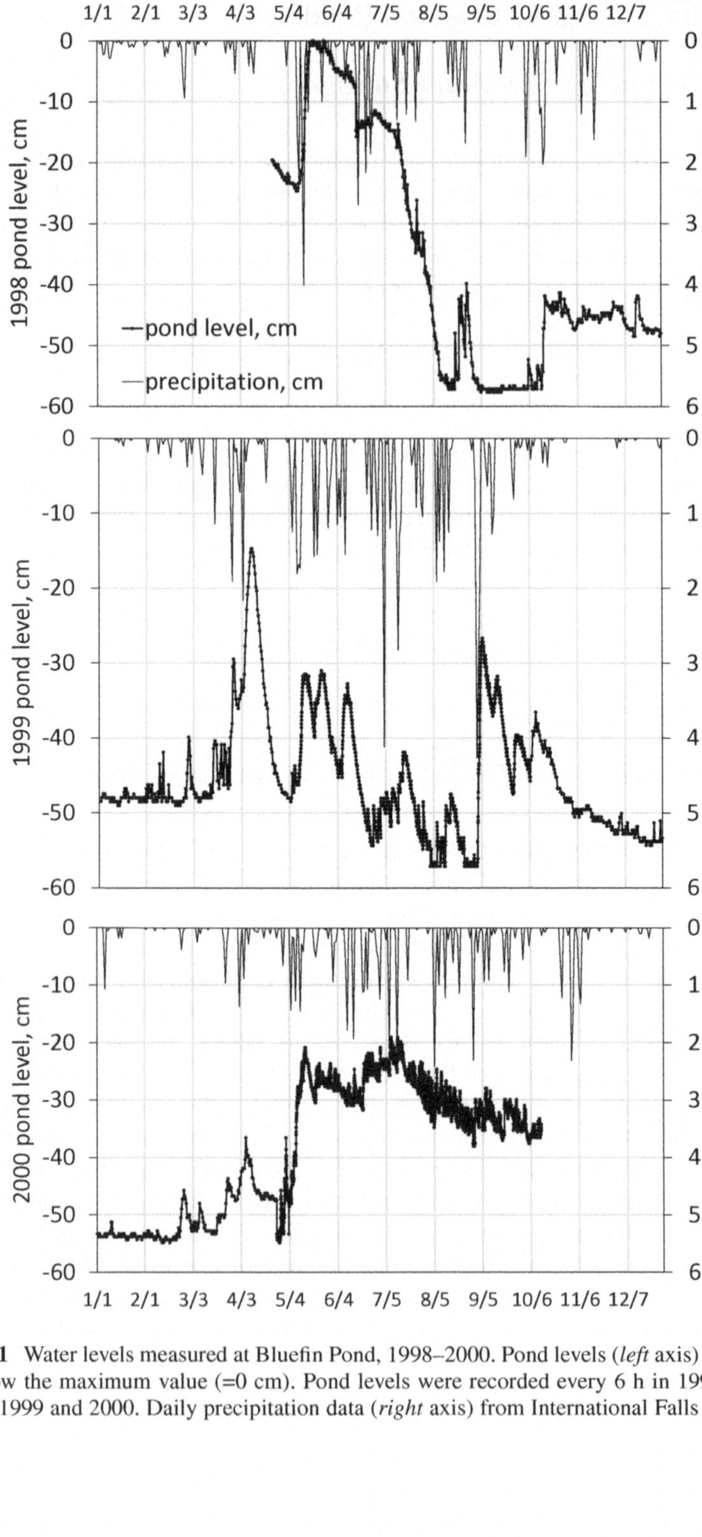

Fig. 3.21 Water levels measured at Bluefin Pond, 1998–2000. Pond levels (*left* axis) are shown in cm below the maximum value (=0 cm). Pond levels were recorded every 6 h in 1998 and every hour in 1999 and 2000. Daily precipitation data (*right* axis) from International Falls MN weather station

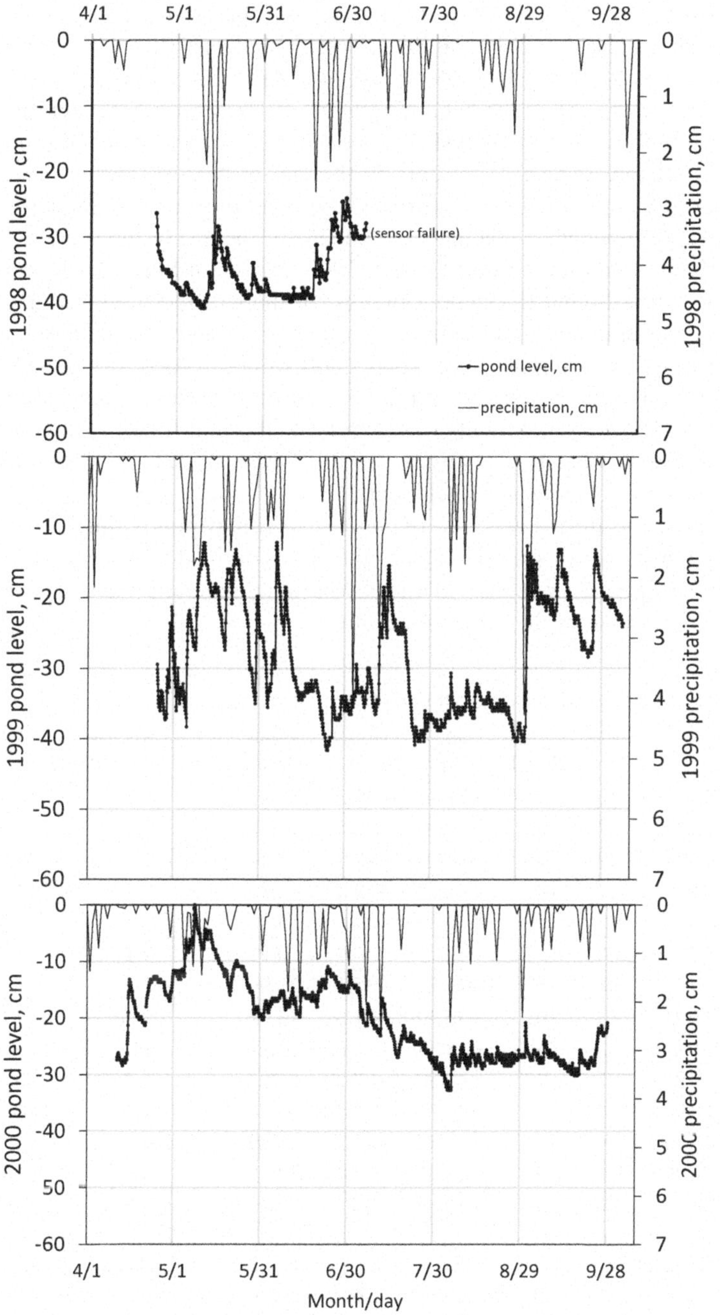

Fig. 3.22 Water levels measured at Found Pond, 1998–2000. Data recorded every 6 h in 1998, every hour in 1999 and 2000. Daily precipitation data from International Falls MN weather station

The winter of 1998–1999 had little snow; the accumulated snowpack at International Falls (15.2 cm) was the lowest on record (Fig. 3.5). The level of Bluefin Pond peaked early in 1999 (−14.7 on 8 April), dropping back to its winter water level (−48.5 cm) by the beginning of May. There were frequent rains during the summer and fall of 1999 that shifted the pond level back up several times, but it dropped to its lowest level (−57 cm) by the end of August. Heavy rains in early September (9.8 cm September 1–4) raised the pond to its second highest peak that year, a level from which it gradually declined over the rest of the year.

The Bluefin Pond level peaked late in 2000 (−20.8 on May 14), but maintained a level of −36 or higher throughout the rest of the ice-free season. Rains in 2000 were light but frequent. The extreme diurnal pond level fluctuations shown in Fig. 3.20 occurred during that year. Pond levels were recorded for Found Pond only during April through September, and records during 1998 were truncated by equipment failure. The temporal trends were similar as those in Bluefin Pond, but August 1999 drawdowns were less severe at Found (which has an inlet stream) than at Bluefin (which is a headwater catchment).

I analyzed the drawdown rate of the two residual ponds during dry periods following rainstorms to see how rapidly the ponds drained (Table 3.2). Drawdowns ranged from 0.6 to 3.1 cm day^{-1}. Drawdowns were generally faster at Found Pond than Bluefin Pond after the same rain event.

Table 3.2 Pond level decline rates after rain events

Rain dates	Rain (cm)	Pond	Peak (cm)	Trough (cm)	Water level decline (cm)	Recovery time (days)	Decline (cm day^{-1})
5/10–5/17/1998	10.7	FP	−28.40	−39.37	10.97	10.0	1.10
8/26–8/27/1998	1.7	BF	−39.90	−57.20	17.30	7.3	2.39
3/27–4/7/1999	6.9	BF	−14.70	−48.50	33.80	25.5	1.33
5/5–5/12/1999	8.2	BF	−31.50	−39.88	8.38	6.3	1.32
5/5–5/12/1999	8.2	FP	−12.20	−27.40	15.20	7.6	1.99
5/20–5/23/1999	3.8	BF	−30.99	−41.91	10.92	6.4	1.70
5/20–5/23/1999	3.8	FP	−13.20	−35.10	21.90	7.1	3.07
6/4–6/9/1999	4.3	BF	−32.80	−51.30	18.50	11.8	1.57
6/4–6/9/1999	4.3	FP	−18.54	−41.66	23.11	14.7	1.58
8/15–8/17/1999	1.5	BF	−48.00	−57.20	9.20	7.7	1.19
8/15–8/17/1999	1.5	FP	−32.26	−40.39	8.13	13.5	0.60
9/11–9/14/1999	2.4	BF	−31.75	−47.24	15.49	9.6	1.62
9/11–9/14/1999	2.4	FP	−13.20	−27.40	14.20	7.5	1.88
4/2–4/6/2000	2.6	BF	−36.60	−46.20	9.60	7.0	1.37

Rain totals based on data collected at the International Falls, MN weather station, and are cumulative over the days listed. "Peak" is the highest pond level recorded in response to the rain event, and "trough" is lowest pond level achieved before the next rain event; readings relative to maximum pond level (Figs. 3.21 and 3.22). *BF* Bluefin Pond, *FP* Found Pond

3.5 Groundwater Fluxes

Groundwater was measured in 42 shallow wells (30–100 cm below ground surface) and seven deeper wells (150–300 cm below ground surface) installed throughout Found meadow (Box 3.5) (Cahow no date). Hydraulic conductivity values calculated from the shallow wells were variable, averaging 1.36×10^{-3} cm/s. With the exception of well FP19D, which was screened in a sand-rich material, hydraulic conductivity of the medium and deep wells was about 3 orders of magnitude slower, averaging 1.39×10^{-6} cm/s (Table 3.3). Hydraulic conductivity was also calculated from grain size analysis of borehole soil samples at the Univ. of Minnesota Limnological Research Laboratory (http://lrc.geo.umn.edu) using several different methods (Hazen 1893; Puckett et al. 1984; Krumbein and Monk 1943; Harleman et al. 1963). With the exception of the Krumbein and Monk method (Krumbein and Monk 1943), these hydraulic conductivity values were much slower than those determined using manual pumping (Table 3.3).

Table 3.3 Hydraulic conductivity values determined by pump tests and particle size analyses

Well	Depth class	Hvorslev (1951)	Bouwer and Rice (1976)	Hazen (1893)	Krumbein and Monk (1943)	Puckett et al. (1984)	Harleman et al. (1963)
FP02	S	8.74E-03	3.87E-03	1.64E-06	1.35E-02	2.48E-04	1.75E-04
FP03	S	1.10E-03	3.35E-03	1.68E-06	4.40E-03	2.73E-04	1.79E-04
FP06	S	–	–	1.15E-06	3.19E-04	1.17E-04	1.23E-04
FP08	S	2.12E-03	2.00E-03	–	–	–	–
FP10	S	2.92E-06	7.26E-06	–	–	–	–
FP12	S	3.34E-04	2.32E-04	–	–	–	–
FP14	S	3.00E-04	2.25E-04	–	–	–	–
FP21	S	9.37E-04	5.30E-04	–	–	–	–
FP22	S	1.59E-05	3.55E-05	–	–	–	–
FP30S	S	1.56E-03	1.51E-03	–	–	–	–
FP33	S	1.59E-05	3.04E-05	–	–	–	–
FP35	S	1.06E-03	2.19E-03	–	–	–	–
FP19S	S	8.11E-05	9.40E-05	–	–	–	–
FP19D	D	4.13E-04	1.34E-04	–	–	–	–
FP29D	D	–	1.09E-06	–	–	–	–
FP30M	M	–	1.47E-06	–	–	–	–
FP31D	D	–	1.45E-06	–	–	–	–
FP32D	D	–	1.55E-06	–	–	–	–

S shallow, *M* medium, *D* deep

Box 3.5: Well Measurements

The shallow wells were 3.2 cm diameter PVC pipes that were drilled with holes and pushed into a hand-augered 3.2 cm hole to the clay-rich Bt or C layer (see Chap. 5). The deeper wells were 5.1 cm diameter PVC pipes screened with slots every 3 mm along the lower 76 cm of the pipe. The deep well pipes were placed in a 6.4 cm diameter hand-augered hole that was backfilled with sand the full length of the screen and then sealed with bentonite above the screen to prevent the entry of surface water.

Manual pumping tests were performed on several groundwater wells installed in Found meadow (Cahow no date). For this technique each well was pumped dry and water level measurements were taken at specific time intervals until the well had recovered. Recovery of shallow wells (20–70 cm below ground surface) usually occurred within 1.5 h, but a few took as long 5 h to recover. Recovery took up to a day for deep wells in piezometer nests (150–300 cm deep). For these wells, water levels were taken, the wells were pumped dry, and the water level was measured again the next day to get the recovered water level, with an estimated 24 h recovery period. Hydraulic conductivity values were calculated from the pumping test data using the Hvorslev (Hvorslev 1951) and Bouwer and Rice (Bouwer and Rice 1976) methods.

Vertical groundwater gradients were measured in 1998 at six piezometer well nests at Found meadow (Figs. 3.23 and 3.24) (Cahow no date). The magnitude and direction of vertical gradients differed in different areas of the meadow and pond complex, depending on their proximity to the stream or pond. Well nests FP19, FP30, and FP32 were at ground elevations of 355.517, 355.938, and 355.686, respectively, 0.9–1.3 m above the level of the residual pond. Vertical hydrologic gradients at these piezometers were consistently positive, indicating that surface water was recharging the groundwater system (Fig. 3.23a). The gradients were especially large at FP19 before its surface well dried up in August 1998, a period of severe drought (Fig. 3.22). Two well nests nearest the stream flowing through Found meadow were sometimes groundwater recharge sites (positive gradients) and sometimes groundwater discharge sites (negative values, groundwater moving up toward the surface) (Fig. 3.23b). This fluctuation was probably due to interactions with the stream flowing out from the pond upstream of Found meadow. Well FP37 was at the edge of the residual pond; it was a groundwater discharge site in spring and late fall 1998 (negative values), but showed virtually no vertical gradients during the summer (Fig. 3.24).

Equipotential maps constructed using water table elevations from the shallow wells showed that groundwater was flowing into the residual pond from two sides of Found meadow: the northern side along the stream descending from the upper dam and the western side centered on well nest 19 (Fig. 3.18). The water table

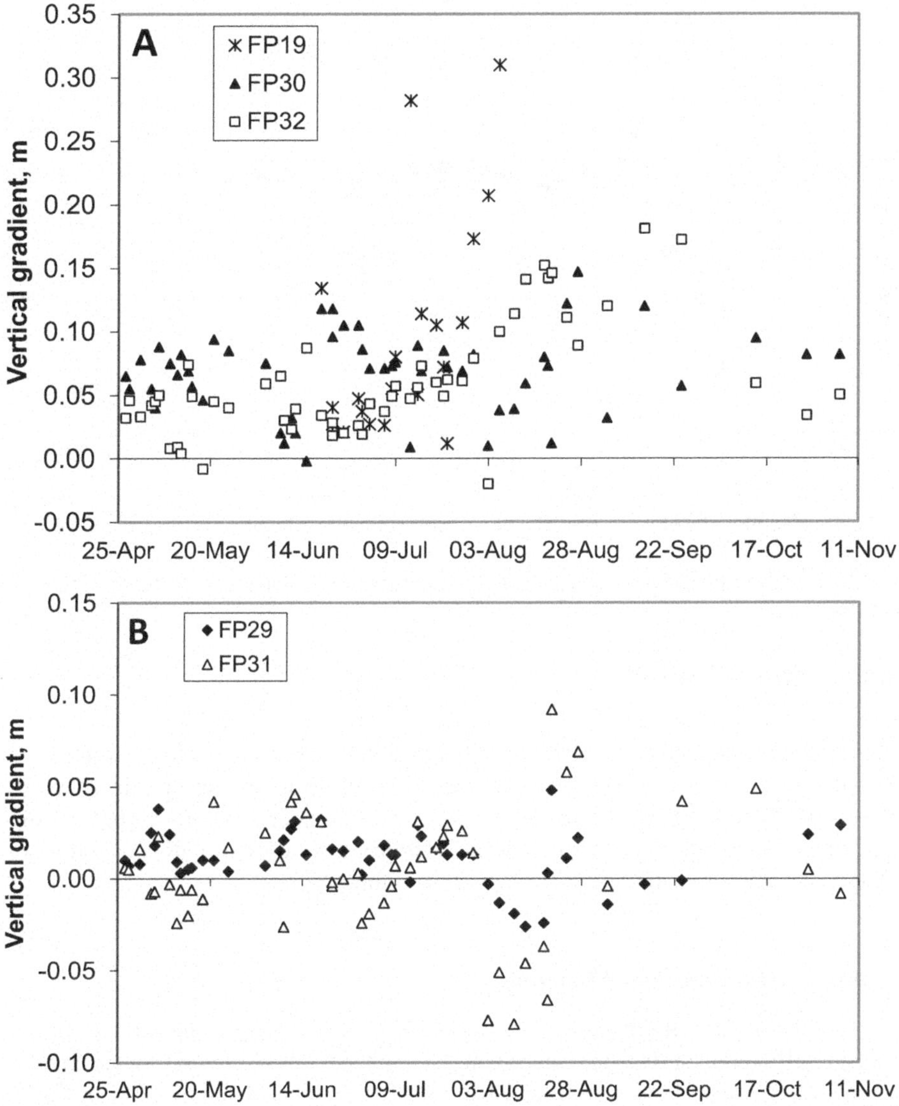

Fig. 3.23 Vertical groundwater gradients (meters) measured at meadow well nests during 1998. (**a**) Wells FP19, FP30, and FP32. (**b**) Wells FP29 and FP31

gradient was steepest in the western side, dropping from an average elevation of 356 m at the western boundary of the meadow to ~354.8 at the residual pond edge, a 1.7% slope. This side also showed a steepening of the water table gradient after significant (>5 mm) precipitation events. The water table slope from the north side was about 0.8%, and was less responsive to precipitation events.

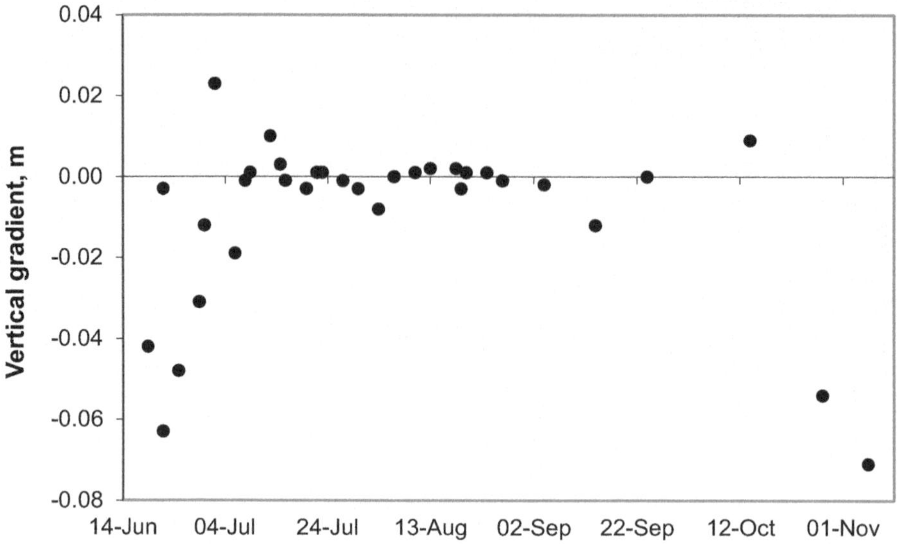

Fig. 3.24 Vertical groundwater gradients (meters) measured at residual pond well nest FP37 during 1998

3.6 Lateral Runoff

Beaver ponds and meadows are surrounded by upland slopes that deliver water from lateral runoff (Fig. 3.1). Because bedrock is so close to the surface in much of Voyageurs National Park (Chap. 5), we expected that lateral runoff might be a significant water source to beaver meadows. Therefore, we used monitoring and modeling approaches to estimate water inputs from this source.

Box 3.6: Lateral Flow Modeling

Lateral flow into the meadow was modeled by routing flow among 2 × 2 m grid cells representing the catchment. Cover type was estimated for all grid cells, and water flow was traced from each cell through other cells in the upland until it reached the meadow boundary. The proportion of water *p* retained by each cell was determined as:

$$p = \left(1 - s/90\right) \sum_{i=1}^{n} k_i P_i \tag{3.8}$$

where *s* is the slope in degrees, k_i (Table 3.4) is the proportion of available flow retained by a cell with 100% coverage of vegetation type *i*, and P_i is the proportion of vegetation type *i* in the cell. Equation (3.8) gives a simple linear decrease from maximum retention with no slope to zero retention for a vertical slope. No water was retained once the cell's storage limit was reached.

Monitoring of lateral runoff inputs was done using "edge collectors," which are basically weirs designed to focus lateral runoff into a collection bucket (detailed edge collector construction described in (Brown et al. 2003)) (Fig. 3.25). Edge collectors were installed at four locations around Found meadow (Fig. 3.18). Edge collector 3 had a gently sloping catchment with ≥0.3 m of soil, whereas edge collectors 1, 2, and 4 had somewhat steeper catchments with less soil and more exposed bedrock. There was measurable flow at all four edge collectors in response to precipitation events of 5 mm or more; values were typically 5–20 liters/day, with peak flows in excess of 100 liters/day (Fig. 3.26). However, the temporal pattern of flow was different for edge collector 3 than the other sites. In July, site 3 exhibited major flow after 1 and 2 are dry; in early October the reverse was true. The catchment draining to site 3 appears to have had greater groundwater storage that prolonged flow after the wet period between June 15 and early July but delayed it after the brief rain events in September and October.

Table 3.4 Cover type parameters used in lateral runoff model

Cover type	Capture proportion	Storage (mm)
Litter, bare ground	0.40	5
Bare rock	0.00	0
Moss	0.40	4
Trees	0.05	2
Forbs	0.30	5

Fig. 3.25 Researcher Kirsten Cahow measures water volume in "edge collector" bucket

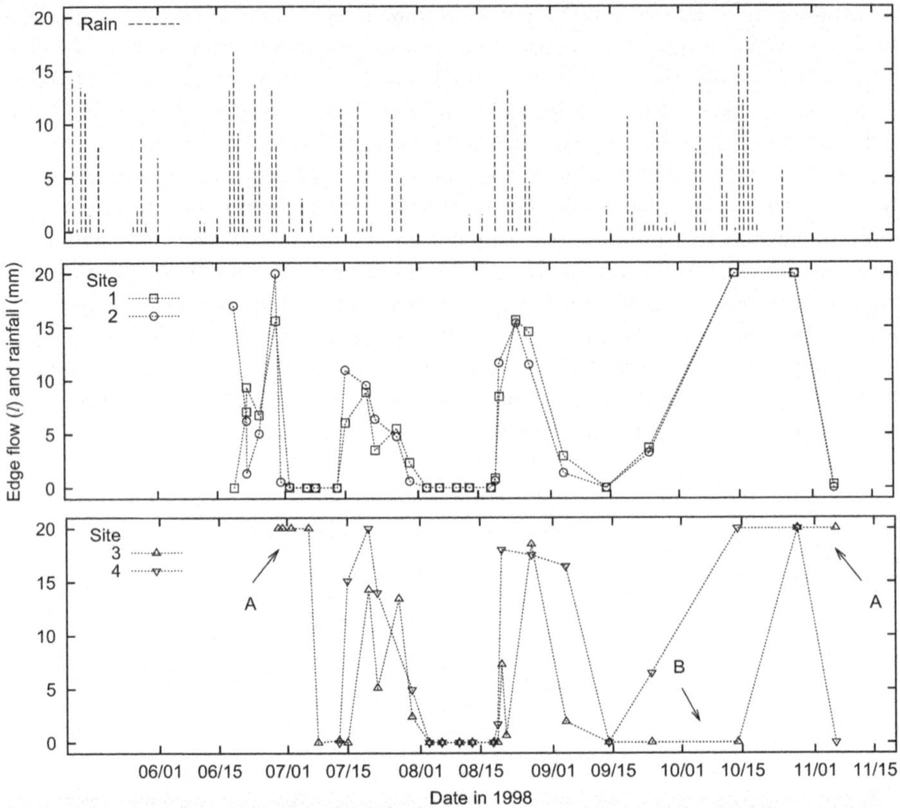

Fig. 3.26 Edge collector readings at four sites in "Found" meadow in 1998. Readings of 20 L indicate bucket overflow. Note A periods where output from site 3 is sustained while other sites are dry, and B period where output from site 3 is delayed. Reprinted from Brown et al. 2003, with permission from Elsevier

The DEMON flow routing algorithm (Costa-Cabral and Burges 1994) provided the most realistic water movement results. Results from this flow simulation were averaged among the grid cells that were within 10 m of each edge collector to estimate the volume of lateral runoff per mm of rainfall (Fig. 3.27). Simulated runoff amounts from all edge collector sites increased quickly as rainfall increased to 4 mm, with amounts received at site 3 being slightly less than those received at the other three sites. At about 5 mm of rainfall, however, the volume of water received at site 3 jumped more dramatically and changed its relative position in the plot, moving site 3 from lowest to highest output, consistent with edge collector measurements. The model successfully represented the change from storage-dominated flow control to catchment-topography-dominated flow control (Brown et al. 2003).

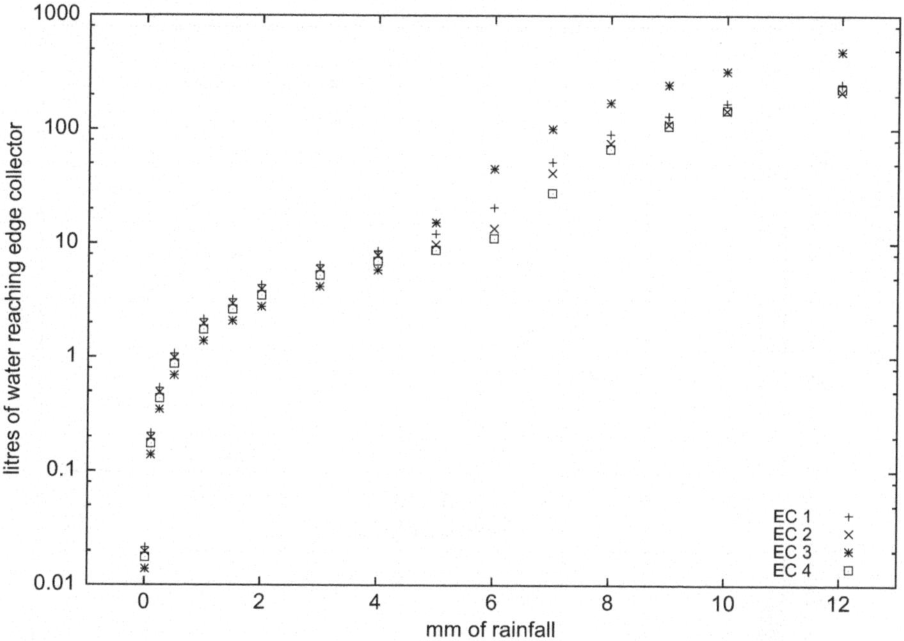

Fig. 3.27 Modeled upland to meadow flow at edge collector positions. Note that edge collector 3 moves from lowest to highest flow at about 5 mm of rain. Reprinted from Brown et al. 2003, with permission from Elsevier

3.7 Stream Inflow and Outflow

Data collected at the inflow and outflow stream weirs at Found meadow (Fig. 3.18, Box 3.3.4) allowed us to calculate the volume of surface water entering and leaving the system for five weeks in 1997 and eight weeks in 1998 (Figs. 3.28 and 3.29). There was little precipitation in the late summer of 1997, and by September the region was experiencing moderate drought (NOAA 2016b). The stream had completely dried up at the outflow weir by the week of 3 August 1997, and at the inflow weir by the week of 24 August 1997 (Fig. 3.28).

The results were quite different in 1998, because stream outputs greatly (and unrealistically) exceeded stream inputs (Fig. 3.29). During the week of June 1998, stream outflow from the residual beaver pond exceeded measured stream inflow by 524 m^3, an amount that cannot be accounted for from other possible sources (e.g., direct precipitation on the pond surface, groundwater fluxes into the pond). The

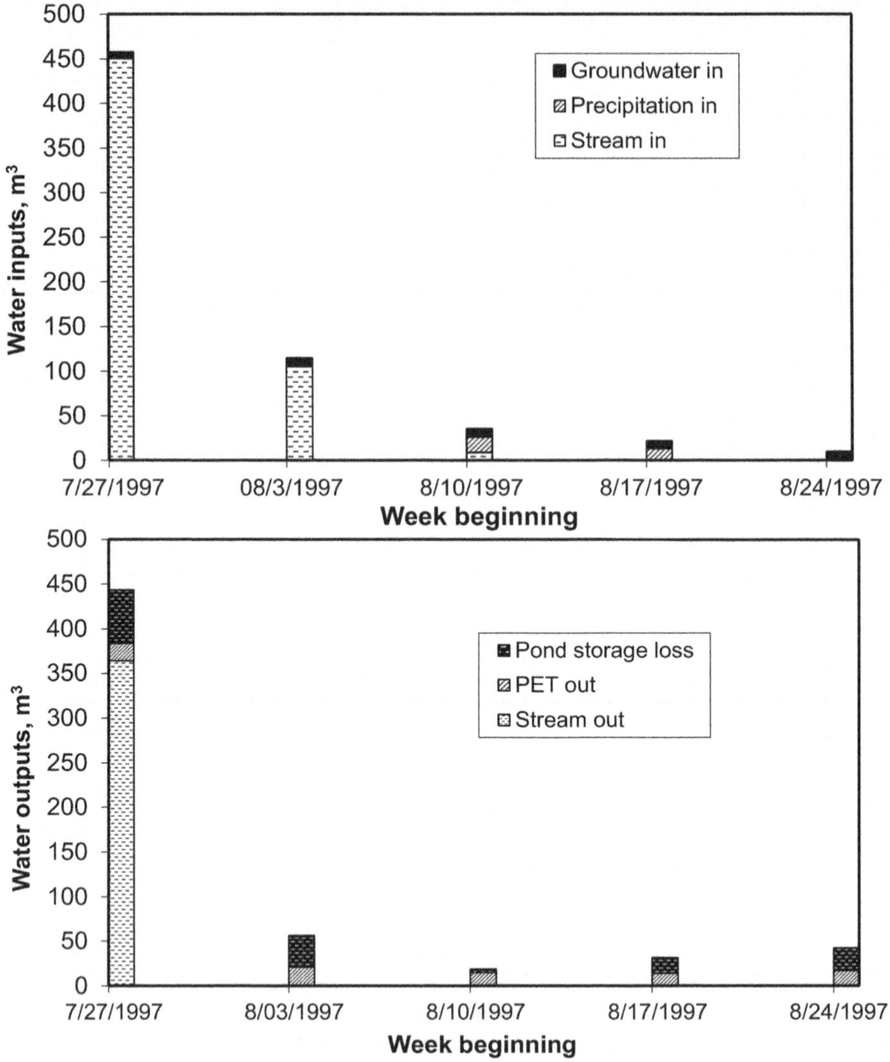

Fig. 3.28 Water balance analysis for Found pond and meadow complex, 1997

pond *gained* storage during that week and the preceding week, so it was not a source of the extra outflow. Possible explanations for the discrepancy are: (1) channelized stream inflow may not have included all seepage from the upper dam above Found meadow, resulting in underestimation of surface water inputs, (2) the lower stream weir may have received groundwater inputs that were underestimated, or (3) the V-notch elevation of the lower weir was lowered relative to the reference elevation

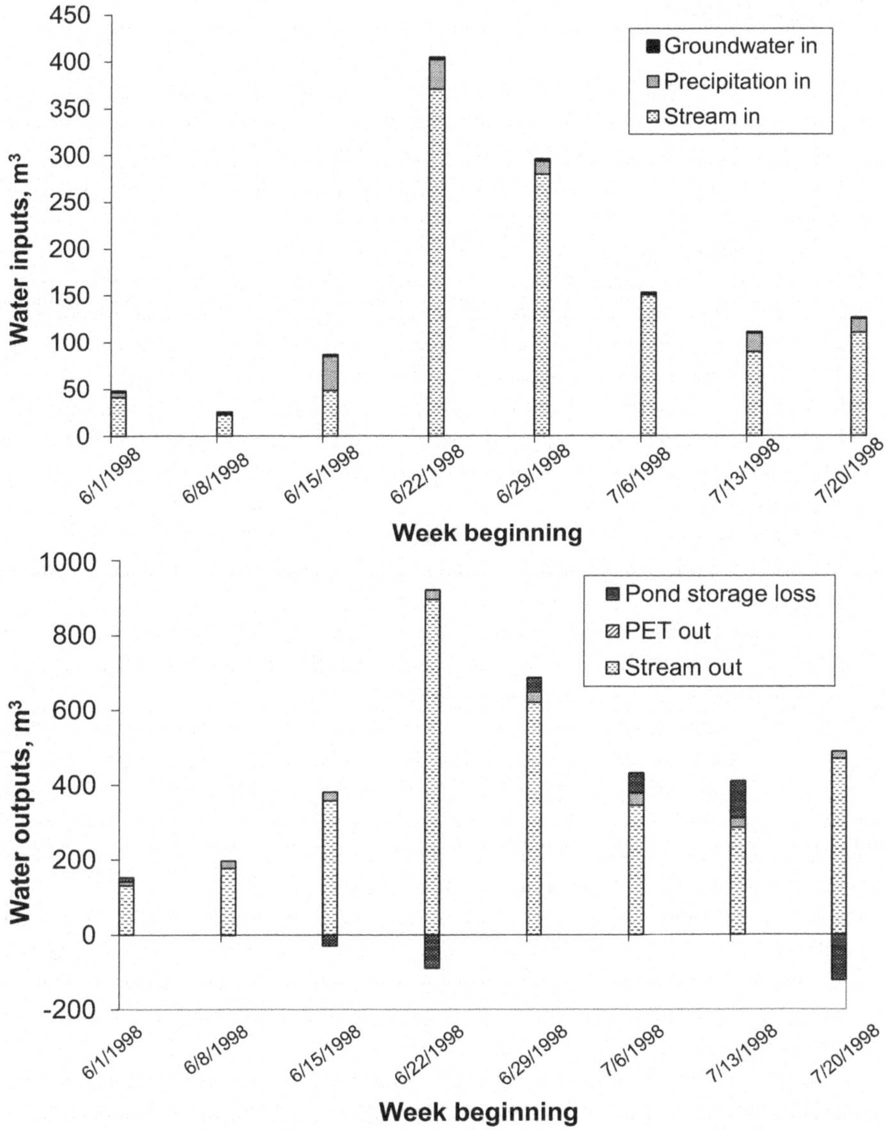

Fig. 3.29 Water balance analysis for Found pond, 1998

of the WL40 due to frost heave, so that the *H* term in Eq. 3.5 (Box 3.3.4) was too high, overestimating outflow (the V-notch elevations were surveyed in 1997 but not 1998). What is clear from the streamflow data in both years is that the Found Pond water balance is dominated by streamflow except during periods when the stream dries up.

3.8 Conclusions

- Two-thirds of precipitation falls during the five summer months of May through September at Voyageurs National Park, but the snowpack that accumulates overwinter provides the most runoff to rivers and streams.
- On average across the landscape, evapotranspiration accounts for 60% of precipitation inputs, but evapotranspiration rates are greater from lakes, ponds, and beaver meadows.
- Potential evapotranspiration from the Found beaver meadow calculated from temperature data was comparable to actual evapotranspiration calculated by a water balance approach except at the end of a long drought, when soil water was depleted and plant water availability was limited.
- Within beaver meadows, evapotranspiration rates were greatest at marshy sites where water inundated the ground surface and herbaceous plant biomass was high. Evapotranspiration was four times greater from wet stands of *Calamagrostis* grass than from dry stands.
- Evapotranspiration caused diurnal fluctuations of as much as 4.5 cm per day in pond and groundwater levels. The diurnal fluctuations were most pronounced during late summer.
- Water levels in residual (inactive) ponds rose quickly in response to large precipitation events but declined more gradually, ranging from 0.6 to 3.1 cm day^{-1}. The highest pond levels usually occurred in the spring.
- Hydraulic conductivity of sediments in the beaver meadows was low, averaging 1.36×10^{-3} cm/s in shallow wells and 1.39×10^{-6} cm/s in deeper wells. Vertical hydrologic gradients at three piezometer nests were consistently positive, indicating that surface water was recharging the groundwater system, but two well nests nearest the stream flowing through Found meadow were sometimes groundwater discharge sites, where water moved up toward the surface.
- Lateral runoff into the beaver meadow from the surrounding forest measured 5–20 liters/day following precipitation events of 5 mm or more. A runoff model successfully represented the change from storage-dominated flow control to catchment-topography-dominated flow control at a site having a relatively flat catchment with soil depth ≥ 0.3 m.
- The water balance of a beaver pond at the terminus of a larger drainage network was dominated by streamflow except during periods when the stream dried up.

References

Abtew W (2012) Wetland evapotranspiration. In: Abtew W, Melesse A (eds) Evaporation and evapotranspiration. Springer Netherlands, Dordrecht, pp 83–108

Allison DW, Mooers HD, Johnston CA, Pastor J, Brown T (1999) Use of a hemispherical chamber for measurement of evapotranspiration in a boreal forest wetland ecosystem (poster). Paper presented at the Geological Society of America Annual Meeting, Denver

Arguez A, Durre I, Applequist S, Squires M, Vose R, Yin X, Bilotta R (2010) NOAA's U.S. Climate Normals (1981–2010), MN INTL FALLS INTL AP (GHCN Daily ID: USW00014918). NOAA National Centers for Environmental Information

Aronson JJ, Mooers HD (1998) Construction and monitoring of weirs in remote locations (poster). Paper presented at the Minnesota Water '98, Collection of Abstracts, Minneapolis, 5–6 May

Bouwer H, Rice RC (1976) A slug test method for determining hydraulic conductivity of unconfined aquifers with completely or partially penetrating wells. Water Resour Res 12(3):423–428

Brown SM, Petrone RM, Mendoza C, Devito KJ (2010) Surface vegetation controls on evapotranspiration from a sub-humid Western Boreal Plain wetland. Hydrol Process 24(8):1072–1085

Brown TN, Johnston CA, Cahow KR (2003) Lateral flow routing into a wetland: field and model perspectives. Geomorphology 53(1–2):11–23

Burns DA, McDonnell JJ (1998) Effects of a beaver pond on runoff processes: comparison of two headwater catchments. J Hydrol 205(3–4):248–264

Butler JJ Jr, Kluitenberg GJ, Whittemore DO, Loheide SP II, Jin W, Billinger MA, Zhan X (2007) A field investigation of phreatophyte-induced fluctuations in the water table. Water Resour Res 43(2):W02404

Cahow KR (no date) Spatial and temporal variations in the groundwater system of boreal beaver meadows. M.S. Thesis (incomplete), University of Minnesota, Duluth

Cahow KR, Mooers HD (eds) (1998) Hydrology of a beaver meadow on the Kabetogama Peninsula, Voyageurs National Park, USA. Minnesota Water '98, Collection of Abstracts. Minnesota Water Resources Center, Minneapolis

Costa-Cabral M, Burges S (1994) Digital elevation model networks (DE-MON): a model of flow over hillslopes for computation of contributing and dispersal areas. Water Resour Res 30(6):1681–1692

Dadaser-Celik F, Stefan HG (2008) Lake evaporation response to climate in Minnesota. St. Anthony Falls Laboratory, University of Minnesota, Minneapolis

Drexler JZ, Snyder RL, Spano D, Paw UKT (2004) A review of models and micrometeorological methods used to estimate wetland evapotranspiration. Hydrol Process 18(11):2071–2101

Fetter CW (1994) Applied hydrogeology, 3rd edn. Prentice-Hall, Englewood Cliffs

Goulden ML, Litvak M, Miller SD (2007) Factors that control Typha marsh evapotranspiration. Aquat Bot 86(2):97–106

Harleman DRF, Melhorn PF, Rumer RR (1963) Dispersion-permeability correlation in porous media. Am Soc Civil Eng 89:67–85

Hazen A (1893) Some physical properties of sand and gravels. Massachusetts State Board of Health, Boston

Hill AR, Duval TP (2009) Beaver dams along an agricultural stream in southern Ontario, Canada: their impact on riparian zone hydrology and nitrogen chemistry. Hydrol Process 23(9):1324–1336

Hood GA, Bayley SE (2008) Beaver (*Castor canadensis*) mitigate the effects of climate on the area of open water in boreal wetlands in western Canada. Biol Conserv 141(2):556–567

Hvorslev MJ (1951) Time lag and soil permeability in ground-water observations. U.S. Army Corps of Engineers, Vicksburg

Karran DJ, Westbrook CJ, Wheaton JM, Johnston CA, Bedard-Haughn A (2017) Rapid surface-water volume estimations in beaver ponds. Hydrol Earth Syst Sci 21:1039–1050

Kool D, Agam N, Lazarovitch N, Heitman J, Sauer T, Ben-Gal A (2014) A review of approaches for evapotranspiration partitioning. Agric For Meteorol 184:56–70

Krumbein WC, Monk GD (1943) Permeability as a function of the size parameters of unconsolidated sand. Trans AIME 151(1):153–163

Loheide SP, Butler JJ, Gorelick SM (2005) Estimation of groundwater consumption by phreatophytes using diurnal water table fluctuations: a saturated-unsaturated flow assessment. Water Resour Res 41(7):W07030

Loheide SP II (2008) A method for estimating subdaily evapotranspiration of shallow groundwater using diurnal water table fluctuations. Ecohydrology 1:59–66

Lowry MM (1993) Groundwater elevations and temperature adjacent to a beaver pond in central Oregon. Oregon State University, Corvallis

Majerova M, Neilson B, Schmadel N, Wheaton J, Snow C (2015) Impacts of beaver dams on hydrologic and temperature regimes in a mountain stream. Hydrol Earth Syst Sci 19(8):3541–3556

Malmström VH (1969) A new approach to the classification of climate. J Geogr 68(6):351–357

Masoner JR, Stannard DI (2010) A comparison of methods for estimating open-water evaporation in small wetlands. Wetlands 30(3):513–524

Meybloom P (1966) Unsteady groundwater flow near a willow ring in hummocky moraine. J Hydrol 4:38–62

Moreo MT, Laczniak RJ, Stannard DI (2007) Evapotranspiration rate measurements of vegetation typical of ground-water discharge areas in the Basin and Range Carbonate-Rock Aquifer System, White Pine County, Nevada, and adjacent areas in Nevada and Utah, September 2005–August 2006. U.S. Geological Survey, Reston

Neff TAM (1996) Mesoscale water balance of the boreal forest using operational evapotranspirational approaches in a distributed hydrological model. University of Waterloo, Waterloo

NOAA (2016a) Climate data online. NOAA National Centers for Environmental Information. http://www.ncdc.noaa.gov/cdo-web

NOAA (2016b) Historical Palmer drought indices. NOAA National Centers for Environmental Information. http://www.ncdc.noaa.gov/temp-and-precip/drought/historical-palmers/

NOAA (2016c) Interactive Snow Information, Station KABM5 - Kabetogama Lake Visitor Center. National Weather Service, National Operational Hydrologic Remote Sensing Center. http://www.nohrsc.noaa.gov/interactive/html/graph.html?station=KABM5

NOAA (2016d) NOWData - NOAA Online Weather Data. National Oceanic and Atmospheric Administration. http://w2.weather.gov/climate/xmacis.php?wfo=dlh. Accessed Jan 2016

Nyssen J, Pontzeele J, Billi P (2011) Effect of beaver dams on the hydrology of small mountain streams: example from the Chevral in the Ourthe Orientale basin, Ardennes, Belgium. J Hydrol 402(1–2):92–102

Puckett WE, Dane JH, Hajek BF (1984) Physical and mineralogical data to determine soil hydraulic properties. Soil Sci Soc Am J 49(4):831–836

Rodionov SN (2006) Sequential Regime Shift Detection, Version 3.2 (software add-in to Excel 2002). NOAA Pacific Marine Environmental Laboratory. http://www.beringclimate.noaa.gov/regimes/

Rosenberry DO, Stannard DI, Winter TC, Martinez MI (2004) Comparison of 13 equations for determining evapotranspiration from a prairie wetland, Cottonwood Lake area, North Dakota, Wetlands 483–497

Sanford WE, Selnick DL (2013) Estimation of evapotranspiration across the conterminous United States using a regression with climate and land-cover data. JAWRA J Am Water Res Assoc 49(1):217–230

Stannard DI (1988) Use of a hemispherical chamber for measurement of evapotranspiration. Denver

Stannard DI, Weltz MA (2006) Partitioning evapotranspiration in sparsely vegetated rangeland using a portable chamber. Water Resour Res 42(2):W02413

Triska FJ, Jackman AP, Duff JH, Avanzino RJ (2000) Subchannel flow velocity and dispersion beneath a relict beaver dam, Shingobee River, Minnesota, USA: Implications for nutrient cycling. Verhandlungen des Internationalen Verein Limnologie, vol 27. pp 463–467

U.S. Geological Survey (2017) National Water Information System: USGS 05131500 Little Fork River at Littlefork, MN. http://waterdata.usgs.gov/nwis/inventory/?site_no=05131500&agency_cd=USGS

Westbrook CJ, Cooper DJ, Baker BW (2006) Beaver dams and overbank floods influence groundwater-surface water interactions of a Rocky Mountain riparian area. Water Resour Res 42(6):W06404
Winter TC, Rosenberry DO, Sturrock AM (1995) Evaluation of 11 equations for determining evaporation for a small lake in north central United States. Water Resour Res 31:983–993
Woo M-K, Waddington JM (1990) Effects of beaver dams on subarctic wetland hydrology. Arctic 43(3):223–230

Chapter 4
Beaver Loggers: Beaver Herbivory Alters Forest Structure

Abstract Beavers engineer forests by cutting trees for dam construction and winter forage. Beavers are the only North American animal besides humans that can fell mature trees. Our studies at Kabetogama and Duluth, Minnesota confirmed beavers' preference for trembling aspen, *Populus tremuloides*. Near the Duluth beaver ponds, beavers traveled more than twice as far to obtain trembling aspen as they did to obtain speckled alder (*Alnus incana*), a species which they avoided relative to its abundance in the riparian zone. We analyzed several boreal tree species for secondary metabolites that might make them unpalatable to beavers. Mature trembling aspen twigs had lower concentrations of phenolics in comparison to most other common boreal tree species, including its sibling, *Populus balsamifera*. The compounds tested did not explain beaver avoidance of speckled alder, however. Selective foraging by beavers within 50 m of their pond edges significantly decreased forest stem density and altered tree species composition, creating a "bathtub ring" of avoided conifers (*Abies balsamea, Picea glauca*) around the pond that was confirmed by GIS analysis of a forest cover map of the Kabetogama Peninsula.

4.1 Introduction

Beavers are nature's loggers (Fig. 4.1), the only boreal animal other than humans that can cut down mature live trees. The largest tree felled by beavers was a cottonwood (*Populus deltoides*) in British Columbia with a diameter of 170 cm and a height of 35 m (Hatt 1944). Beavers fell trees in order to access the tree canopy for its branches, leaves, and twigs that are consumed or used in construction.

Beavers are also gardeners, creating shallow ponds conducive to the aquatic vegetation that is important in their diet (Ray et al. 2001; Milligan and Humphries 2010; Belovsky 1984). Water lilies (*Nuphar* and *Nymphaea*) are particularly favored in the summer, and their fleshy submerged rhizomes are also eaten by beavers in the winter (Northcott 1972; Doucet and Fryxell 1993; Shelton 1966). Other floating-leaf species common in Kabetogama beaver ponds include water shield (*Brasenia schreberi*) and floating pondweed (*Potamogeton natans*) (Morris 2014). A stable isotope analysis of Kabetogama beaver tissues showed that floating-leaf and

C.A. Johnston, *Beavers: Boreal Ecosystem Engineers*,
DOI 10.1007/978-3-319-61533-2_4

Fig. 4.1 An aspen tree that has been partially cut by beavers at Voyageurs National Park. Note wood chips at the base

emergent vegetation constituted about 55% percent of the diet in *both* summer and winter (Severud et al. 2013a), and that beavers assimilated more floating-leaf vegetation where more was available (Severud et al. 2013b). The effects of beaver herbivory on the residual aquatic plant community are unknown for boreal beaver ponds, but a study in Georgia reported that beaver herbivory reduced aquatic plant biomass by 60%, plant litter by 75%, and dramatically shifted plant species composition (Parker et al. 2007).

Previous chapters focused on beaver engineering by dam building, but this chapter examines how beavers engineer forests with their sharp orange incisors, a wood chip at a time (Stefen et al. 2016). Although beaver herbivory may also alter aquatic plant communities, this chapter is primarily about how beaver browsing alters the boreal forests surrounding their ponds. Information is presented from the literature and from my own research, both published and unpublished.

4.2 Why Cut Wood?

Dam and lodge construction is the most visible purpose for beaver wood cutting. The average beaver dam contains about 7 Mg of wood, and a large (80 m long) beaver dam could contain as much as 52 Mg of wood (Johnston and Naiman 1990). However, the majority of wood harvested for this purpose is probably acquired during the first few years of beaver occupancy, since dam maintenance requires much less wood than initial dam construction.

Fig. 4.2 A beaver food cache partially submerged in a beaver pond

Another purpose of beaver wood cutting is to obtain forage. The boreal beaver's overwinter survival depends on its storage of woody branches in a submerged food pile (cache) that is placed near the lodge and accessed underwater when the pond ices over (Fig. 4.2). Cache preparation causes a flurry of beaver logging activity in the fall, although beavers may continue to cut fresh trees even in winter (Jenkins and Busher 1979). The presence of a food cache in the autumn is used as evidence of an active beaver colony (Johnston and Windels 2015). Cache size is proportional to colony size (Easter-Pilcher 1990).

Some of the wood cut by beavers serves a dual purpose for food *and* dam construction. In a North Dakota study, 43.6% of the *Populus tremuloides* logs and sticks used in dam construction were peeled, but only 13.3% of *Fraxinus pennsylvanica* sticks were peeled (Pinkowski 1983). Peeled *Populus tremuloides* logs are commonly used in dam construction at Voyageurs National Park (Fig. 2.5).

4.3 Beaver Morticulture and the Fate of Beaver-Cut Wood

The term "morticulture" has recently been coined to describe the culturing of woody detritus in forests, and acknowledges the importance of dead wood in forest ecosystems (Harmon 2001). A number of wildlife species benefit from downed wood for food and shelter, and decomposition of the felled trees returns carbon and nutrients to the forest floor (Hassinger and Payne 2008). The 4 July 1999 blowdown in the Boundary Waters Canoe Area that lies east of Voyageurs National Park caused foresters to rethink some of the benefits of downed wood (Harmon 2001). Beavers are masters of morticulture.

Fig. 4.3 Debarked aspen sticks floating on the water surface at a beaver feeding station

Unlike other boreal browsers, beavers ingest only a fraction of the wood that they cut (Fig. 4.3). An individual beaver cuts at least 1 Mg yr^{-1} of woody vegetation (Howard 1982; Johnston and Naiman 1990), but feeding studies show that it ingests only 0.2–0.3 Mg yr^{-1} (Aldous 1938). Thus, the majority of the biomass cut by beavers is not consumed.

Beavers divide felled trees into chunks that they can move into their ponds. Large tree boles are left on the ground, although the bark may be consumed from the fallen tree trunks (Fig. 4.4). Some trees become hung up in the canopy of dense forests after being cut by beavers, and cannot be used at all. A Minnesota study at five active beaver dams showed that 27% of the 456 trembling aspen (*Populus tremuloides*) stems cut by beavers were completely utilized (i.e., moved off site or consumed), 44% were partially utilized, and 29% were "completely wasted" (Aldous 1938). Beavers transported or consumed only 36% of the computed food value of these aspen trees, determined as the mass of tree bark plus twigs <1.3 cm in diameter. Other studies have reported 1–10% wastage of aspens <10 cm in diameter, but 50–64% wastage of larger aspens (Northcott 1971; Gibson 1957). The term "wastage" reflects a utilitarian bias, however, that is outdated.

4.4 Forage Selectivity

Beavers have been called "choosy generalist" herbivores, showing clear preferences in cutting certain tree species for food, but also cutting substantial numbers of trees of non-preferred species (Jenkins 1975; Gerwing et al. 2013;

Fig. 4.4 Aspen boles left on the ground surrounding a beaver pond at Voyageurs National Park

Barnes and Dibble 1988; Busher 1996). The predominance of a given plant species in a beaver's diet does not in itself indicate preference unless availability of that species is taken into account. I applied the electivity method (Box 4.4) to data that I collected along transects in the beaver-browsed forests surrounding two ponds north of Duluth, Minnesota, called Arnold Pond and Ash Pond, that were previously described in Chap. 2 (Table 2.4). Beavers showed a highly significant preference for trembling aspen (*Populus tremuloides*) at both ponds (Table 4.1), a finding that is consistent with other beaver foraging studies (Barnes and Mallik 2001; Donkor and Fryxell 1999; Fryxell and Doucet 1993; Hall 1960; Northcott 1971). There was also a preference for chokecherry (*Prunus virginiana*) at Ash Pond. Ironically, beavers significantly *avoided* a different species of poplar, *Populus balsamifera*, as indicated by the negative sign on the electivity index (Table 4.1). They also avoided *Alnus incana*, a wetland shrub species commonly found near boreal beaver ponds. The only conifer species present, *Picea glauca*, was unbrowsed. Other taxa (*Amelanchier* spp., *Acer spicatum*, *Betula papyrifera*, *Corylus cornuta*, *Salix* spp.) were cut in proportion to their availability or unbrowsed.

A similar analysis of beaver browse on Isle Royale National Park, Michigan showed a significant preference for *Acer saccharum* and *Betula alleghaniensis*, and a significant avoidance of *Betula papyrifera*, *Corylus cornuta*, and all conifers present (*Abies balsamea*, *Thuja occidentalis*, *Picea* spp.) (Belovsky 1984). Use of *Acer rubrum*, *A. spicatum*, *Alnus incana*, *Cornus stolonifera*, and *Sorbus americana* was proportional to their abundance at Isle Royale. Other authors confirm that conifers are avoided except when nothing else is available (Smith and Tyers 2008; Müller-Schwarze et al. 1994).

Table 4.1 Beaver electivity (E_{ij}) for riparian forest species surrounding two beaver ponds north of Duluth, MN

Species	E_{ij}	X^2
Arnold Pond		
Alnus incana ssp. rugosa	−0.50	2.16
Amelanchier spp.	0.70	2.04
Betula papyrifera	Unbrowsed	na
Corylus cornuta	Unbrowsed	na
Picea glauca	Unbrowsed	na
Populus balsamifera	−1.69	2.79
Populus tremuloides	1.50	24.25**
Salix spp.	Unbrowsed	na
Ash Pond		
Acer spicatum	−0.09	0.35
Alnus incana ssp. rugosa	−0.42	5.64*
Betula papyrifera	0.01	0.00
Corylus cornuta	0.45	2.44
Fraxinus nigra	−0.41	1.75
Populus balsamifera	−1.67	7.72**
Populus tremuloides	0.80	22.98**
Prunus virginiana	0.77	6.68**

*$P < 0.05$
**$P < 0.01$

Box 4.4. Calculating Electivity

Beaver preference for woody plant species can be measured by an electivity index, E_{ij} (Jenkins 1979):

$$E_{ij} = \ln\left[\frac{(r_{ij})(1-p_{ij})}{(p_{ij})(1-r_{ij})}\right] \tag{4.1}$$

where r_{ij} is the proportion of tree species i cut at time j and p_{ij} is the proportion of tree species i present at time j. This is equivalent to the natural logarithm of the cross-product ratio, a frequently used measure of association in contingency table analysis. Significance of each E_{ij} was tested by computing X^2 values:

$$\chi^2 = E_{ij}^2 / \left[1/x_{ij} + 1/(m_j - x_{ij}) + 1/y_{ij} + 1/(n_j - y_{ij})\right] \tag{4.2}$$

where x_{ij} and y_{ij} are the number of trees of species i cut and available, respectively, at time j, and m_j and n_j are the total number of trees cut and available, respectively, at time j. These computed values were compared with a χ^2 distribution with one degree of freedom (Fienberg 1977).

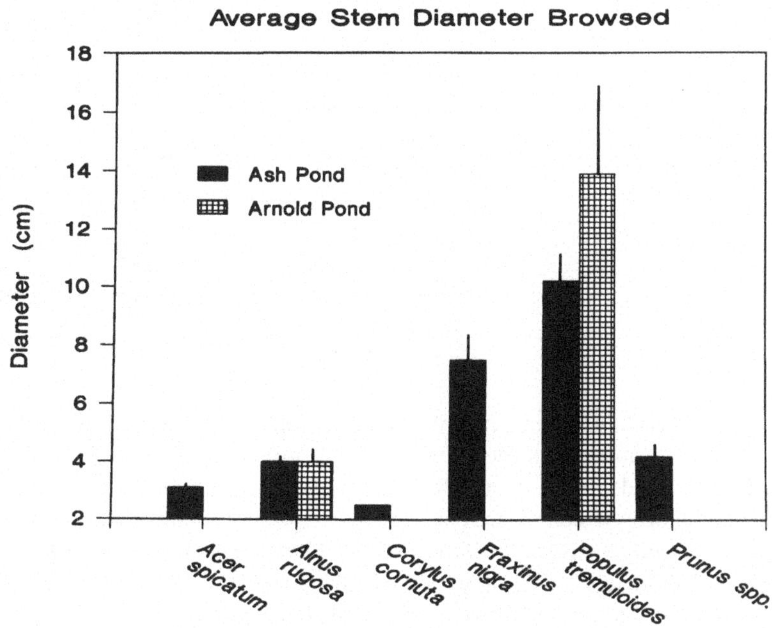

Fig. 4.5 Average stem diameter browsed, by plant species, at two Duluth MN beaver ponds

The average stem diameter browsed also varied by species (Fig. 4.5). Average stem diameter browsed was <5 cm for the shrubby species present (*Acer spicatum, Alnus incana* ssp. *rugosa*, *Corylus cornuta*, and *Prunus virginiana*), consistent with their life-form. Average stem diameters browsed of *Fraxinus nigra* and *Populus tremuloides* were larger, but these trees grow to larger sizes. The largest tree browsed was a 43.5 cm diameter *Populus tremuloides*.

The time required for a beaver to cut a woody stem (t) increases exponentially with its diameter (D) (Belovsky 1984):

$$t = 0.63e^{0.24D} \tag{4.3}$$

Therefore, if browse selection were based purely on effort expended, we would expect beavers to preferentially select for small stems, which are more quickly harvested and result in less waste. In order to assess size preference by beavers, I applied the electivity analysis to determine if the Duluth beavers were selecting for small diameter stems (<5 cm) rather than larger stems. I found that smaller diameter stems were preferred over large stems for *Alnus incana* (positive electivity index, Table 4.2). This is consistent with a study from northern Ontario, where beavers highly preferred small diameter *Alnus* stems (1.5–3.5 cm) for dam construction, placing them intact on the dam without peeling (Barnes and Mallik 1996).

Table 4.2 Beaver electivity (E_{ij}) for small (<5 cm) versus large (≥5 cm) stems of riparian forest species surrounding two beaver ponds north of Duluth, MN

Species	E_{ij}	X^2
Arnold Pond		
Alnus incana ssp. rugosa	0.61	0.60
Populus tremuloides	−1.39	13.00**
Ash Pond		
Acer spicatum	−0.30	0.40
Alnus incana ssp. rugosa	0.99	4.49*
Fraxinus nigra	0.39	0.38
Populus tremuloides	−0.44	2.20
Prunus virginiana	−0.13	0.04

* = $P < 0.05$
** = $P < 0.01$

However, the Duluth beavers preferentially selected larger stems of *Populus tremuloides* over small ones (negative electivity index, Table 4.2). At both of the Duluth beaver ponds, more than half of the trees 35–40 cm in diameter that were available in the browse zone were cut by beavers (Johnston and Naiman 1990). In a southern Utah study, beavers similarly preferred large branches of *Populus fremontii* over small branches (McGinley and Whitham 1985). Avoidance of toxins (see Sect. 4.5 on Plant Metabolites) may explain this behavior.

4.5 Beavers and Plant Metabolites

Trees cannot flee herbivores, but they have developed other mechanisms to avoid being eaten. One tree defense strategy is to grow quickly, raising leaf canopies out of reach (Coley et al. 1985). This strategy works well for earth-bound herbivores such as moose and deer, but beavers have outsmarted the trees by bringing their canopies crashing back down to earth. An alternate tree defense strategy is to defend themselves from being eaten by manufacturing secondary metabolites that are unpalatable to herbivores (Bryant et al. 1991; Rosenthal and Janzen 1979; Müller-Schwarze et al. 1994). For example, beavers have been shown to avoid a poplar species that contains condensed tannins (*Populus angustifolia*) in preference to a species that does not (*P. fremontii*) (Bailey et al. 2004). In a cafeteria feeding experiment with different sizes of *Populus tremuloides*, beavers preferred branches from adult trees and adult-form suckers over juvenile-form suckers that contained significantly greater concentrations of a deterrent phenolic compound (Basey et al. 1988; Basey et al. 1990). Another study found comparable concentrations of total phenolics in mature trembling aspens sampled from beaver-browsed versus unbrowsed stands, so secondary metabolites apparently do not affect spatial patterns of mature stand selection (Masslich et al. 1988).

My colleagues and I analyzed the secondary metabolite chemistry of six woody plant species commonly found in the riparian zones of boreal beaver ponds to determine if it could explain beaver forage selectivity (Box 4.5). We also compared the

Box 4.5. Plant Metabolite Methods
Twig samples were collected from 3 to 4 different plants at the Duluth beaver ponds and shipped on dry ice for analysis in the laboratory of Rex Cates at Brigham Young University, Provo, Utah. Terpene identification was done on a Gas Chromatography Mass Spectrophotometer (GC-MS) with a library of 70,000 compounds. Tannin, a polyphenolic, was analyzed by the radial diffusion method using quebracho tannin as the standard (Hagerman 1987). Salicortin, a phenolic glycoside, was identified by high pressure liquid chromatography (HPLC), using purified salicortin as a standard (Clausen et al. 1989). Monomeric phenolics were extracted using a butanol/acetic acid/water solvent with a C_{18} reverse phase column and analyzed by HPLC.

concentrations of secondary metabolites in mature versus juvenile *Populus tremuloides*, because other authors have reported that higher concentrations of phytochemical deterrents in younger trees have deterred beaver herbivory (Basey et al. 1988).

Two tree species that were avoided or unbrowsed, balsam poplar (POBA2) and white spruce (PIGL), had the greatest amount of chemical defenses. Both contained numerous terpenes and sesquiterpenes (Table 4.3), whereas the other plant species had one (10-demethylsqualene in *Betula papyrifera*) or none (all other species). White spruce (PIGL) contained the highest concentrations of four phenolic compounds analyzed: catechin, caffeic acid, benzoic acid, and tannin (Table 4.4). Balsam poplar (BOPA2) contained benzoic acid and tannin, compounds that were absent from trembling aspen (POTR5). Both of the poplar species (POTR5, POBA2) contained comparable catechin, chlorogenic acid, and salicortin concentrations.

Table 4.3 Terpenes and sesquiterpenes identified by GC-MS in balsam poplar and white spruce twigs sampled from the Duluth beaver ponds

Plant species	Terpenes	Sesquiterpenes
Balsam poplar (*Populus balsamifera*)	1,8 cineole a- copaene a- yalangene g- gurjunene valencene b- maalene camphor	delta-cadinene gamma-cadinene
White spruce with needles (*Picea glauca*)	alpha-pinene camphene beta-pinene delta-3-carene p-cymene limonene cis-ocimene alpha-terpinolene neophytadiene	beta-elemene trans-caryophyllene b-cubebene delta-cadinene gamma-cadinene alpha-humulene

Table 4.4 Mean concentrations of various phenolic compounds in twigs of woody plants sampled from the Duluth beaver ponds

Species	Catechin (mg/g)		Chlorogenic acid (μg/g)		Caffeic acid (μg/g)		P-coumaric acid (μg/g)		Benzoic acid (μg/g)		Salicortin (mg/g)		Tannin (mg/g)	
POTR5 (mature)	2.8	ab	540	c	139	cd	85	d	0	a	21	c	0	a
POTR5 (sapling)	3.3	ab	410	c	379	ef	106	d	0	a	26	c	0	a
ALINR	1.2	a	29	a	0	a	0	a	0	a	3	ab	0	a
BEPA	4.6	bc	44	ab	16	ab	11	b	0	a	0	a	24	b
FRNI	4.2	bc	311	bc	195	de	1097	e	0	a	2	ab	0	a
PIGL	6.5	c	502	bc	592	f	0	a	1704	c	3	ab	29	c
POBA2	3.6	bc	396	c	83	bc	54	c	268	b	16	bc	21	b

Values expressed as mass per gram fresh weight of twig tissue. Within each column, values followed by the same letter are not significantly different (Tukey HSD, $\alpha = 0.05$)
POTR5 Populus tremuloides, *ALINR* Alnus incana, *BEPA* Betula papyrifera, *FRNI* Fraxinus nigra, *PIGL* Picea glauca, *POBA2* Populus balsamifera

Twigs from mature trembling aspen (POTR5) had much lower concentrations of caffeic acid than did those from trembling aspen saplings, which may explain the beaver's preference for larger stem sizes. Mature and juvenile samples of trembling aspen had comparable concentrations of other secondary metabolites (Table 4.4).

The measured secondary metabolites did *not* explain beaver avoidance of speckled alder (*Alnus rugosa*), which had the lowest concentrations of all the compounds analyzed. Alder is unique among these species in its ability to symbiotically fix nitrogen, so it may contain toxic nitrogen-based secondary metabolites such as alkaloids that were not analyzed.

4.6 Central Place Foraging

Beavers are central place foragers, rarely venturing more than 50 m from the safety of their ponds to access food items in the riparian zone (Jenkins 1980; Donkor and Fryxell 2000; Pinkowski 1983; Basey and Jenkins 1995), but traveling as far as 200 m when there is little predation risk (Smith et al. 1994; Anderson et al. 2009). At the Duluth beaver ponds, foraging extended to 45 m from the pond edges, but was most intense within 20 m (Johnston and Naiman 1990). At the intensively browsed Ash Pond, 83% of the sample plots within 20 m of the pond contained evidence of beaver foraging (Johnston and Naiman 1990).

The central place foraging theory predicts that, because provisioning costs increase with distance, animals should forage more selectively farther from the central place (McGinley and Whitham 1985). Browse patterns at the Duluth beaver

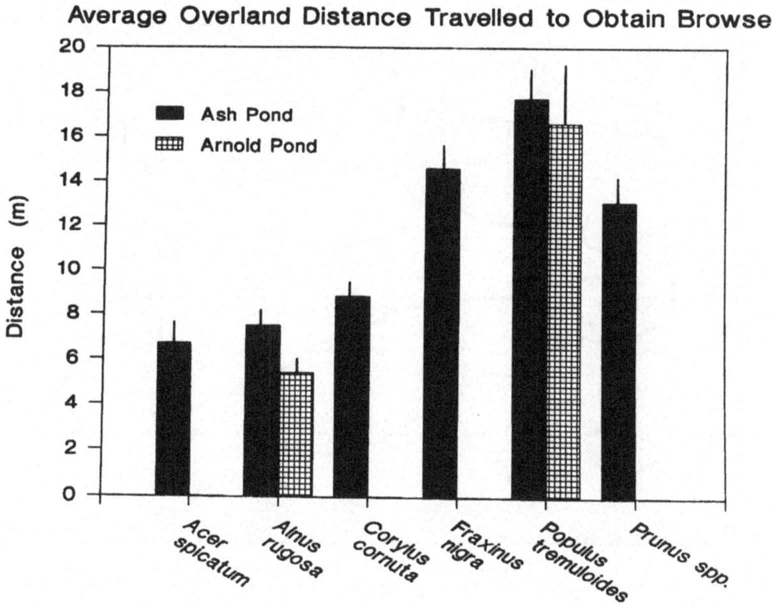

Fig. 4.6 Average overland distance traveled by beavers to obtain browse, by plant species, at two Duluth MN beaver ponds

ponds were consistent with this principle, because beavers traveled greater distances to access preferred species (Fig. 4.6). On average, beavers traveled 17 m to obtain *Populus tremuloides*, more than twice the distance traveled to obtain *Acer spicatum* or *Alnus* (~6 m). Beavers also traveled relatively far to obtain the other two tree species, *Prunus virginiana* and *Fraxinus nigra*.

Another prediction of central place foraging is that the size of stems selected should generally decrease with distance from the pond, a relationship that was found in several beaver foraging studies (Belovsky 1984; Jenkins 1980; Pinkowski 1983). However, the opposite trend (larger stems harvested with increasing distance from the water body) has also been shown (Raffel et al. 2009; Gallant et al. 2004). This relationship is consistent with central place foraging theory provided that the food value of a branch increases with branch size and provisioning time is independent of branch size (McGinley and Whitham 1985).

4.7 Beaver Silviculture: How Does Beaver Logging Affect the Remnant Forest?

Beaver logging can substantially alter forest stem density and structure. At Ash Pond, a single beaver colony cut 8.42 Mg ha^{-1} yr^{-1} of woody stems ≥5 cm, reducing the average stem density and basal area of the forest by 43% within a 6-year period

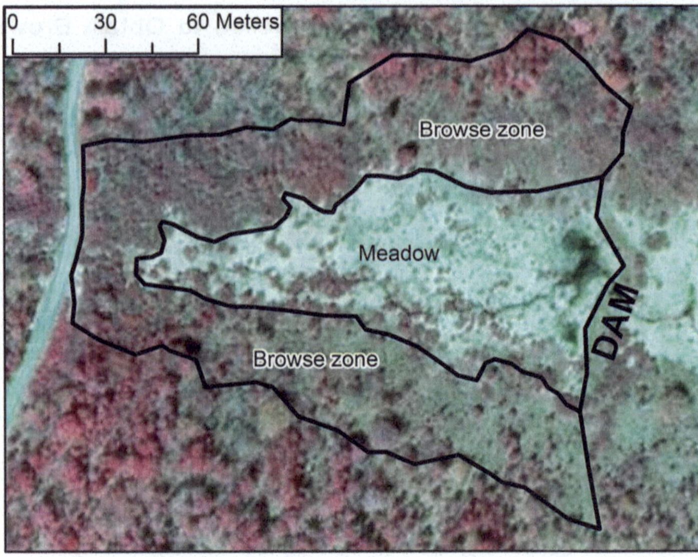

Fig. 4.7 *Color* infrared image of the clear-cut browse zone surrounding Ash Pond near Duluth MN. "Meadow" is the footprint of the former pond, which was drained as of this image date (2009)

(Johnston and Naiman 1990). The majority of the woody plant basal area in the pre-browse forest was *Populus tremuloides* (60%), so the beavers essentially clear-cut the riparian forest within a 35 m radius of the pond edge, leaving behind shrubs and tree suckers (Fig. 4.7). The site was occupied by beavers from 1980 through at least 1991, but was abandoned until 2013, when beavers recolonized the site. Beavers typically abandon ponds that cease to provide adequate forage, moving to new locations that can supply their forage needs.

Clear-cutting is the recommended silvicultural practice to promote vegetative reproduction ("suckering") of *Populus tremuloides* in the north-central United States. Managers are advised to remove as much of the overstory as possible so as to allow heat and light to reach the forest floor (Perala 1977). Therefore, when the pre-browse forest is dominated by aspen, the intense browsing pressure exerted by beaver tends to perpetuate that species, thereby promoting site reoccupancy by beavers when the aspen suckers have re-grown and can once again supply sufficient forage.

An ecosystem model was used to better understand how beaver foraging would affect the structure and function of a willow (*Salix*) community, a plant species that also grows suckers in response to browsing. Foraging by beaver increased aboveground net primary production, decreased biomass, and increased structural heterogeneity in the simulated willow community (Peinetti et al. 2009).

Beavers can also act as "molecular geneticists" through their browsing. Selective foraging by beavers was shown to promote hybridization of two *Populus* species, *P. fremontii* and *P. angustifolia*. Beaver selectively browsed *P. fremontii* because its

Fig. 4.8 The browse zone next to this beaver pond contains only conifers, but taller deciduous trees are visible beyond (Photo courtesy of Steve Windels)

bark contained fewer condensed tannins than *P. angustifolia*, and the foraging pressure promoted the transmission of this ecologically functional genetic trait (i.e., tannin production) via hybridization of the two species (Bailey et al. 2004).

When the pre-browse forest contains a mix of avoided and preferred tree species, selective browsing by beavers can alter forest composition as avoided species increase in relative abundance (Wright et al. 2002; Barnes and Dibble 1988; Donkor and Fryxell 1999). On the Kabetogama Peninsula, beaver browsing of the overstory aspen and birch releases the slower growing conifers beneath from light competition. The conifers then shade out deciduous reproduction in the riparian zone around the pond, creating a "bathtub ring" of conifers within the beaver browse zone (Fig. 4.8). As noted in Chap. 2, beavers also inhabit lakes and rivers, and beaver herbivory affects those riparian zones, too (Herbison and Rood 2015; Mortenson et al. 2008; Raffel et al. 2009; Barnes and Dibble 1988).

The spatial distribution of beaver browse may not be uniform around a pond (Jenkins 1979). On the Kabetogama Peninsula, bedrock bluffs adjacent to some ponds create barriers to beaver movement. Wetlands with shallow water and dense vegetation may also impede beaver movements unless surrounded by a water moat (Fig. 2.9) or traversed by a beaver-dug canal (Fig. 2.28). Beavers canals and trails tend to focus their access to the upland, concentrating their herbivory (Fig. 4.9). Such patchy herbivory would result in a mix of riparian zone vegetation patches rather than a homogeneous clear-cut or bathtub ring.

Fig. 4.9 Beaver trail leading into a beaver canal

4.8 Beaver Browse on the Kabetogama Peninsula

So far, this chapter has summarized information from the Duluth beaver ponds and other literature to describe beaver foraging and its effects. The following section describes beaver foraging specific to the Kabetogama Peninsula (Box 4.8, Fig. 4.10).

Box 4.8. Beaver Forage Study

Woody plant stems were measured in plots at 20, 40, and 80 m along transects perpendicular to the pond edge from starting points that were randomly chosen; the number of transects sampled per pond (3–12) was proportional to the pond perimeter. Two concentric circular plot sizes were used: stems ≥5 cm were sampled within 50 m^2 plots (4 m radius), and stems <5 cm were sampled within 10 m^2 plots (1.78 m radius). Stems were measured by species at stump height (~30 cm from ground level) using a gauge with incremental 0.5 cm slots for stems 2.5–4.9 cm and a diameter tape for stems ≥5 cm (Fig. 4.10). All trees were identified to species; the shrubs *Amelanchier* and *Salix* were identified only to genus (Table 4.6). Measurements of both browsed and unbrowsed stems were made in 1989, and measurements of new browse were made in 1990–1994. Browsed stumps were marked with fluorescent paint so that they would not be recounted in future revisits.

The community composition of browsed and unbrowsed stems at the 11 beaver pond sites was analyzed using PRIMER version 6 statistical software (Clarke and Gorley 2001). Square-root transformation was done to down-weight high abundance species. Similarity was computed after Bray and Curtis (Bray and Curtis 1957), and used to classify sites by nonmetric multidimensional scaling (MDS) and by agglomerative hierarchical clustering with group-average linking. The SIMPER procedure (Clarke and Gorley 2001) was used to determine species contributions to the similarity of a cluster.

Fig. 4.10 Measuring a beaver-browsed stump using a diameter tape

Table 4.5 Characteristics of ponds used in beaver forage studies

Site No.	Location	First impounded	Activity	No. of Transects
1	Lost Ponds	1961	Inactive	12
2	Lost Ponds	1961	Active	8
3	Lost Ponds	1961	Active	3
4	Lost Ponds	1961	Active	3
7	Lost Ponds	1981	Inactive	3
8	Shoepack Ponds	1981	Active	3
10	Nashata Point	1961	Active	8
11	Cutover Island	1961	Active	8
12	Bluefin Bay	1961	Inactive	5
13	Clyde Creek	1981	Active	4
14	Bluefin Bay	1961	Active	5

See Fig. 4.11 for location of pond numbers. "First impounded" is the date of historical aerial photography on which the ponds were first observed; 1961 = old, 1981 = new

4.8.1 Characteristics of the Riparian Forest

Riparian forests surrounding 11 beaver ponds on the Kabetogama Peninsula were selected for study that represented two categories, beaver pond activity and beaver pond age (Table 4.5, Fig. 4.11). A total of 1790 live stems representing 20 different taxa were measured within the 186 plots observed (Table 4.6). Trembling aspen (*Populus tremuloides*) was the most abundant species, occurring at all 11 sites and constituting 45% of basal area and 27% of stems (Fig. 4.12). Balsam fir (*Abies balsamea*) also occurred at all 11 sites and was comparably abundant (26% of stems), but made up only 17% of basal area. Paper birch (*Betula papyrifera*) was the next most important species, sampled at ten of the 11 sites. Together these three species made up three-fourths of all basal area and two-thirds of all stems in the forest (Fig. 4.12). Two shrubs, *Amelanchier* and *Acer spicatum*, each constituted 9% of the stems, but represented a small amount of basal area because of their small stem diameters. The fourth and fifth most important species by basal area were white spruce (*Picea glauca*) and red maple (*Acer rubrum*).

The MDS and cluster analysis of the unbrowsed stem density separated the 11 beaver pond sites into four groups at a similarity of 57% (Fig. 4.13). The two largest groups, c and d, were distinguished primarily by their differing proportions of mountain maple and balsam fir (Table 4.7). Group c is comparable to the aspen–birch/boreal conifer forest and group d is comparable to the balsam fir–paper birch forest defined as community types for Voyageurs National Park (Faber-Langendoen et al. 2007). Group a (site 4 only) had no birch, the least aspen and most red maple of any site, and had the lowest stem density (1511 stems/ha) and basal area (1040 dm^2/ha). Group b (site 3 only) had the most serviceberry of any site and the least balsam fir.

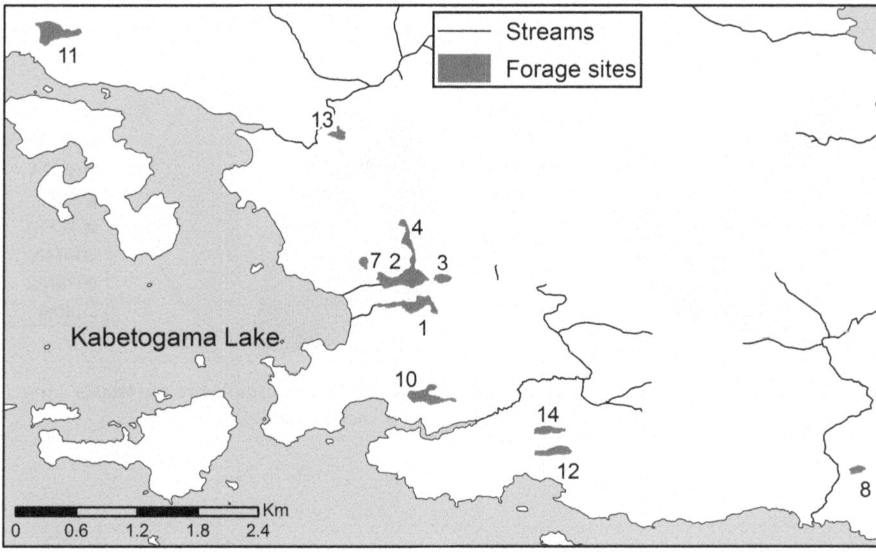

Fig. 4.11 Locations and pond numbers of sites used for forage studies on the southwestern Kabetogama Peninsula

Table 4.6 Woody species that occurred in forage transects. Codes are from http://plants.usda.gov

Species	Common name	Code	Life-form	Leaf type
Abies balsamea	Balsam fir	ABBA	Tree	Evergreen
**Acer rubrum*	Red maple	ACRU	Tree	Deciduous
**Acer spicatum*	Mountain maple	ACSP2	Shrub	Deciduous
**Alnus incana*	Speckled alder	ALINR	Shrub	Deciduous
**Amelanchier* spp.	Serviceberry	AMELA	Shrub	Deciduous
**Betula papyrifera*	Paper birch	BEPA	Tree	Deciduous
Corylus cornuta	Beaked hazelnut	COCO6	Shrub	Deciduous
**Fraxinus nigra*	Black ash	FRNI	Tree	Deciduous
Picea glauca	White spruce	PIGL	Tree	Evergreen
Picea mariana	Black spruce	PIMA	Tree	Evergreen
Pinus banksiana	Jack pine	PIBA2	Tree	Evergreen
Pinus resinosa	Red pine	PIRE	Tree	Evergreen
Pinus strobus	White pine	PIST	Tree	Evergreen
Populus grandidentata	Bigtooth aspen	POGR4	Tree	Deciduous
**Populus tremuloides*	Trembling aspen	POTR5	Tree	Deciduous
Prunus virginiana	Chokecherry	PRVI	Shrub	Deciduous
Quercus macrocarpa	Burr oak	QUMA2	Tree	Deciduous
Quercus rubra	Northern red oak	QURU	Tree	Deciduous
**Salix* spp.	Willow	SALIX	Shrub	Deciduous
Thuja occidentalis	Eastern white cedar	THOC2	Tree	Evergreen

*Woody species that were browsed by beavers

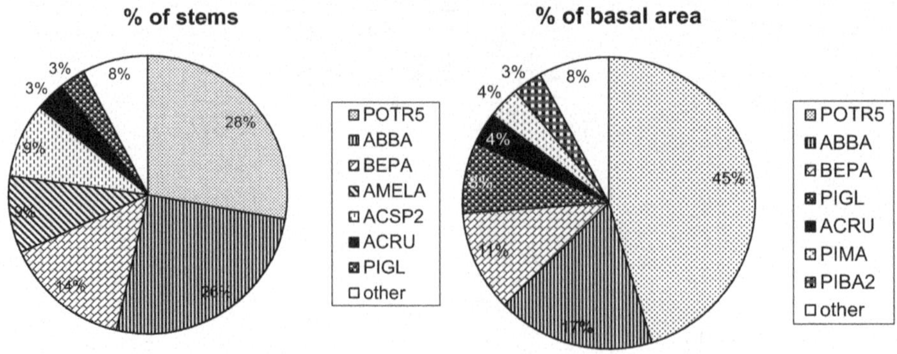

Fig. 4.12 Species composition of the standing live woody plants sampled in the beaver forage study, % of stems and % of basal area. See Table 4.6 for species codes

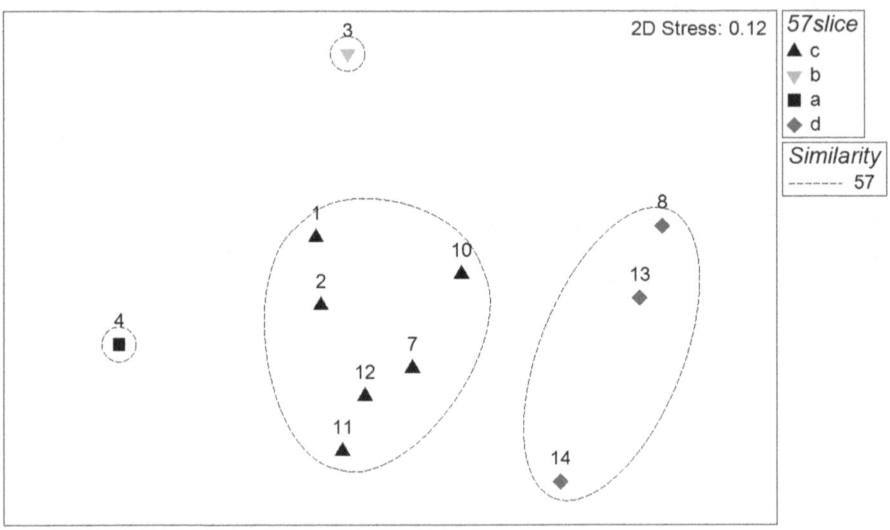

Fig. 4.13 Nonmetric multidimensional scaling plot of beaver riparian zones based on live woody stems/ha (Table 4.5). *Dashed line* encompasses sites at 57% similarity, identified by hierarchical clustering

These four community groups differed significantly in basal area/ha ($F = 5.54$, $df = 3$, $P = 0.029$). Group a (site 4) had the least basal area, and group d had the most (Fig. 4.14). There was no significant difference in unbrowsed stems/ha, browsed stems/ha, or browsed basal area/ha across the four groups.

Table 4.7 Relative contribution (%) of woody plants within the four plant community clusters identified for the unbrowsed forest, based on stem density

Species	a	b	c	d
Abies balsamea	19.6	3.5	20.0	33.2
Acer rubrum	23.0	10.6	5.2	
Acer spicatum			13.4	
Alnus incana	11.0			
Amelanchier spp.	15.5	29.5	8.4	
Betula papyrifera		16.9	15.3	23.4
Picea glauca	13.0		6.6	
Picea mariana				10.5
Populus tremuloides	13.0	31.7	25.0	26.3
Quercus rubra	4.9			
Salix spp.		7.9		

Data for groups c and d are percent contribution to group similarity from SIMPER analysis; data for a and b (single sites) are relative stem abundance after square root transformation

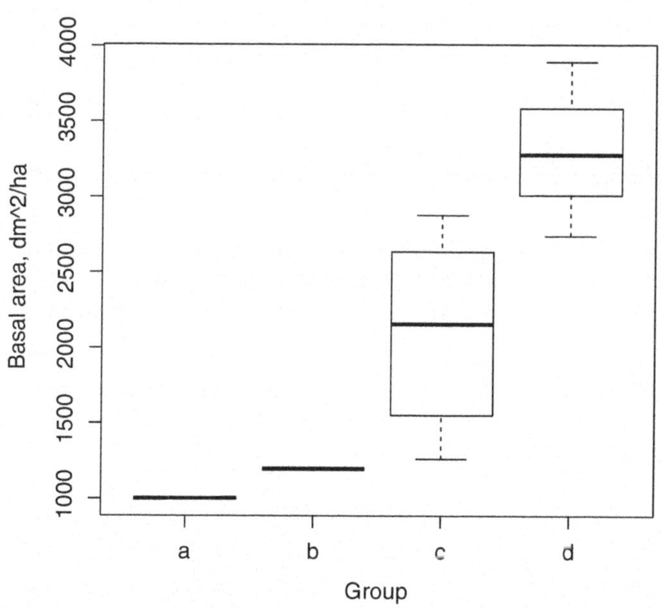

Fig. 4.14 Basal area (dm²/ha) by community group (see Fig. 4.13). In each box, the heavy *horizontal line* crossing the box is the median, the *bottom* and *top* of the box are the *lower* and *upper* quartiles, the *lower* whisker is the minimum value, the *upper* whisker is 1.5 times the interquartile range, and *circles* are outliers larger than 1.5 times the interquartile range

4.8.2 Browsed Stems

Eight deciduous species had been browsed within the 186 plots as of the initial sampling date (Table 4.6). Trembling aspen was by far the preferred browse species, accounting for 56% of the stems/ha and 81% of the basal area/ha browsed (Fig. 4.15). Trembling aspen was the only species that was browsed at the 80-m plot distance. The maximum stem diameter browsed was 34.7 cm (a trembling aspen). Birch and alder made up 32% of browsed stems but only 13% of the basal area browsed due to their smaller diameters. No evergreen species was browsed. There was a tendency for beavers to browse more small stems closer to the pond ($F = 2.80$, df = 2, $P = 0.062$) (Fig. 4.16).

Follow-up measurements of new browse in five subsequent years showed that browsing was spatially and temporally variable (Table 4.8). The lack of new browse was partially due to inactivity (ponds #1, #7, #12) and abandonment (ponds #4 and #14 became inactive after 1989 and 1991, respectively). At the active ponds, browse tended to be localized. For example, the extremely high basal area harvested at pond #8 in 1991 was due to the removal of 15 trembling aspen stems from two plots closest to the pond; this was the only browse that occurred along the transects at this pond, which became abandoned after 1991. Unlike the evenly distributed browsing witnessed at Ash Pond in Duluth (Fig. 4.7), browsing at the Kabetogama beaver ponds was more patchy, and our transects did not always intersect those patches. In 1994, there was no new browse at *any* of the 114 plots observed.

4.8.3 Effect of Browsing on Residual Forest: Field Study

I used generalized allometric regressions to convert both browsed and unbrowsed stem diameter to biomass (Freedman 1984), adding the two values to estimate the pre-browsed forest biomass at each site. Based on these calculations,

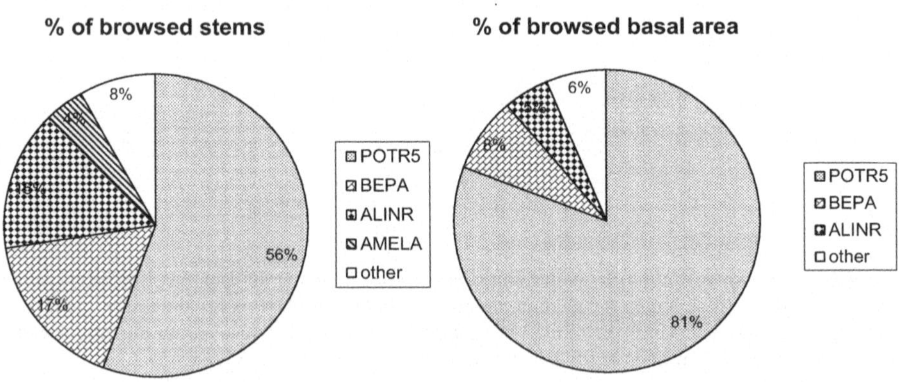

Fig. 4.15 Species composition of browsed woody plants sampled in the beaver forage study, % of stems and % of basal area. "Other" category includes ACSP2, ACRU, FRNI, and SALIX. See Table 4.6 for species codes

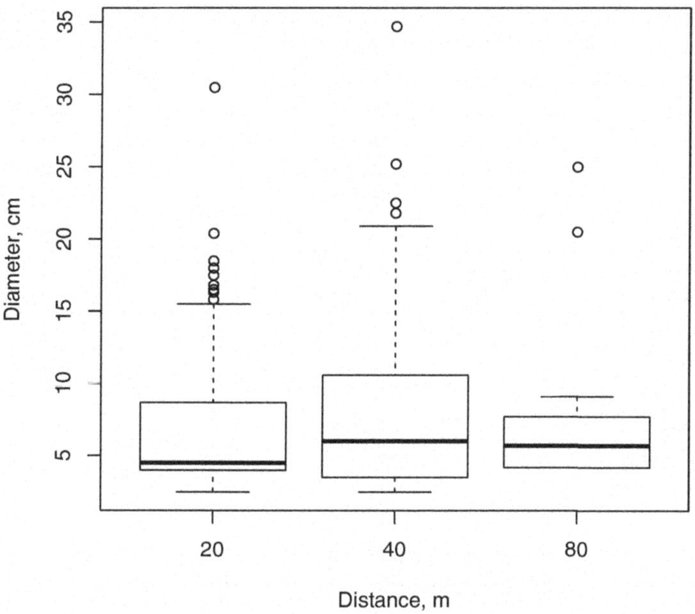

Fig. 4.16 Boxplot of average stem diameter browsed at 20, 40, and 80 m distances from the pond edge. Boxplots as in Fig. 4.14

Table 4.8 Basal area (dm^2/ha) browsed in sample plots, 1990–1994

Pond No.	1990	1991	1992	1993	1994
1	31	12.3	dis.	dis.	dis.
2	99.3	48.3	0	17.4	0
3	10.1	0	0	60.5	0
4	0	0	0	0	0
7	0	0	dis.	dis.	dis.
8	0	385.2	0	0	0
10	0	31.5	34.4	51.6	0
11	63.5	8.9	0	50.3	0
12	0	0	dis.	dis.	dis.
13	104.9	57.4	0	63.2	0
14	0	0	0	0	0

dis = sampling discontinued

beavers had cut ~32% of the pre-browse biomass within 80 m of their ponds at sites #7 and #13 (Table 4.9). Sites #3 and #4, which were distinct community types based on the MDS and cluster analyses, also had very high rates of biomass removal (18–20%); beaver herbivory could explain their lower residual basal area (Fig. 4.14).

Table 4.9 Residual (unbrowsed) biomass, browsed biomass, and the percentage of the pre-browse forest (browsed plus unbrowsed biomass) cut by beavers as of 1989 in 186 sample plots

Pond	Residual biomass as of 1989, kg/ha	Browsed biomass as of 1989, kg/ha	Percent browsed
1	59,925	4000	6.3%
2	87,713	8090	8.4%
3	53,136	12,957	19.6%
4	42,413	8511	16.7%
7	61,139	29,000	32.2%
8	128,590	0	0.0%
10	117,018	1168	1.0%
11	166,104	5020	2.9%
12	135,760	8020	5.6%
13	98,861	44,514	31.0%
14	195,880	0	0.0%

I expected that pond age and beaver activity (i.e., colony presence or absence) would affect the residual forest, hypothesizing that the forest near older ponds and inactive ponds would have fewer stems/ha and less basal area/ha due to prolonged beaver herbivory. A 2-way analysis of variance using the groups in Table 4.5 showed that there was a significant effect of age on stem density: older ponds (established by 1961) had significantly fewer stems/ha than newer ponds (established by 1981) ($F = 13.95$, $P = 0.0073$). However, there was no significant effect of beaver activity (active versus inactive ponds, Table 4.5) nor the interaction of terms (Fig. 4.17). There was no significant effect of either factor on residual basal area/ha.

4.8.4 *Effect of Browsing on Residual Forest: GIS Study*

The "bathtub ring" effect, bands of conifer forests around ponds due to beavers' selective central place foraging of deciduous trees, has already been noted (Fig. 4.8). Could this alter the characteristics of the forest as a whole? A GIS analysis was used to determine the landscape-scale effects of beaver browsing on forest cover types of the Kabetogama Peninsula. A 50-m buffer was created surrounding each interior lake and beaver pond that existed as of 1988, and used to clip a 1988 forest cover map (Allen et al. 1993). Summary statistics compared the areal proportion of cover types within and outside of the buffer.

This GIS analysis showed large differences in two forest cover categories: aspen and balsam fir–white spruce (Allen et al. 1993). Aspen cover was 13.5% less within the 50-m browse zone than it was in the unbrowsed forest, and balsam–spruce cover was 11.9% greater (Table 4.10). Thus, selective herbivory of aspens by beavers has reduced their abundance in the landscape, favoring balsam–spruce and other evergreens that are avoided by beavers.

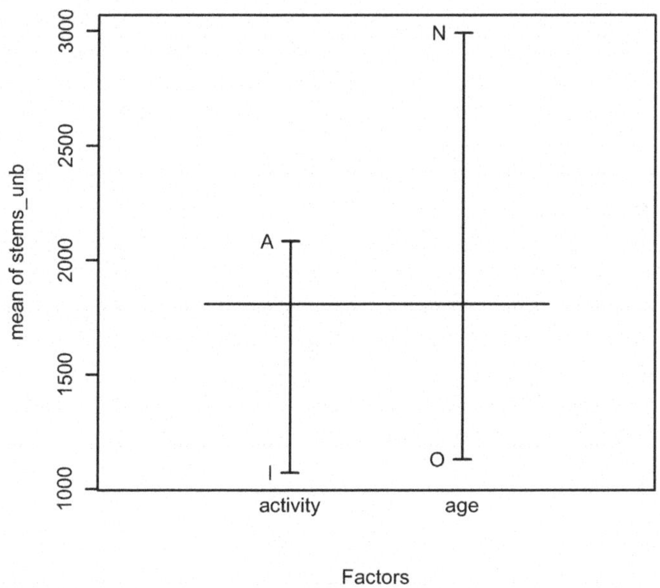

Fig. 4.17 Effects of pond age (N = new, O = old) and activity (A = active, I = inactive) on stems/ha (mean of stems_unb) in the residual forest around beaver ponds

Table 4.10 1988 land cover percentages of upland forest types within a 50-m buffer of beaver ponds

Cover type	Percentage of unbrowsed upland	Percentage of browsed upland
Aspen	39.2%	25.7%
Aspen w/ conifer understory	28.2%	26.2%
Mixed aspen—conifer	4.1%	2.5%
Balsam fir—white spruce	4.7%	16.6%
Red and white pine	4.6%	3.0%
Jack pine	14.7%	17.9%
Brush—shrub	4.6%	6.8%
Oak	0.7%	1.4%

4.9 Conclusions

- Beaver engineering includes horticulture, silviculture, and morticulture (i.e., the cultivation of woody debris).
- Beavers cut wood primarily for forage, but also to acquire dam and lodge building materials.
- Beavers fell mature trees, cutting off branches to transport back to their ponds.
- The trembling aspen (*Populus tremuloides*) is by far the beaver's preferred woody forage species. Conifers are avoided.

- Plant secondary metabolites (e.g., phenols, terpenes, tannins) deter beaver browsing of some woody species. High caffeic acid concentrations in juvenile trembling aspen stems may explain beavers' preference for larger trembling aspen trees.
- The plant secondary metabolites examined did not explain beavers' avoidance of speckled alder (*Alnus incana*).
- Beavers are central place foragers, rarely traveling more than 50 m from the safety of their ponds.
- Beaver foraging can clear-cut the forests near their ponds. Beaver selective herbivory can create distinct plant assemblages.
- Beavers have a sustained need for forage acquisition, but the pond riparian zone may become depleted of suitable forage when browsing is intensive and/or prolonged.
- Selective beaver herbivory near beaver ponds and beaver-occupied lakes can create a "bathtub ring" of avoided conifers around their periphery.

References

Aldous SE (1938) Beaver food utilization studies. J Wildl Manag 2(4):215–222

Allen B, Salés J, Bonde J, Meysembourg P, Johnston CA (1993) Effects of past disturbance on present-day forests at Voyageurs National Park. In: ARC/INFO maps 1992. Environmental Systems Research Institute, Redlands, p 63

Anderson CB, Pastur GM, Lencinas MV, Wallem PK, Moorman MC, Rosemond AD (2009) Do introduced North American beavers Castor canadensis engineer differently in southern South America? An overview with implications for restoration. Mammal Rev 39(1):33–52

Bailey JK, Schweitzer JA, Rehill BJ, Lindroth RL, Martinsen GD, Whitham TG (2004) Beavers as molecular geneticists: a genetic basis to the foraging of an ecosystem engineer. Ecology 85(3):603–608

Barnes DM, Mallik AU (1996) Use of woody plants in construction of beaver dams in northern Ontario. Can J Zool 74(9):1781–1786

Barnes DM, Mallik AU (2001) Effects of beaver, *Castor canadensis*, herbivory on streamside vegetation in a northern Ontario watershed. Can Field-Nat 115(1):9–21

Barnes WJ, Dibble E (1988) The effects of beaver in riverbank forest succession. Can J Bot 66:40–44

Basey JM, Jenkins SH (1995) Influences of predation risk and energy maximization on food selection by beavers (*Castor canadensis*). Can J Zool 73(12):2197–2208

Basey JM, Jenkins SH, Busher PE (1988) Optimal central-place foraging by beavers: tree-size selection in relation to defensive chemicals of quaking aspen. Oecologia 76(2):278–282

Basey JM, Jenkins SH, Miller GC (1990) Food selection by beavers in relation to inducible defenses of *Populus tremuloides*. Oikos 59(1):57–62

Belovsky GE (1984) Summer diet optimization of beaver. Am Midl Nat 111(2):209–222

Bray JR, Curtis JT (1957) An ordination of the upland forest communities in southern Wisconsin. Ecol Monogr 27:325–349

Bryant JP, Provenza FD, Pastor J, Reichardt PB, Clausen TP, du Toit JT (1991) Interactions between Woody plants and browsing mammals mediated by secondary metabolites. Annu Rev Ecol Syst 22:431–446

Busher PE (1996) Food caching behavior of beavers (*Castor canadensis*): selection and use of woody species. Am Midl Nat 135(2):343–348

Clarke KR, Gorley RN (2001) PRIMER v6: user manual/tutorial. PRIMER-E, Plymouth

Clausen TP, Evans TP, Reichardt PB, Bryant JP (1989) A simple method for the isolation of salicortin, tremulacin, and tremuloiden from quaking aspen (*Populus tremuloides*). J Nat Prod 52(1):207–209

Coley PD, Bryant JP, Chapin FS (1985) Resource availability and plant antiherbivore defense. Science 230(4728):895–899

Donkor NT, Fryxell JM (1999) Impact of beaver foraging on structure of lowland boreal forests of Algonquin Provincial Park, Ontario. For Ecol Manag 118(1–3):83–92

Donkor NT, Fryxell JM (2000) Lowland boreal forests characterization in Algonquin Provincial Park relative to beaver (Castor canadensis) foraging and edaphic factors. Plant Ecol 148(1):1–12

Doucet CM, Fryxell JM (1993) The effect of nutritional quality on forage preference by beavers. Oikos 67(2):201–208

Easter-Pilcher A (1990) Cache size as an index to beaver colony size in northwestern Montana. Wildl Soc Bull 18(2):110–113

Faber-Langendoen D, Aaseng N, Hop K, Lew-Smith M (2007) Field guide to the plant community types of Voyageurs National Park. U.S. Geological Survey, Reston

Fienberg SE (1977) The analysis of cross-classified categorical data. MIT Press, Cambridge

Freedman B (1984) The relationship between the aboveground dry weight and diameter for a wide size range of erect land plants. Can J Bot 62(11):2370–2374

Fryxell JM, Doucet CM (1993) Diet choice and the functional-response of beavers. Ecology 74(5):1297–1306

Gallant D, Berube CH, Tremblay E, Vasseur L (2004) An extensive study of the foraging ecology of beavers (Castor canadensis) in relation to habitat quality. Can J Zool 82(6):922–933

Gerwing TG, Johnson CJ, Alström-Rapaport C (2013) Factors influencing forage selection by the North American beaver (Castor canadensis). Mamm Biol Z Säugetierkd 78(2):79–86

Gibson GG (1957) A study of beaver colonies in southern Algonquin Park, Ontario, with particular reference to the available food. University of Toronto, Toronto

Hagerman AE (1987) Radial diffusion method for determining tannin in plant extracts. J Chem Ecol 13(3):437–449

Hall JG (1960) Willow and aspen in the ecology of beaver on Sagehen Creek, California. Ecology 41(3):484–494

Harmon ME (2001) Moving towards a new paradigm for Woody detritus management. Ecol Bull 49:269–278

Hassinger J, Payne J (2008) Dead wood for wildlife. Pennsylvania woodlands no. 7. Pennsylvania State University, University Park

Hatt RT (1944) A large beaver-felled tree. J Mammal 25(3):313

Herbison B, Rood SB (2015) Compound influences of river damming and beavers on riparian cottonwoods: a comparative study along the Lardeau and Duncan Rivers, British Columbia, Canada. Wetlands 35(5):945–954

Howard RJ (1982) Beaver habitat classification in Massachusetts. University of Massachusetts, Amhearst

Jenkins SH (1975) Food selection by beavers: a multidimensional contingency table analysis. Oecologia 21:157–173

Jenkins SH (1979) Seasonal and year-to-year differences in food selection by beavers. Oecologia 44(1):112–116

Jenkins SH (1980) A size-distance relation in food selection by beavers. Ecology 61(4):740–746

Jenkins SH, Busher PE (1979) Castor canadensis. Mamm Species 120:1–8

Johnston CA, Naiman RJ (1990) Browse selection by beaver: effects on riparian forest composition. Can J For Res 20:1036–1043

Johnston CA, Windels SK (2015) Using beaver works to estimate colony activity in boreal landscapes. J Wildl Manag 79(7):1072–1080

Masslich W, Brotherson J, Cates R (1988) Relationships of aspen (Populus tremuloides) to foraging patterns of beaver (Castor canadensis) in the Strawberry Valley of central Utah. West North Am Nat 48(2):250–263

McGinley MA, Whitham TG (1985) Central place foraging by beavers (*Castor canadensis*): a test of foraging predictions and the impact of selective feeding on the growth form of cottonwoods (*Populus fremontii*). Oecologia 66:558–562
Milligan HE, Humphries MM (2010) The importance of aquatic vegetation in beaver diets and the seasonal and habitat specificity of aquatic-terrestrial ecosystem linkages in a subarctic environment. Oikos 119(12):1877–1886
Morris DM (2014) Aquatic habitat use by North American moose (Alces alces) and associated richness and biomass of submersed and floating-leaved aquatic vegetation in north-central Minnesota. Lakehead University, Thunder Bay
Mortenson SG, Weisberg PJ, Ralston BE (2008) Do beavers promote the invasion of non-native Tamarix in the Grand Canyon riparian zone? Wetlands 28(3):666–675
Müller-Schwarze D, Schulte BA, Sun L, Müller-Schwarze A, Müller-Schwarze C (1994) Red maple (Acer rubrum) inhibits feeding by beaver (Castor canadensis). J Chem Ecol 20(8):2021–2034
Northcott TH (1971) Feeding habits of beaver in Newfoundland. Oikos 22:407–410
Northcott TH (1972) Water lilies as beaver food. Oikos 23(3):408–409
Parker JD, Caudill CC, Hay ME (2007) Beaver herbivory on aquatic plants. Oecologia 151(4):616–625
Peinetti HR, Baker BW, Coughenour MB (2009) Simulation modeling to understand how selective foraging by beaver can drive the structure and function of a willow community. Ecol Model 220(7):998–1012
Perala DA (1977) Managers handbook for aspen in the north central states. North Central Forest Experiment Station, St Paul
Pinkowski B (1983) Foraging behavior of beavers (Castor canadensis) in North Dakota. J Mammal 64(2):312–314
Raffel TR, Smith N, Cortright C, Gatz AJ (2009) Central place foraging by beavers (*Castor canadensis*) in a complex lake habitat. Am Midl Nat 162(1):62–73
Ray AM, Rebertus AJ, Ray HL (2001) Macrophyte succession in Minnesota beaver ponds. Can J Bot 79(4):487–499
Rosenthal GA, Janzen DH (eds) (1979) Herbivores: their interaction with secondary plant metabolites. Academic Press, Orlando
Severud WJ, Belant JL, Windels SK, Bruggink JG (2013a) Seasonal variation in assimilated diets of American beavers. Am Midl Nat 169(1):30–42
Severud WJ, Windels SK, Belant JL, Bruggink JG (2013b) The role of forage availability on diet choice and body condition in American beavers (*Castor canadensis*). Mamm Biol Z Säugetierkd 78(2):87–93
Shelton PC (1966) Ecological studies of beavers, wolves, and moose in Isle Royale National Park, Michigan. Purdue University, Lafayette
Smith DW, Trauba DR, Anderson RK, Peterson RO (1994) Black bear predation on beavers on an Island in Lake superior. Am Midl Nat 132(2):248–255
Smith DW, Tyers DB (2008) The beavers of Yellowstone. Yellowstone Sci 16(3):4–15
Stefen C, Habersetzer J, Witzel U (2016) Biomechanical aspects of incisor action of beavers (*Castor fiber* L.) J Mammal 97(2):619–630
Wright JP, Jones CG, Flecker AS (2002) An ecosystem engineer, the beaver, increases species richness at the landscape scale. Oecologia 132(1):96–101

Chapter 5
Soils of Beaver Meadows

Abstract The Kabetogama Peninsula is underlain by dipping beds of biotite schist bedrock that form ridges and valleys which focus land surface runoff. A variety of glacial deposits that influence soil properties are superimposed upon this bedrock base. Soils derived from glaciolacustrine deposits have pedogenic carbonates and a high clay content (>30% clay), whereas soils derived from glaciofluvial deposits have a low carbonate content and are sandy. Analysis of soils from well boreholes within a single beaver meadow showed textures ranging from sandy loam (mean phi size = 3.59) to clay (mean phi size = 7.31). The soil survey for the Kabetogama Peninsula maps beaver impoundments primarily as Sax-Tacoosh complex, ponded.

Hans Jenny's classical treatise on soil development lists the five soil forming factors as: parent material, topography, climate, biota, and time (Jenny 1941). By changing the moisture regime of soils that are impounded, beaver engineering alters soil climate and biota. By changing organic matter accretion, beaver engineering alters parent material. By digging canals and promoting erosion/deposition, beaver engineering alters topography. Thus, beavers affect four of the five soil forming factors, and the soils of beaver ponds and meadows differ from the unimpounded soils of the boreal upland. Each of these soil forming factors is discussed below.

5.1 Parent Material

Parent material is the underlying geologic material in which soil horizons form. Parent materials on the Kabetogama Peninsula include disintegrated rock, various glacial deposits, alluvium, and organic matter. The soil survey for Voyageurs National Park lists nine generic types of parent material, some of which are combinations of two materials (Table 5.1). Two-thirds of the Kabetogama Peninsula is mapped as having soils derived from "loamy drift over igneous rock" (USDA/NRCS 2015).

C.A. Johnston, *Beavers: Boreal Ecosystem Engineers*,
DOI 10.1007/978-3-319-61533-2_5

Table 5.1 Parent materials of soils throughout the Kabetogama Peninsula

Parent material	Area (km^2)
Loamy drift over igneous rock	211.3
Herbaceous organic material over silty glaciolacustrine deposits	22.5
Herbaceous organic material	16.2
Mossy organic material over herbaceous organic material	16.1
Silty glaciolacustrine deposits	9.4
Clayey glaciolacustrine deposits	8.9
Organic material over loamy drift	3.4
Sandy colluvium and/or sandy glaciolacustrine deposits and/or sandy outwash over clayey glaciolacustrine deposits	2.1
Herbaceous organic material over stratified loamy herbaceous organic material over loamy alluvium	1.5
Water	*10.6*
Grand total	301.9

Data derived from Soil Survey Geographic Database (SSURGO) for Voyageurs National Park (USDA/NRCS 2015)

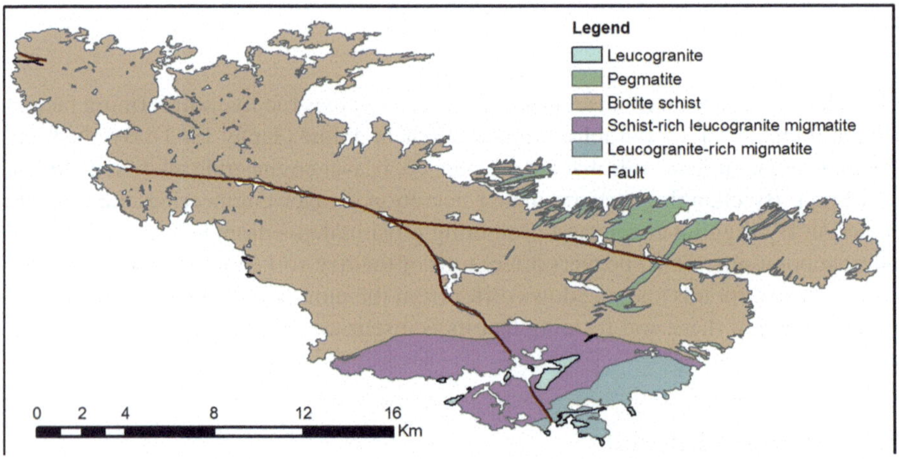

Fig. 5.1 Bedrock of the Kabetogama Peninsula, extracted from *Bedrock geologic map of Voyageurs National Park and vicinity, Minnesota* (Hemstad et al. 2002)

5.1.1 Bedrock

The main bedrock type underlying the Kabetogama Peninsula is biotite schist, a metamorphic rock with sheet-like, roughly parallel grains of biotite mica interleaved with other minerals (Fig. 5.1) (Hemstad et al. 2002). At the landscape scale, the dipping beds of biotite schist form parallel ridges and valleys (see exposed bedrock on Fig. 1.5) that contribute to the peninsula's complex topography (Fig. 2.1).

The southern portion of the peninsula contains migmatite, consisting of biotite schist interlayered with leucogranite, a light colored granitic rock with almost no dark minerals. Pegmatite, a granite with coarse crystals (larger than 2.5 cm), also occurs in some areas of the peninsula. All of the bedrock is Early Precambrian, about 2700 million years old (Ojakangas and Matsch 1982).

A major fault line runs east-west across the peninsula from Black Bay to Mica Bay, passing under a number of the permanent lakes: Locator, War Club, Quill, Loiten, Oslo, Brown, Beast (Fig. 5.1). A secondary fault extends southeast from Loiten Lake, past Shoepack and Little Shoepack Lakes, to the southern edge of the peninsula. Topographic depressions that mark the faults are often filled with beaver ponds.

Flooding by beavers exposes the crystalline structure of the bedrock by killing the dark-colored lichens that grow on bedrock, as evidenced by high water marks that are visible on exposed bedrock shorelines along abandoned beaver ponds (Fig. 5.2). Beaver ponds may also affect rock weathering rates. Disintegrated schist bedrock residuum that was distinct in color and texture from overlying glacial deposits was found at the base of a beaver meadow soil, indicating that beaver pond waters contributed to its weathering (Johnston 2000).

5.1.2 Glacial Till (Drift)

Glaciers moved over northern Minnesota throughout the Wisconsin glaciation period, fed by the Laurentide Ice Sheet. The low topography encountered by the Laurentide Ice Sheet along its southern border influenced the geometry of the glacier margin, and distinctive ice bulges ("lobes") developed (Ojakangas and Matsch 1982). The Rainy lobe originated northeast of the region, and last advanced about 15,500 years ago (Clayton and Moran 1982). Its retreat allowed the Koochiching lobe (formerly the St. Louis sublobe) to advance from the northwest about 13,500 years ago (Mooers 1988). The Kabetogama Peninsula was mapped as the limit of glacial readvance of the Koochiching lobe (Sado et al. 1995), and contains deposits from both the Rainy and Koochiching lobes (Fig. 5.3).

The two glacial lobes have distinctive till (also called "drift"), which is the poorly sorted mix of earth and rocks that is deposited when the glacier retreats. The till from the Rainy lobe is pale reddish-brown, sandy, noncalcareous, and incorporates diverse rock fragments from various locations in Ontario and northeast Minnesota (Sado et al. 1995; Ojakangas and Matsch 1982) (Fig. 5.4). The till from the Koochiching lobe is yellowish-brown, finer textured (loam and silt loam), very calcareous, and contains primarily shale, limestone, and dolomite rock fragments (Sado et al. 1995).

Fig. 5.2 Mark on bedrock denotes high water level of former beaver pond. Ski pole in photo for scale

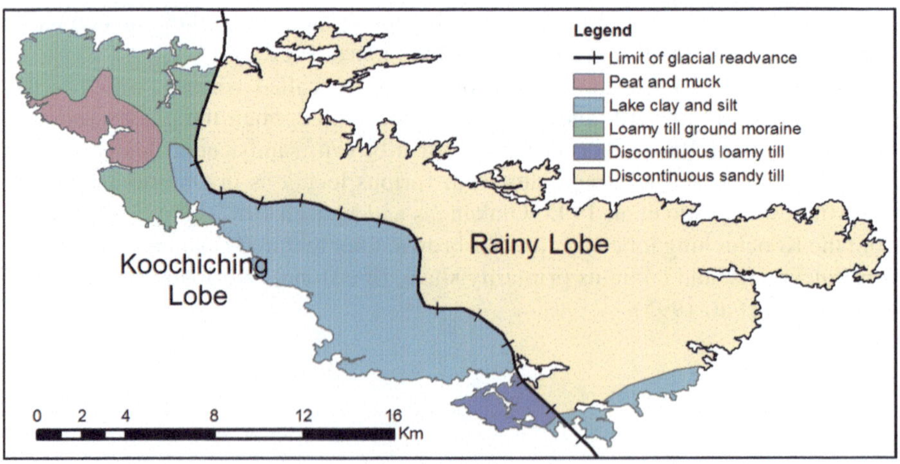

Fig. 5.3 Glacial geology of the Kabetogama Peninsula, extracted from *Quaternary geologic map of the Lake of the Woods 4° × 6° quadrangle, United States and Canada* (Sado et al. 1995)

5.1.3 Glaciofluvial Deposits

The Kabetogama Peninsula has local glaciofluvial deposits (also known as outwash), which are sands and gravels that were sorted and deposited by flowing glacial meltwaters (Fig. 5.5). Glacial outwash is located in drainageways, and its cumulative area is small (Table 5.1). Soils formed in outwash are sandy. Some beaver ponds have impounded glaciofluvial soils.

Fig. 5.4 Rocks embedded in glacial till exposed in bottom of drained beaver pond

Fig. 5.5 Sand and gravel outwash deposited in the Clyde Creek valley. Shovel head in picture for scale

5.1.4 *Glaciolacustrine Deposits*

As noted in Chap. 2, the Kabetogama Peninsula was submerged under the waters of early Glacial Lake Agassiz 11,000 years ago, which left behind silty and clayey glaciolacustrine deposits (Teller 1985). These deposits are most extensive on the flat western end of the peninsula, but also occur in drainageways throughout (Fig. 2.4). These glaciolacustrine deposits are one of the factors that makes the Kabetogama Peninsula such a haven for beavers.

5.1.5 *Alluvium and Colluvium*

Alluvium is sediment deposited by the running water of streams during the Holocene (i.e., 9700 years ago to the present). The particle size of alluvium varies from silt to gravel and cobble, depending on the streamflow. Floodplains that experience infrequent overbank flow have silty alluvium, whereas streambeds have coarser deposits.

In regions where streams transport substantial amounts of sediment, beaver dams can greatly increase sedimentation rates. Published sedimentation rates range from 0.4 cm $yr.^{-1}$ in a headwater beaver pond in central Ontario (Devito and Dillon 1993) to 27.9 cm $yr.^{-1}$ in a beaver pond in the mountains of Montana (Butler and Malanson 1995). Beavers have been called geomorphic agents due to the long-term accumulation of sediments in valley bottoms impounded by their dams (Ives 1942; Rudemann and Schoonmaker 1938; Butler and Malanson 1994; Gurnell 1998). However, the interpretation of these landforms is confounded by their presence in glaciated terrain, where U-shaped valleys may have been caused by glacial and post-glacial geomorphic processes rather than beaver activity (Retzer et al. 1956).

Catastrophic failure of beaver dams can cause erosion, a process that is less well documented in the literature (Butler 1989). A stream in Alberta was drastically altered when a flood wave from a breached beaver dam eroded and redeposited large amounts of debris (Hillman 1998). Alluvium that remains downstream of a dam breach is usually gravel, cobbles, and boulders (Johnston 2012).

The streams of the Kabetogama Peninsula have low discharge and transport little sediment, so alluvium is an uncommon parent material there. Only four areas are mapped on the soil survey as having alluvial soils, all along creeks: Cranberry Creek downstream of Locator Lake (catchment #3), the Shoepack Lake outlet creek (catchment #8), the creek connecting Cruiser, Mud, and Weir Lakes (catchment #13), and the creek flowing from Agnes Lake out to Namakan Lake (catchment #16). Drained beaver ponds with bedrock bottoms show slight evidence of sedimentation (Fig. 5.6).

Colluvium is material on or at the foot of a slope that was moved there chiefly by gravity. Colluvial deposits occur around the edges of beaver ponds where wave action has caused slumping of the shoreline. Two soil profiles that I studied in Kabetogama Peninsula beaver meadows contained a layer of colluvium overlying the pre-beaver soil surface (Johnston et al. 1995; Johnston 2000). Because the colluvium is sandy it contains much less carbon than the soil it buried (which used to be the soil surface) or the soil on top of it (which is the *current* soil surface) (Fig. 5.7).

Fig. 5.6 Drained beaver pond bottom showing cracked patches of alluvial silt deposited over the bedrock bottom

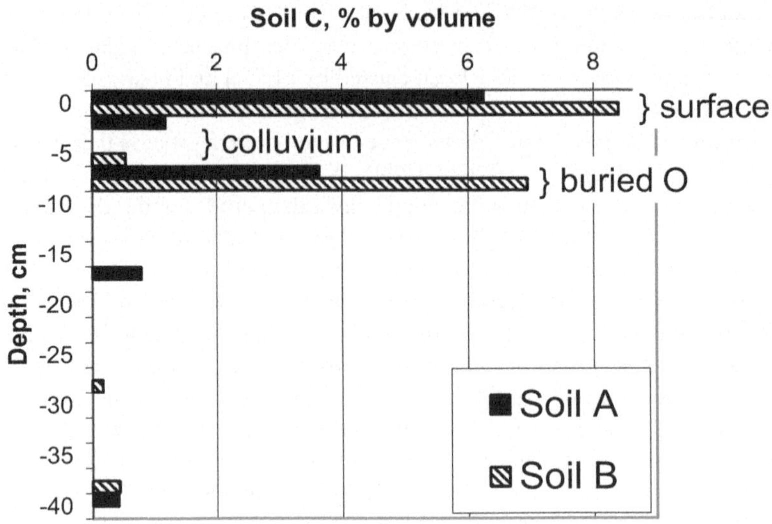

Fig. 5.7 Colluvium layers (low carbon content) overlying the pre-beaver soil surface (buried O, high carbon content)

5.1.6 Organic Matter

Organic matter has been accumulating in boreal wetlands since the retreat of the glaciers. Organic parent material is the partially decomposed remains of plants that have been growing on site. Dark, finely divided organic matter in which individual plant parts are indiscernible to the naked eye is generically called "muck," and coarser, less decomposed organic matter is called "peat." I referred to organic soil wetlands as "peatlands" in Chap. 2. Soil scientists refer to organic soils as Histosols.

Organic soils provide a harsh growing environment for plants due to the lack of oxygen in the rooting zone, nutrient limitations, and cold soil temperatures. Therefore, boreal organic soils contain special plant communities, bogs and fens, that are adapted to these conditions. My colleagues and I used the 1940 aerial photography to identify and map peatlands of the Kabetogama Peninsula by looking for flat terrain that supported these special plant communities (Fig. 5.8). We did this to distinguish peatlands that existed prior to beaver population expansion from wetlands that were *created* by beavers.

The largest peatlands occur on the low-lying western end of the Kabetogama Peninsula (Fig. 5.8). The vegetation of these areas is primarily low shrubs such as leatherleaf, sweet gale, and bog-birch (Faber-Langendoen et al. 2007). Elsewhere, such as on Soldier Point and near Shoepack Lake, peatland vegetation types are bogs and northern conifer swamps. The vegetation of the peatland shown in Fig. 2.12 is a "black spruce/labrador tea poor swamp" (Faber-Langendoen et al. 2007).

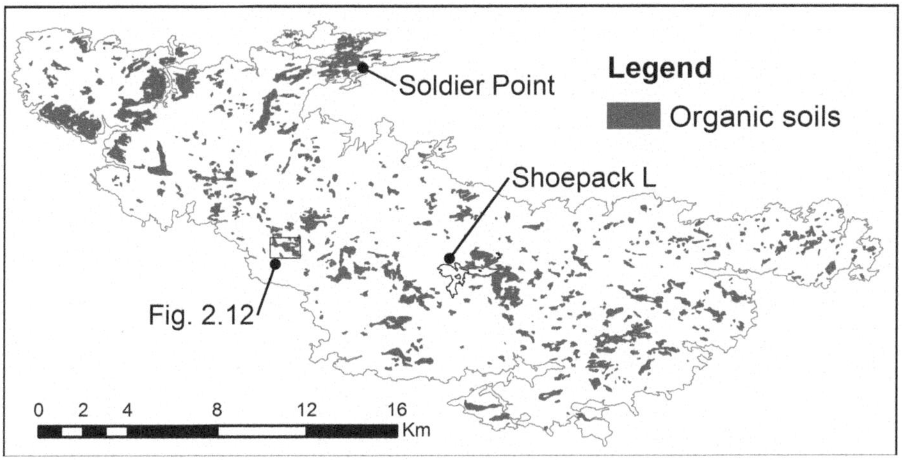

Fig. 5.8 Organic soils of the Kabetogama Peninsula, mapped using 1940 aerial photos. Rectangle denotes the footprint of the peatland aerial photo image shown in Fig. 2.12

5.2 Particle Size Distribution of Beaver Meadow Soils

The individual grains that make up a mineral soil include clay (<0.002 mm), silt (0.002–0.05 mm), and sand (0.05–2 mm). The U.S. Department of Agriculture uses ternary (3-axis) graphs to display the relative proportions of these three components, and to group them into soil texture classes (Soil Survey Staff 1951). Sedimentologists often use a finer Phi scale to distinguish particle size distribution, with a Phi value of 0 representing 1 mm sand and a Phi value greater than 9 representing clay (<0.002 mm) (Krumbein and Aberdeen 1937). The particle size distribution of soils affects properties such as bulk density (i.e., mass per unit volume), hydraulic conductivity, and wilting point (Saxton et al. 1986).

Particle size distribution was measured on samples collected during the field study of beaver meadow soil profiles (Johnston et al. 1995), and on samples collected during well installation at Found and Bluefin meadows (Cahow n.d.). The latter samples were analyzed at the Limnological Research Center laboratory of the University of Minnesota-Twin Cities, which also analyzed for organic matter (loss on ignition at 550°, $LOI_{550°}$) and $CaCO_3$ content (calculated from loss on ignition at 1000° (Heiri et al. 2001)).

The Bluefin borehole samples were mostly fine textured (mean Phi sizes ranging from 5.6 to 7.8, Table 5.2). The deepest Bluefin sample, extracted from 5.2 m, had a relatively high organic content ($LOI_{550°} = 28.4\%$) and a mean Phi size of 4.9. Some of the Bluefin samples had $CaCO_3$ contents as high as 30%, but there was no clear pattern of change with sample depth.

The Found borehole samples were collected from much shallower strata, the deepest only 70 cm (Table 5.2). Within each of the three Found boreholes, organic

Table 5.2 Physical properties of samples collected from well boreholes

Pond	Borehole	Depth (cm)	LOI 550° (%)	$CaCO_3$ (%)	Mean size (phi)	USDA class
BlueFin	LB78m	46	15.0	5.25	5.57	sil
BlueFin	LB78m	61	5.1	0.87	6.32	sil
BlueFin	LB78m	122	3.9	30.09	6.94	sicl
BlueFin	LB78m	137	11.7	11.62	7.11	sicl
BlueFin	LB78m	152	5.0	21.58	6.66	sil
BlueFin	LB78m	183	4.9	22.58	6.62	sil
BlueFin	LB78m	213	7.6	8.18	6.64	sil
BlueFin	LB78m	244	7.2	6.19	7.49	sicl
BlueFin	LB78m	259	6.5	7.24	7.85	sicl
BlueFin	LB78m	274	5.1	10.50	7.84	sic
BlueFin	UB79s	61	12.5	2.10	6.74	sil
BlueFin	UB79s	76	17.7	2.52	6.60	sil
BlueFin	UB80d	457	16.8	29.41	6.81	sil
BlueFin	UB80d	518	28.4	8.80	4.91	loam
Found	FP02	10	10.7	0.17	3.59	sl
Found	FP02	30	10.6	0.35	5.40	loam
Found	FP02	55	7.8	0.47	6.15	sil
Found	FP03	8	41.5	0.61	4.58	loam
Found	FP03	23	22.8	0.69	4.43	loam
Found	FP03	40	16.0	0.53	7.65	sil
Found	FP03	55	18.9	0.88	7.31	clay
Found	FP03	70	28.2	1.28	6.40	cl
Found	FP06	8	27.6	0.37	6.11	sil
Found	FP06	40	17.4	0.10	6.97	sicl

LOI 550° loss on ignition at 550° (Heiri et al. 2001). Phi scale as described by Krumbein (Krumbein and Aberdeen 1937). *cl* clay loaml; *sic* silty clay; *sicl* silty clay loam; *sil* silt loam; *sl* sandy loam (Soil Survey Staff 1951)

matter decreased and mean Phi increased with depth (i.e., deeper soils were finer textured). Borehole FP02 was located adjacent to the stream, and had surface soils that would classify as sandy loam in the USDA classification system, indicative of alluvium.

When all particle size analyses were plotted together on a ternary graph, their diversity was apparent (Fig. 5.9). The glaciofluvial soils were very sandy, and clustered at the lower left axis of the ternary graph. Most of the glaciolacustrine soils contained 30% or more clay, and tended to fall in the upper 2/3 of the ternary graph. Samples collected from the boreholes contained mixtures ranging from sandy loam to clay (Table 5.2).

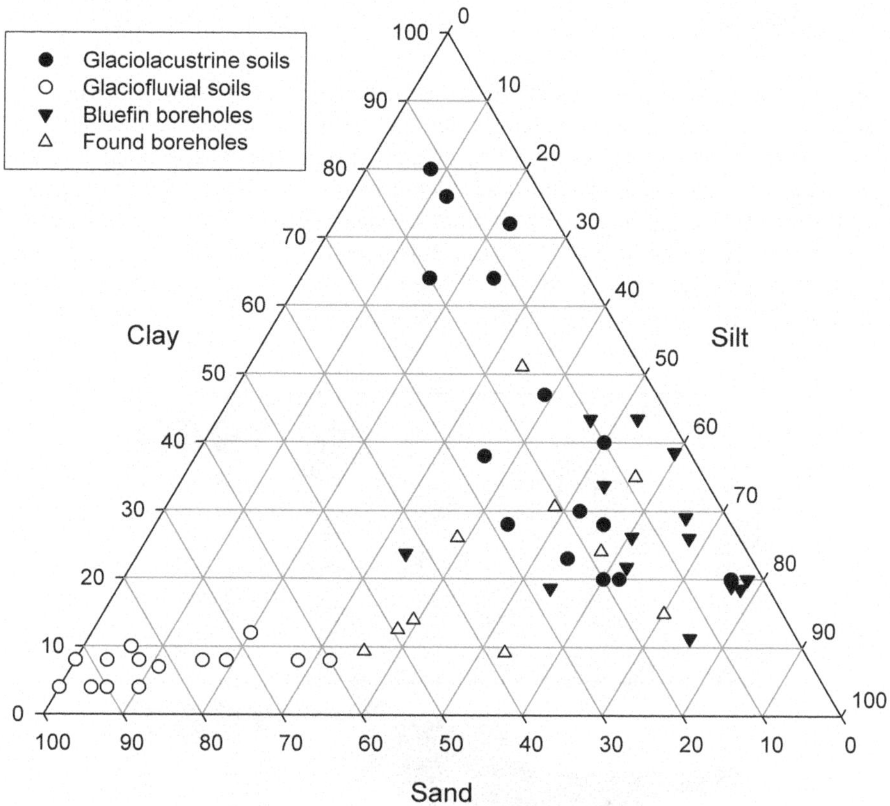

Fig. 5.9 Ternary graph plotting the relative proportion of sand, silt, and clay in soils sampled from well boreholes and soil profiles

5.3 Field Study of Beaver Meadow Soil Profiles

Soil profiles were described and sampled at several drained beaver ponds (Fig. 3.2). Soil pits were hand-dug in hydrosequences extending from dry (never impounded) to wet (formerly submerged, now saturated). The hydrosequences had two different parent materials: glaciofluvial outwash and glaciolacustrine sediments. An organic soil ("Histosol") was also sampled using a 3 × 3 cm box corer (Cuttle and Malcolm 1979) to remove an intact core of material that was 70 cm long. Soil colors (e.g., 10YR 4/6) were described with a Munsell color chart, and soil texture was determined by hand texturing in the field supplemented by laboratory analysis. Detailed methods and results are previously published (Johnston et al. 1995; Johnston 2014; Johnston 2000).

5.3.1 Histosol

Soils with surface peat layers 40 cm or more thick are called “Histosols.” A histosol was sampled from the Lost Pond site, which had been a small peatland with woody vegetation prior to inundation by a beaver dam (Fig. 5.10). The organic soil was hemic (intermediate decomposition) to a depth of 60 cm, and sapric (very decomposed) from 60–70 cm, the maximum depth sampled. Soil color was very dark grayish brown (10YR 3/2) to black (5YR 2/1). Bulk density of all horizons was very low, from 0.09 Mg m^{-3} at the surface to 0.35 Mg m^{-3} at 60–70 cm. The organic parent material in the profile was judged to be of pre-beaver origin (Johnston 2014).

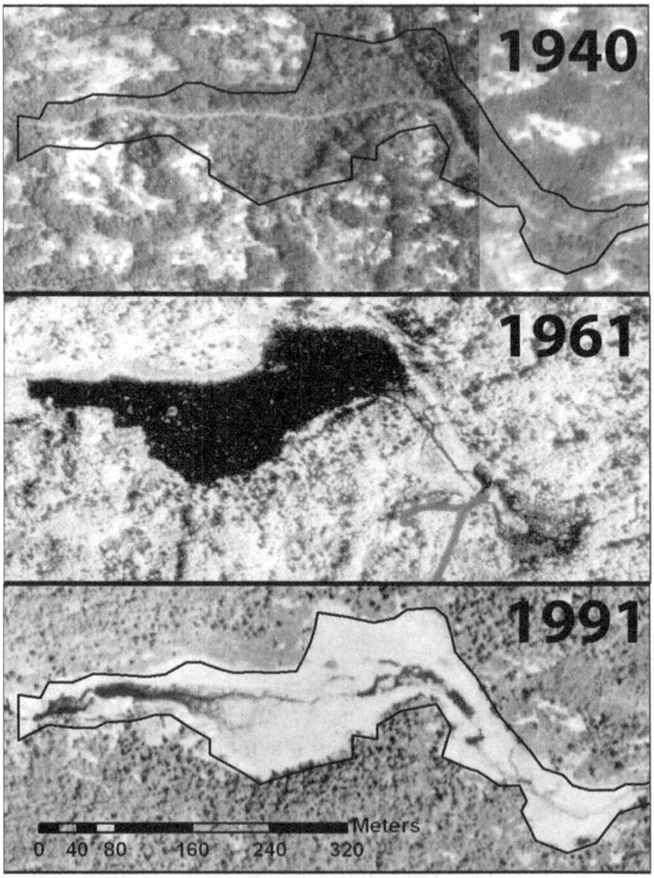

Fig. 5.10 Aerial photographs illustrating changes in vegetation and ponding of the Lost Pond site. Black boundary on 1940 and 1991 photos shows the 1991 beaver meadow extent superimposed upon the same area in 1940. Reprinted from Johnston CA (2014) Beaver pond effects on carbon storage in soils. Geoderma 213:371–378, with permission from Elsevier

5.3.2 Argiaquoll

The Argiaquoll was sampled from the edge of a former cedar swamp (trees killed by beaver flooding) at the Clyde Creek field site. The soil taxon "Argiaquoll" refers to a water-saturated mineral soil that contains illuviated (argillic) clay and a relatively thick, dark surface (A horizon) containing at least 5.8 g kg^{-1} organic carbon. This type of soil is commonly associated with wet prairies in southwest Minnesota, but is not common in northern Minnesota. The soil was formed in glaciolacustrine parent material, and its A horizon was very dark gray (10YR 3/1) to a depth of 39 cm, the dark color resulting from the incorporation of organic carbon (Johnston et al. 1995). The implications of this thick A horizon for carbon storage are discussed in Chap. 8.

5.3.3 Other Mineral Soils

Soils developed in glacial outwash were quite different than those developed in glaciolacustrine deposits (Table 5.3). The most obvious difference was soil texture, which was sandy loam to sand in the glaciofluvial soils versus clay to silty clay loam in the glaciolacustrine soils. This difference in soil texture affected soil structure, which was much more developed in the glaciolacustrine soils than in the glaciofluvial soils.

In the glaciofluvial hydrosequence, an E horizon occurred only in the driest soil as an albic (bleached) horizon, caused by downward leaching of sesquioxides (Fig. 5.11).

Table 5.3 Comparison of beaver meadow soils with glaciolacustrine versus glaciofluvial parent material

Soil characteristic	Glaciolacustrine hydrosequence	Glaciofluvial hydrosequence
O horizon	Increasing thickness with increasing wetness	Increasing thickness with increasing wetness
E horizon	Eluviation of clay, platy structure, bulk density ≥ 1.6 Mg/m^3	Albic E horizon in driest soil of hydrosequence
Bt horizon	Clay content 3.5–4× that of E horizon	None
Bk horizon	$CaCO_3$ masses present in drier soils, effervescent	None
Bg horizon	Present in wetter soils, greenish gray (5GY 5/1) to dark greenish gray (5G 4/1)	Present in wetter soils, dark grayish brown (2.5YR 4/2) to olive gray (5Y 5/2)
C horizon	Clay content ≥40%	Clay content ≤12%
Oxalate-extractable Fe	≥6.2 mg/kg in B horizons	≤1.5 mg/kg in B horizons
Redox concentrations	Yellowish brown (10YR 5/6) pore linings	Strong brown (7.5YR 4/6) masses
Structure	Moderate fine subangular blocky	Weak fine granular to subangular blocky

Fig. 5.11 This upland soil profile in loamy fine sand outwash consists of well-decomposed black (10YR 2/1) surface litter, a grayish brown (10YR 5/2) E horizon, and brown (7.5YR 3/4) to dark yellowish brown (10YR 4/6) Bw horizons

There were no E horizons in the wetter soils of the glaciofluvial hydrosequence, which were gleyed (Bg horizon) due to the reduction of iron compounds. The side-by-side comparison of a dry and wet soil from the glaciofluvial hydrosequence illustrates the color change and increase in surface organic matter that accompanied beaver impoundment (Fig. 5.12).

The wetter soils in both hydrosequences were gleyed, indicating prolonged reducing conditions. The color of the glaciolacustrine Bg layers was greenish gray (5GY 5/1) to dark greenish gray (5G 4/1). The saturated glaciofluvial soils were also gleyed, but their Bg horizons had chromas of 2 and hues that were more yellow. Rust-colored mottles (redox concentrations) occurred in both soils, indicating fluctuating water tables. The redox concentrations occurred as strong brown (7.5YR 4/6) masses in the weakly structured glaciofluvial soils and as prominent yellowish brown and dark yellowish brown (10YR 4/6–5/6) pore linings in the moderately structured glaciolacustrine soils (Fig. 5.13).

The eluvial horizons (E horizons) in the glaciolacustrine soils were quite dense, consistently denser than the B horizons below them. The Bt horizons sampled

Fig. 5.12 This side-by-side comparison shows the reddish-brown colors and thin surface organic layer of an upland glaciofluvial soil (*left*), versus the gleyed colors and 10 cm thick surface litter of a beaver meadow glaciofluvial soil (*right*)

contained 3.5–4 times the clay content of their overlying E horizon, indicating the downward translocation of clay particles from the E into the argillic Bt horizon (Johnston et al. 1995). The glaciofluvial soils lacked Bt horizons.

The glaciolacustrine soils were also calcareous, and the drier members of the hydrosequence contained Bk horizons. Pedogenic carbonates, which were visible as white masses in the Bk horizons, effervesced violently upon treatment with 2.87 N HCl. The glaciofluvial soils did not contain Bk horizons nor exhibit effervescence.

Soil horizons were analyzed for oxalate-extractable iron, which migrates with clay during interhorizon translocation (Richardson and Hole 1979). There were pronounced subsurface peaks of oxalate-extractable iron ($\geq$6.2 mg/kg) in the glaciolacustrine soils (Table 5.3). Depth to these peak concentrations decreased with increasing wetness (Johnston et al. 1995). The source of this iron was probably the overlying E horizons, which were depleted in iron. There was no comparable illuviation in soils developed in glaciofluvial soils, which contained little oxalate-extractable iron.

Fig. 5.13 Soil profile in a drained beaver pond having glaciolacustrine parent material. Note the rust-colored mottles following the root channels

5.4 Beaver Ponds and Meadows on Soil Surveys

How are beaver ponds depicted on contemporary soil surveys? I determined this by intersecting the Kabetogama beaver pond site maps (Johnston and Windels 2015) with the soil survey for Voyageurs National Park that was completed in 2015 (USDA/NRCS 2015). The map distinguished 23 mapping units within the Kabetogama Peninsula (Table 5.4). Of these, the "Sax-Tacoosh complex" was almost exclusively applied to beaver-impounded soils: 90% of this mapping unit area within the Kabetogama Peninsula occurred in beaver impoundments, and about one half of all beaver impoundments had this soil. The "Bowstring and Fluvaquents" soil, which was the only soil with alluvial parent material, was almost entirely flooded by beaver impoundments, although it covered a much smaller area than did the Sax-Tacoosh complex. Other soils that frequently occurred in beaver impoundments (but were also found in other settings) were Rifle mucky peat, Greenwood peat, and Cathro and Tacoosh soils, ponded. The "water" designation was not applied to beaver ponds on the soil survey; it was used only for permanent lakes and ponds of the Kabetogama Peninsula.

Table 5.4 Soils of the Kabetogama Peninsula as used in the SSURGO map for Voyageurs National Park (USDA/NRCS 2015), and soil type (*U* upland; *M* wet mineral; *O* organic)

Type	Beaver pond area (ha)	% of Total mapped	Mapping unit
U	4	0.8	Baudette, till substratum-Conic, very stony-Littleswan complex, 0–12% slopes
M	132	86.6	Bowstring and Fluvaquents loamy, soils, 0–2% slopes, frequently flooded
M	26	5.2	Brickton-Dalbo complex, 0–3% slopes
M	55	14.0	Brickton-Hassman complex, 0–2% slopes
U	5	2.2	Canthook-Durkeelake complex, 0–12% slopes
O	148	43.2	Cathro and Tacoosh soils, 0–1% slopes, ponded
O	184	14.6	Greenwood peat, 0–1% slopes
O	108	31.1	Greenwood peat, 0–1% slopes, seasonally ponded
U	20	1.7	Insula, very stony-Voyageurs-Wahlsten, very stony complex, 0–12% slopes
U	116	2.4	Insula-Conic-Rock Outcrop complex, 8–25% slopes, very bouldery
U	53	1.1	Insula-Conic-Wahlsten complex, 0–12% slopes, very stony
U	64	1.3	Insula-Conic-Wahlsten complex, 0–25% slopes, very stony
M	19	4.4	Littleswan-Spooner complex, 0–3% slopes
U	1	0.4	Quetico-Insula, very bouldery-Greenwood complex, 0–12% slopes
U	22	1.1	Quetico-Insula-Rock Outcrop complex, 3–18% slopes, very bouldery
U	34	2.2	Quetico-Insula-Rock Outcrop complex, 35–60% slopes, very bouldery
O	217	36.9	Rifle mucky peat, 0–1% slopes
O	573	55.6	Rifle mucky peat, 0–1% slopes, seasonally ponded
O	1573	89.5	Sax-Tacoosh complex, 0–1% slopes, ponded
M	35	7.1	Spooner-Sax complex, 0–2% slopes
O	38	20.3	Tacoosh and Sax soils, 0–1% slopes, frequently flooded
O	61	19.7	Tacoosh and Sax soils, 0–1% slopes, occasionally flooded
U	14	1.2	Wahlsten, very stony-Spooner-Voyageurs complex, 0–8% slopes

% of Total Mapped = beaver-impounded area as a percent of total area of this soil on the Kabetogama Peninsula

Of the three beaver meadow field sites at which we had mapped the soils, only Lost Pond was correctly classified by the soil survey as Sax-Tacoosh complex. The Clyde Creek site was also classified as Sax-Tacoosh complex, but based on its glaciofluvial parent material should probably have been classified as Canthook-

Durkeelake complex. The Shoepack site (Fig. 5.13) was incorrectly classified as a Histosol (Rifle mucky peat, seasonally ponded).

Field soil scientists cannot visit every square meter of a landscape, especially in wilderness terrain as rugged as the Kabetogama peninsula. Furthermore, soil scientists generally do not inspect subaqueous soils, such as those submerged beneath the water of beaver ponds (Johnston 1977). Also, traversing a beaver meadow can be difficult and even dangerous due to tussocks, boulders, and the sharp spears of beaver-cut stumps hidden in the grass. Based on this knowledge and the map evidence, it seems that the contemporary soil scientists generally applied the Sax-Tacoosh complex to beaver-impounded soils, except where the impoundments were clearly part of a larger peatland.

5.5 Conclusions

- The dipping beds of biotite schist bedrock that underlie most of the Kabetogama Peninsula form parallel valleys that are conducive to beaver pond construction.
- Beaver ponds also occur in valleys formed over bedrock faults.
- Material left by two glacial lobes and Glacial Lake Agassiz forms a veneer of Quaternary deposits over the bedrock.
- The most common soil parent materials in beaver meadows of Kabetogama Peninsula beaver meadows are glaciolacustrine deposits and organic matter (Histosols). Less extensive parent materials of beaver meadow soils include glacial till, glaciofluvial deposits, alluvium, and colluvium.
- The texture of samples from soil profiles and well boreholes varied from loamy sand in glaciofluvial soils to clay in glaciolacustrine soils.
- Soils described in the field included a Histosol, an Argiaquoll, and hydrosequences in glaciolacustrine and glaciofluvial parent materials.
- E horizons in the glaciolacustrine soils were very dense, had platy structure, and exhibited clay eluviation, whereas the glaciofluvial soils exhibited little eluviation
- B horizons in the glaciolacustrine soils exhibited properties that were absent from the glaciofluvial soils, including argillic clay illuviation (Bt) and carbonate masses (Bk).
- The wettest soils in the hydrosequences were gleyed, regardless of parent material
- The thickness of the surface O horizon increased with wetness, regardless of parent material
- On a contemporary soil survey, beaver-impounded soils were usually designated as the Sax-Tacoosh complex, or as various types of organic soil.

References

Butler DR (1989) The failure of beaver dams and resulting outburst flooding: a geomorphic hazard of the Southeastern Piedmont. Geogr Bull 31(1):29–38

Butler DR, Malanson GP (1994) Beaver landforms. Can Geogr 38(1):76–79

Butler DR, Malanson GP (1995) Sedimentation rates and patterns in beaver ponds in a mountain environment. Geomorphology 13(1–4):255–269

Cahow KR. Spatial and temporal variations in the groundwater system of boreal beaver meadows. M.S. thesis (incomplete), University of Minnesota, Duluth

Clayton L, Moran SR (1982) Chronology of late Wisconsinan glaciation in middle North America. Quat Sci Rev 1(1):55–82

Cuttle SP, Malcolm DC (1979) A corer for taking undisturbed peat samples. Plant Soil 51:297–300

Devito KJ, Dillon PJ (1993) Importance of runoff and winter anoxia to the P and N dynamics of a beaver pond. Can J Fish Aquat Sci 50(10):2222–2234

Faber-Langendoen D, Aaseng N, Hop K, Lew-Smith M (2007) Field guide to the plant community types of Voyageurs National Park. U.S. Geological Survey, Reston

Gurnell AM (1998) The hydrogeomorphological effects of beaver dam-building activity. Prog Phys Geogr 22(2):167–189

Heiri O, Lotter AF, Lemcke G (2001) Loss on ignition as a method for estimating organic and carbonate content in sediments: reproducibility and comparability of results. J Paleolimnol 25(1):101–110

Hemstad CB, Southwick DL, Ojakangas RW (2002) M-125 Bedrock geologic map of Voyageurs National Park and vicinity, Minnesota. Minnesota Geological Survey, St. Paul

Hillman GR (1998) Flood wave attenuation by a wetland following a beaver dam failure on a second order boreal stream. Wetlands 18(1):21–34

Ives RL (1942) The beaver-meadow complex. J Geomorphol 5:191–203

Jenny H (1941) Factors of soil formation. McGraw-Hill, New York

Johnston CA (1977) Considerations for wetland inventories in Wisconsin. University of Wisconsin, Madison

Johnston CA (2000) Wetland soil and landscape alteration by beavers. In: Richardson JL, Vepraskas MJ (eds) Wetland soils: their genesis, morphology, hydrology, landscapes, and classification. CRC Press, Boca Raton, pp 391–408

Johnston CA (2012) Beaver wetlands. In: Batzer DP, Baldwin A (eds) Wetland habitats of North America: ecology and conservation concerns. University of California Press, Berkeley, pp 161–171

Johnston CA (2014) Beaver pond effects on carbon storage in soils. Geoderma 213:371–378

Johnston CA, Windels SK (2015) Using beaver works to estimate colony activity in boreal landscapes (online spatial data). CUAHSI Spatial Data Service, Consortium of Universities for the Advancement of Hydrologic Science Inc., Medford. http://spatial.cuahsi.org/johnstonC01/

Johnston CA, Pinay G, Arens C, Naiman RJ (1995) Influence of soil properties on the biogeochemistry of a beaver meadow hydrosequence. Soil Sci Soc Am J 59(6):1789–1799

Krumbein WC, Aberdeen E (1937) The sediments of Barataria Bay. J Sediment Res 7(1):3–17

Mooers HD (1988) Quaternary history and ice dynamics of the St. Croix phase of late Wisconsin glaciation, Central Minnesota. University of Minnesota, Minneapolis

Ojakangas RW, Matsch CL (1982) Minnesota's geology. University of Minnesota Press, Minneapolis

Retzer JL, Swope HM, Remington JD, Rutherford WH (1956) Suitability of physical factors for beaver management in the Rocky Mountains of Colorado. Colorado Department of Game, Fish and Parks, Denver

Richardson JL, Hole FD (1979) Mottling and iron distribution in a Glossoboralf-Haplaquoll hydrosequence in a glacial moraine in Northwestern Wisconsin. Soil Sci Soc Am J 43:552–558

Rudemann R, Schoonmaker WJ (1938) Beaver dams as geologic agents. Science 88:523–525

Sado EV, Fullerton DS, Goebel JE, Ringrose SM (1995) Quaternary geologic map of the Lake of the Woods 4° × 6° quadrangle, United States and Canada, U.S. Geological Survey Miscellaneous Investigations Series Map I-1420 (NM-15). U.S. Geological Survey, Denver

Saxton KE, Rawls WJ, Romberger JS, Papendick RI (1986) Estimating generalized soil-water characteristics from texture. Soil Sci Soc Am J 50(4):1031–1036

Soil Survey Staff (1951) Soil Survey Manual. U.S. Department of Agriculture, Washington

Teller JT (1985) Glacial Lake Agassiz and its influence on the Great Lakes. In: Karrow PF, Calkin PE (eds) Quaternary evolution of the Great Lakes, Special Paper 30. Geological Association of Canada, Toronto, pp 1–16

USDA/NRCS (2015) Soil Survey Geographic (SSURGO) database for Voyagers National Park, Minnesota. http://websoilsurvey.nrcs.usda.gov

Chapter 6
Vegetation of Beaver Impoundments

Abstract We investigated the vegetation of beaver meadows and other beaver-flooded sites using a combination of field, greenhouse, model, and GIS studies. The beaver meadows were dominated by two graminoid plant species, *Calamagrostis canadensis* and *Carex lacustris*. *Carex* dominated in wetter sites along streams and the edges of residual beaver ponds which had lower nitrate availability (determined using in situ resin bags), whereas the ubiquitous *Calamagrostis* dominated in mesic to dry portions of beaver meadows that had higher nitrate availability. Greenhouse studies with these two species plus *Scirpus cyperinus* and *Carex rostrata* investigated the effects of fertilization and wetness: increased N supply increased plant height, biomass, leaf nitrogen, and the vertical distribution of photosynthetic machinery, effects that were accentuated at field moisture capacity (as opposed to saturation). In the field, light attenuation in various beaver meadow communities was primarily a function of canopy height. A model that combined the results of the greenhouse and light attenuation studies correctly predicted that *Calamagrostis* had a competitive advantage over *Carex lacustris* under high levels of nitrogen availability, whereas the reverse was true under low nitrogen, saturated conditions. Field observations of plant cover at 108 sample points in two Kabetogama beaver meadows were analyzed using agglomerative hierarchical clustering and non-metric multidimensional scaling to distinguish eight plant assemblages. The most diverse of these assemblages occurred along the upland edges of the beaver meadows, where forest species were beginning to encroach upon the meadows. The GIS analysis showed that "wet meadow/fen mosaic complex" was the most common beaver impoundment vegetation within the Kabetogama Peninsula, followed by deep marsh mosaic complex, speckled alder swamp, Canada bluejoint eastern meadow, and northern water lily aquatic wetland. Voles may affect the recolonization of beaver meadows by conifers due to their influence on ectomycorrhizae (symbiotic root-associated fungi).

Chapter 2 introduced the vegetation associated with beaver-engineered impoundments, describing general wetland types that occur in active and inactive pond sites. Chapter 6 expands on that introduction by exploring specific plant communities and species

C.A. Johnston, *Beavers: Boreal Ecosystem Engineers*,
DOI 10.1007/978-3-319-61533-2_6

that grow in beaver impoundments. Vegetation is described at several scales: individual plants that have key roles in beaver meadows, plant assemblages within beaver meadows, and plant communities that occur throughout beaver impoundments on the Kabetogama Peninsula. Where information in this chapter is previously unpublished, the methods used are described in boxes that accompany the appropriate section.

6.1 Beaver Meadow Plant Species

Beaver meadows are common features of the Kabetogama Peninsula. In contrast to abandoned beaver ponds elsewhere in the U.S. that quickly succeed to shrub wetlands (Remillard et al. 1987; Westbrook et al. 2011; Bartel et al. 2010), the graminoid vegetation of Kabetogama beaver meadows persists long after pond abandonment and drainage. The four most abundant plant species in beaver meadows have special characteristics that contribute to their persistence.

6.1.1 Bluejoint Grass

Calamagrostis canadensis (common name bluejoint grass or marsh reed grass) is the beaver's main plant partner in establishing beaver meadows as persistent landscape features (Fig. 6.1). *Calamagrostis* was the most common species in VNP beaver meadows, growing in 98% of 108 plots sampled in 1997 (see Box 6.1). The exposed soils of a drained beaver pond provide the perfect seed bed for windblown *Calamagrostis* seeds, which remain viable for as long as 5 years (Conn 1990; MacDonald and Lieffers 1991). Foresters often consider *Calamagrostis* to be a weed due to its aggressive colonization of logged and burned boreal forests, preventing tree seedling establishment by *Populus tremuloides* and several conifers (Landhäusser et al. 2007; Landhäusser and Wein 1993; Lieffers and Stadt 1994; Lieffers et al. 1993; Rivard et al. 1990). A mechanism by which *Calamagrostis* excludes competitors is the formation of a litter thatch that shades the soil and cools soil temperatures, delaying vernal soil thaw (Hogg and Lieffers 1991a; Landhäusser et al. 2007; Lieffers et al. 1993). Aboveground net primary production was 4.1 Mg ha^{-1} y^{-1} in our beaver meadows (Erickson 1994), but values as high as 7.9 Mg ha^{-1} y^{-1} have been reported (Stout 1912). Once established, *Calamagrostis* spreads slowly (5–15 cm/year) by rhizomes (Darris 2005). Belowground biomass of *Calamagrostis* roots and rhizomes approximately equals aboveground biomass (Wetzel and van der Valk 2005).

Calamagrostis grows in a range of environments. *Calamagrostis canadensis* occurs in beaver meadows throughout North America (Bonner et al. 2009; Hood and Bayley 2009; Longcore et al. 2006; Pollock et al. 1998; Townsend 1953; Westbrook et al. 2011; McMaster and McMaster 2000) and the genus *Calamagrostis* also grows in European beaver meadows (Simonavičiūtė and Ulevičius 2007). The U.S. National Wetland Plant List classifies *Calamagrostis* as an obligate wet-

Fig. 6.1 Two prominent beaver meadow plant species. (*top*) *Calamagrostis canadensis*. (*bottom*) *Scirpus cyperinus*)

Box 6.1 Plant Communities of Kabetogama Beaver Meadows Methods
Vegetation cover data were collected at 108 grid points in Blue Fin and Found Pond meadows during 1997 through 2000 (Fig. 3.2, Fig. 6.3). Cover estimates were made by visual observation in 1 × 1 m plots, and plants were identified to the lowest possible taxonomic division by trained field botanists. A data matrix was constructed of taxa (153 columns) by plot-year (432 rows). Log(x + 1) transformation was done to downweight high abundance species, and similarity was computed after Bray and Curtis (Bray and Curtis 1957). Plant communities were classified by agglomerative hierarchical clustering with group-average linking based on Bray-Curtis similarities, and the SIMPER procedure in PRIMER software (Clarke and Gorley 2006) was used to determine taxa contributions to the average similarity within each cluster. Average plot diversity metrics were determined for each of the hierarchical clusters identified. Nonmetric multidimensional scaling (MDS) was used with the Bray-Curtis similarity data to ordinate plots, using 25 restarts and a minimum stress of 0.01 (Clarke and Gorley 2006).

land plant in the Northcentral and Northeast Region but facultative (i.e., equally likely to occur in wetlands and non-wetlands) in Alaska (Lichvar et al. 2014). It can grow in waters that are fresh to brackish (0–10 ppt), and in acidic to alkaline soils (pH 3.5–8) (Wynia 2006; Darris 2005). *Calamagrostis* survives best in moist to saturated soils; the average depth of overlying water in Great Lakes coastal marshes containing *Calamagrostis* was 2.6 ± 0.4 cm (Johnston et al. 2008). Container-grown *Calamagrostis* plants had greatest biomass, height, and shoot density when the water table was 10 cm below the surface, decreasing with drier 20 and 40 cm water depth treatments.

Calamagrostis is suitable forage for wild and domestic ungulates (McKendrick et al. 1977; Hogg and Lieffers 1991b), but its value is limited by high silica content and a rapid loss of nutritional quality following seed head formation (Darris 2005). It has low food and cover value for small mammals and terrestrial birds (Quinlan and Cuccarese 2004).

One reason that *Calamagrostis* is so widespread in beaver meadows is its phenotypic plasticity, the ability of an individual organism to alter its physiology/morphology in response to changes in environmental conditions (Schlichting 1986). A study by MacDonald and Lieffers (MacDonald and Lieffers 1993) showed that when *Calamagrostis* was provided the opportunity to "choose" different soil environments in a greenhouse experiment, its rhizomes preferentially grew into the most favorable microhabitats (i.e., warm soil, unshaded, free of competition) (MacDonald and Lieffers 1993). The few rhizomes that grew into the less favorable "cool soil + competition" area were shorter, showed less branching, had shorter internodes, and produced fewer shoots. Rhizomes entering the "cool soil" area showed reduced growth compared to rhizomes from the same individual which grew into the control side. This clonal foraging was dubbed an "opportunistic guerilla" strategy by the authors (MacDonald and Lieffers 1993).

Table 6.1 Cover, height, and density characteristics of Found Pond beaver meadow vegetation types (mean ± SD) (Bourdaghs 1998)

	Cover, %	Canopy height, cm	Litter height, cm	Density of primary species, stems/m^2	Density of other species, stems/m^2
Calamagrostis, wet sites	99 ± 2	108 ± 11	18 ± 11	500 ± 110	29 ± 31
Calamagrostis, dry sites	47 ± 17	44 ± 11	6 ± 3	250 ± 71	43 ± 33
Carex lacustris	96 ± 7	109 ± 5	9 ± 7	186 ± 69	38 ± 29

Cover and height measurements are for primary species

We studied phenotypic plasticity of aboveground *Calamagrostis* stands in relation to soil wetness at Voyageurs National Park (Bourdaghs 1998). Dry *Calamagrostis* stands located near the edges of beaver meadows (depth to groundwater = 51.6 ± 20 cm) were shorter and had half as much cover and stem density as wet *Calamagrostis* stands located closer to the central stream (depth to groundwater = 26.5 ± 22 cm, Table 6.1). This phenotypic plasticity of *Calamagrostis* allowed it to occupy sites with diverse environmental conditions.

6.1.2 Lake Sedge

Carex lacustris (common name lake sedge or hairy sedge) was the second most common species in Kabetogama Peninsula beaver meadows, growing in 83% of plots sampled. *Carex lacustris* was the dominant species at the wettest sites along streams and pond edges, and was mixed with *Calamagrostis* at drier sites (Fig. 6.2). *Carex lacustris* is an obligate wetland plant throughout its range (Lichvar et al. 2014), and tolerates wetter soils than *Calamagrostis*: the average depth of overlying water in Great Lakes coastal marshes containing *C. lacustris* was 3.7 ± 0.6 cm (Johnston et al. 2008). When *Carex lacustris* rhizomes were planted in experimental basins with constant or falling water depths, survival rates and plant production (7.4 and 4.9 Mg ha^{-1} y^{-1}) were greatest when the rhizomes were planted near the water's edge, suggesting that shoots are sensitive to early season water levels (Yetka and Galatowitsch 1999).

Carex lacustris and *Calamagrostis* both have a graminoid growth form, but because the former is a sedge and the latter is a grass they have different plant architecture. *Carex lacustris* leaves emerge from the base of the plant, whereas *Calamagrostis* leaves are attached at regular intervals along culm nodes. *C. lacustris* can reproduce vegetatively by long, creeping rhizomes, and extensive stands of *C. lacustris* can occur without fertile culms (Flora of North America Editorial Committee 2002). Within Kabetogama beaver meadows, the average height of the *C. lacustris* canopy was comparable to that of wet *Calamagrostis* stands (Table 6.1), but the height of standing dead litter in the *C. lacustris* stands (9 ± 7 cm) was only half that of the wet *Calamagrostis* stands (18 ± 11 cm: Table 6.1).

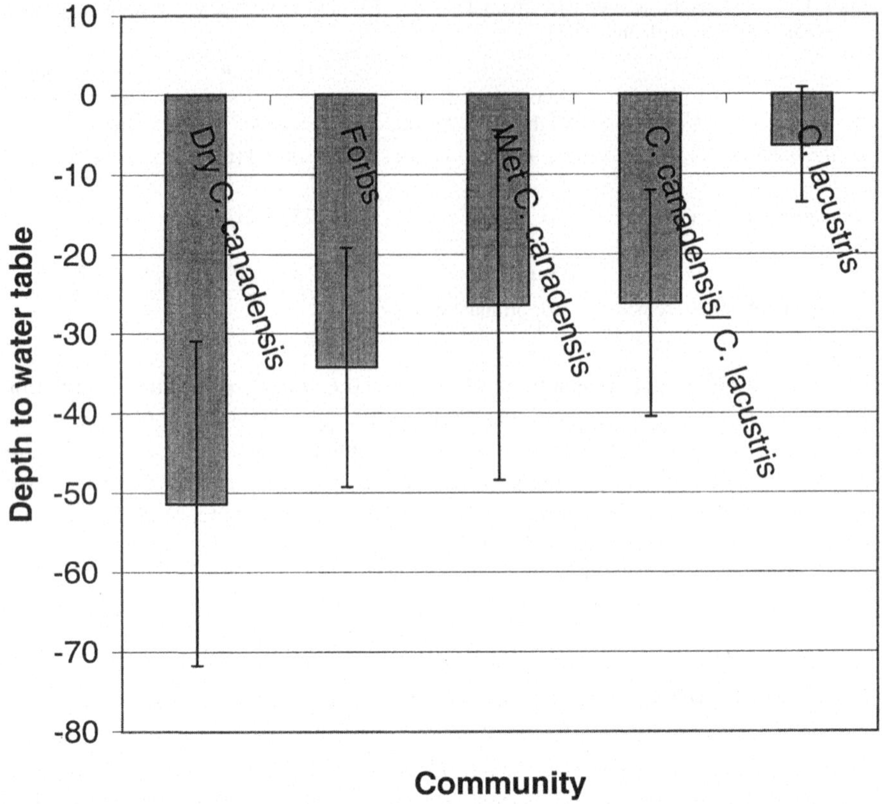

Fig. 6.2 Average depth to groundwater for Found beaver meadow plant assemblages during the sampling period 6/29/98–8/26/98 (Bourdaghs 1998)

Average aboveground net primary production of *C. lacustris* stands from multiple venues was 10.8 ± 2.8 Mg ha^{-1} y^{-1}, but belowground production was only 1.6 ± 0.3 Mg ha^{-1} y^{-1}, 13% of total net primary production (Johnston 1988). In comparison to *Calamagrostis canadensis* and *Typha latifolia*, *C. lacustris* is the most efficient at both N and P uptake, acquiring as much as 93% of the N and 63% of the P made available to it (Shaver and Melillo 1984).

6.1.3 Scirpus cyperinus

Scirpus cyperinus (common name woolgrass) is a member of the sedge family that often invades disturbed wetlands where the soil has been exposed by drawdown. *Scirpus cyperinus* is a prolific seed producer, and soil seed banks may contain 1500–2600 seeds/m^2 (Larson and Stearns 1987; Keddy and Reznicek 1982; Wilcox et al. 1985). Overwinter shoot survival is high, typically 97%, and the mean life

Table 6.2 Aboveground biomass and depth to water table of *S. cyperinus* and *Calamagrostis*-dominated beaver meadows (Erickson 1994)

		Biomass, g/m² (mean ± SE)		Median depth to water table, cm	
Type	Site	1989	1991	1990	1991
S. cyperinus meadows	Lower Bluefin	420 ± 26	385 ± 129	−17.4	5.6
	Lower Shoepack	515 ± 44	478 ± 152	−42.2	−27.1
Calamagrostis meadows	Upper Bluefin	411 ± 37	210 ± 20	−50.4	−37.2
	Upper Shoepack	410 ± 31	296 ± 53	−51.8	−40.3
	Found	400 ± 28	188 ± 21	−42.6	−33.7

Meadow locations shown on Fig. 3.2. Biomass data from 1991 were used as controls for the fertilization experiment, Sect. 6.6 below. Positive value indicates water above the soil surface

span of shoots is 18 months (Larson 1999). On the Kabetogama Peninsula, *S. cyperinus* occurred in 62% of beaver meadow plots sampled.

Scirpus cyperinus typically grows where the water table is 6–12 inches below the surface for most of the growing season, but can tolerate a wide range of water levels because it forms raised tussocks (Larson 1999). The tolerance of *S. cyperinus* for dry conditions and its ability to survive reflooding contribute to its success in impoundments with fluctuating water levels (Kadlec 1958, 1961). Tussocks of other sedge species have been shown to promote biodiversity (Johnston and Zedler 2012; Peach and Zedler 2006), and the same may be true of *S. cyperinus*, because species richness was significantly greater in *S. cyperinus* meadows than in *Calamagrostis* meadows (Erickson 1994). Although *S. cyperinus* was not as abundant as *Calamagrostis* or *C. lacustris* in Kabetogama beaver meadows, the structure provided by its tussocks may diversify and help perpetuate the meadow plant community.

The aboveground biomass of Kabetogama Peninsula *S. cyperinus* stands was compared with that of *Calamagrostis* stands in 1989 and 1991, determined by clipping three subplots per site (30 × 30 cm subplots in 1989, 30 × 60 cm subplots in 1991), oven-drying the samples at 70°C, and weighing (Erickson 1994). Average aboveground biomass was greater in the *S. cyperinus* meadows than the grass meadows in both 1989 (ANOVA, $F_{1,65} = 3.83$, $P = 0.055$) and 1991 (ANOVA, $F_{1,65} = 5.47$, $P = 0.067$) (Table 6.2). The water table was significantly deeper below the surface of the grass meadows than in the *S. cyperinus* meadows in both years (MANOVA, $P < 0.001$).

6.1.4 Rubus idaeus

Grayleaf red raspberry (*Rubus idaeus* spp. *strigosus*) occurred in 70% of the beaver meadow plots sampled at Kabetogama. Following large-scale disturbances, raspberry commonly forms a dense layer that competes with tree seedlings for light, soil moisture, and nutrients (Ruel 1992; Hughes and Fahey 1991; Mou et al. 1993). This rapid invasion is possible due to an abundant seed bank that can remain viable for more than 50 years (Whitney 1982).

The presence of red raspberry may delay woody regrowth in beaver meadows due to allelopathy, which has been shown to affect red pine seedlings (*Pinus resinosa* Ait.) (Norby and Kozlowski 1980). Similarly, raspberry foliar leachates decreased radial growth of black spruce (*Picea mariana* Mill.) seedlings by reducing ectomycorrhizal fungal activity (Coté and Thibault 1988). Red raspberry is intolerant of shade by taller species (Ricard and Messier 1996), so aboveground stems rarely persist longer than 10 years (Whitney 1986).

6.1.5 Cattail Is Notably Absent from Beaver Meadows

Cattail (genus *Typha*) is an iconic and often invasive wetland plant, but it was uncommon in our Kabetogama beaver meadows, occurring at only 13 of 108 plots sampled in 1997. Cattail marshes mapped in Voyageurs National Park by NatureServe occurred almost exclusively in the littoral zones of the three major lakes surrounding the Kabetogama Peninsula (Faber-Langendoen et al. 2007a), and *Typha* was classified as "occasional" in abundance in those littoral zones by an earlier field study (Monson 1986). We classified the cattail we observed in beaver meadows as the native *Typha latifolia*, but genetic studies have confirmed the presence of F1 hybrids between *Typha latifolia* and *Typha angustifolia* in Rainy and Kabetogama Lake marshes (Travis et al. 2010).

6.2 Plant Communities of Kabetogama Beaver Meadows Field Study

Analysis of beaver meadow vegetation data (Box 6.1) identified eight different plant communities at a resemblance level of 36%. Because the four plant species described in the previous section are so common, they contributed to the similarity of multiple groups. *Calamagrostis* contributed to the similarity of all eight communities, *Carex lacustris* contributed to seven, *Rubus idaeus* contributed to five, and *Scirpus cyperinus* contributed to four (Table 6.3).

The eight groups were defined on the basis of vegetation alone, but plotting the points on maps of the study sites showed them to be spatially clustered (Fig. 6.3). Mapping of the groups also provided insight into their association with environmental conditions.

6.2.1 Wet Groups: A, B, C

Groups A, B, and C represented the wettest sites at the two beaver meadows. The A group consisted of a single point dominated by *Calamagrostis* and *Typha latifolia* that was located along the stream flowing into the remnant Found Pond. The B

Table 6.3 Species average similarity contribution percentage to beaver meadow plant groups identified by cluster analysis

Scientific name	Hydrophyte status	Plant group							
		A	B	C	D	E	F	G	H
Abies balsamea[a]	FAC				6.17				
Alisma triviale	OBL			5.15					
Aralia nudicaulis	FACU								3.49
Athyrium filix-femina	FAC								3.06
Betula papyrifera[a]	FACU					6.65			
Calamagrostis canadensis	OBL	25.32	5.63	19.41	13.48	23.3	35.16	43.96	22.07
Carex lacustris	OBL	6.71	44.88	29.61	3.65		4.71	13.67	3.58
Carex utriculata	OBL		15.17						
Cirsium arvense	FACU								5.24
Comptonia peregrina[a]	UPL				5.09				
Cornus canadensis	FAC					7.78			
Dryopteris cristata	OBL	3.89							
Dulichium arundinaceum	OBL			3.12					
Equisetum sylvaticum	FACW								8.25
Eupatorium maculatum	OBL	12.8							
Fragaria vesca	UPL				8.13	19.37	15.48		
Fraxinus pennsylvanica	FACW								3.13
Impatiens capensis	FACW	10.9		12.03					
Lycopus americanus	OBL	6.23							3.02
Lysimachia thyrsiflora	OBL		9.43	6.41				5.98	5.53
Pteridium aquilinum	FACU						6.24		
Rubus idaeus	FACU				8.42	13.62	5.17	16.03	19.46
Rubus pubescens	FACW								5.3
Salix discolor[a]	FACW				11.06				
Salix petiolaris[a]	FACW				11.78				
Scirpus cyperinus	OBL			5.6	3.1		3.61	11.1	
Scutellaria galericulata	OBL	8.29							
Solidago gigantea	FACW				4.55				
Triadenum fraseri	OBL	3.2							
Typha latifolia	OBL	12.95	15.99						

Species shown contributed >3% to average similarity among sites in a cluster

[a]Tree and shrub species

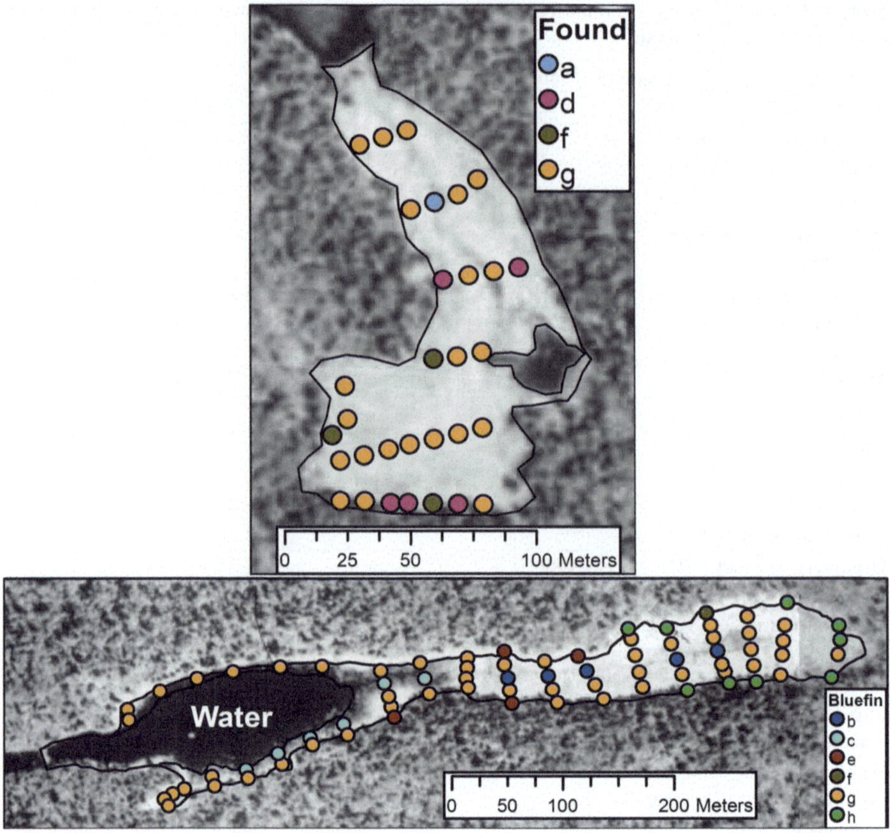

Fig. 6.3 Classification of vegetation sample plots at Found (*upper Fig.*) and Bluefin (*lower Fig.*) beaver meadows. Letter codes refer to classes in Table 6.3. Displayed on *black* and *white* orthophoto taken May 1992

group consisted of five points along the stream flowing into the remnant Bluefin Pond that were dominated by *Carex lacustris* and *Typha latifolia*. The C group consisted of six points along the edge of the remnant Bluefin Pond that were dominated by *Carex lacustris*, *Calamagrostis*, and the annual forb *Impatiens capensis*.

6.2.2 Calamagrostis *Core: Group G*

Calamagrostis canadensis was the dominant cover of this group, contributing 44% of similarity. Associated species included *Carex lacustris*, *Rubus idaeus*, *Scirpus cyperinus*, and *Lysimachia thyrsiflora*. This was also the most common community, occurring at about two-thirds of the grid points in each meadow (Fig. 6.3).

6.2.3 Upland Edge Groups: D, E, F, H

Grid points classified in the remaining groups were located along the upland edges of the beaver meadows (Fig. 6.3). They were all dominated by *Calamagrostis canadensis* and *Rubus idaeus*, but their associated species differed. Group F occurred at both Found and Bluefin meadows, and contained woodland strawberry (*Fragaria vesca*) and bracken fern (*Pteridium aquilinum*). Group E occurred along the edges of the Bluefin meadow, and supported paper birch (*Betula papyrifera*) and bunchberry dogwood (*Cornus canadensis*) as well as *Fragaria vesca*. Group H occurred in the upper Bluefin meadow and contained woodland horsetail (*Equisetum sylvaticum*) and Canada thistle (*Cirsium arvense*). Group D occurred at five points near the upland edges of Found Pond meadow, and was distinguished by supporting giant goldenrod (*Solidago gigantea*) and four woody species: *Abies balsamea, Salix discolor, Salix petiolaris,* and *Comptonia peregrina*. Note that the plants associated with these edge groups are classified as facultative rather than obligate wetland plants (Table 6.3), consistent with their position near the upland edge.

6.2.4 Beaver Meadow Diversity and Succession

Mean cover per plot was greater than 100% for all groups due to layering of vegetation, and was greatest for edge group H (Table 6.4). Group H also had a high proportion of woody cover (23%), as did edge groups D and E. Species richness was greatest for the four edge groups (D, E, F, H), which averaged 19 or more species per m^2 plot. Groups D and H had the highest values for evenness, a measure of how evenly the cover is distributed among the different species (Pielou 1975), and the highest values for the Shannon-Weiner index, which incorporates both richness and evenness (MacArthur and MacArthur 1961). Group B, the streamside vegetation at Bluefin meadow, had the lowest diversity and evenness measures (Table 6.4), being dominated by *Carex lacustris* (Table 6.3).

I created a dominance plot showing the increase in cumulative dominance with species rank for plant communities occupying wet (groups A, B, C), core (group G), and edge environments (groups D, E, F, H) (Fig. 6.4). Core plots achieved nearly

Table 6.4 Mean cover and diversity measures for hierarchical cluster groups identified

Group	Mean cover/plot	Woody cover/plot	Species richness	Evenness (J')	Shannon (H')
A	190%	0%	16.2	0.60	1.7
B	131%	4%	8.5	0.53	1.1
C	188%	0%	13.8	0.64	1.7
D	186%	49%	22.2	0.71	2.2
E	185%	18%	21.4	0.68	2.0
F	138%	4%	19.5	0.59	1.8
G	153%	3%	11.3	0.59	1.4
H	216%	23%	18.8	0.71	2.1

Mean cover per plot exceeds 100% due to layering of vegetation

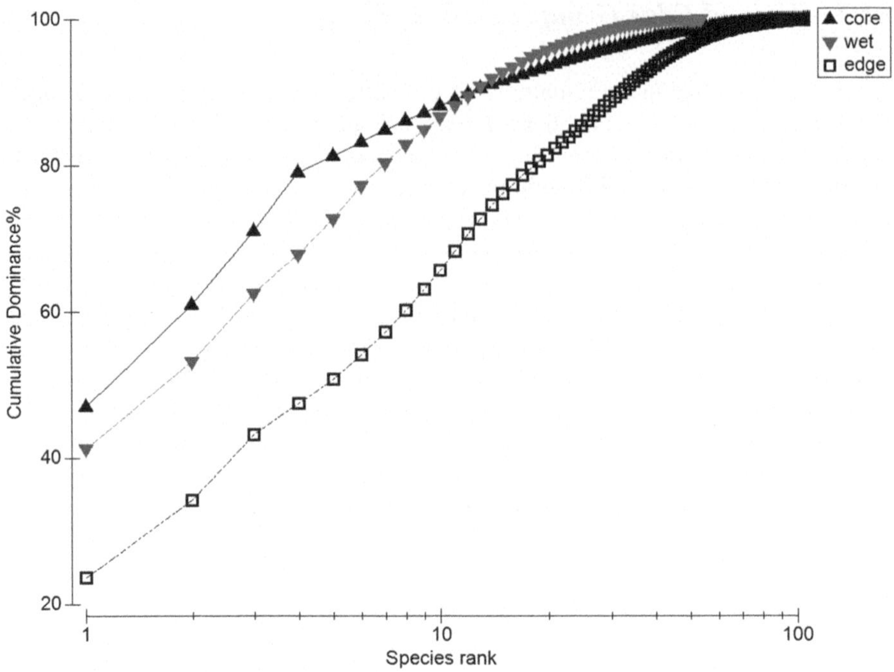

Fig. 6.4 Dominance plots for wet (groups A, B, C), core (group G), and edge (groups D, E, F, H) plant communities in Bluefin and Found beaver meadows

80% dominance by the 4th species, whereas edge plots did not reach 80% dominance until the 18th species, despite nearly equal number of species present in each environmental grouping (112 for core, 115 for edge). Wet plots started at 41% dominance by the top ranking species (*Carex lacustris*), achieved 80% dominance by the 7th species, and reached 100% dominance by the 54th species.

The greater diversity of the edge groups is an indication of ongoing succession. These edge environments are more elevated above the water table, and therefore able to return more quickly to the upland vegetation communities that existed there prior to inundation by beavers. The edge sites are also closer to seed sources in the adjacent unimpounded forest. The succession rate is slow, however, because no edge sites changed to a different cluster group during the 4 years studied.

I looked for changes in the classification of individual plots to try to identify plant community shifts within the 1997–2000 time period. However, only two shifts occurred over the 4-year period, both associated with water level declines between 1997 and 1998: (1) a point near the upstream edge of the residual pond at Bluefin was classified with group C in 1997 and group G the remaining years, (2) a point near the stream flowing into Found Pond was classified with group A in 1997 and group G the remaining years. Both of these changes are from wetter to drier plant assemblages; the remaining 106 grid points did not change over time. Therefore, although succession was inferred by the diversity of the edge groups, it was not detected within the 4-year timeframe of the study.

Pastor and colleagues (1996) generated species-area curves for twelve vegetation plots within the Bluefin and Shoepack beaver meadows (Fig. 3.2), quantifying species richness as a function of area sampled (sample plots = 0.25, 1, 4, and 9 m^2). Species richness increased with increasing sampling area, following the power law function, $S = cA^z$, where S is the expected number of species found in area A, and z is the instantaneous rate by which species richness increases with an incremental increase in area (Arrhenius 1921). The regressions of log(species) against log(area) were significant for each plot ($r > 0.918$ in each case, $P < 0.05$). The z value for the twelve plots averaged 0.25, which is the theoretical value that can be derived from species abundance models (Sugihara 1981).

6.3 Plant Communities of All Kabetogama Beaver Ponds and Meadows GIS Analysis

A comprehensive vegetation map for Voyageurs National Park was prepared by NatureServe from 1995 aerial photography using the U.S. National Vegetation Classification (USNVC) (Faber-Langendoen et al. 2007a, b). This map showed all plant associations, but was not designed to identify beaver impoundments, *per se*. Two of the NatureServe land cover classes were specific to beaver impoundments, the "Beaver Basin Break-up Complex" (NatureServe code BBX) and "Water - Beaver Pond" (NatureServe code WBP), but these classes were applied to only 23 and 18 ha, respectively, of the Kabetogama Peninsula. The remaining 3356 ha of beaver impoundments on the Kabetogama Peninsula encompassed a variety of vegetation types (Table 6.5). I intersected my beaver impoundment maps (Johnston and Windels 2015) with the NatureServe vegetation map to determine their correspondence (Box 6.2, Table 6.6).

In addition to the BBX and WBP classes, other plant communities that were strongly associated with beaver impoundments were Canada Bluejoint Eastern Meadow (code BJ, 78% occurring in beaver ponds), Midwest Pondweed Submerged Aquatic Wetland (code PW, 75% occurring in beaver ponds), and Northern Water Lily Aquatic Wetland (code WL, 82% occurring in beaver ponds).

Box 6.2 Plant Communities of All Kabetogama Beaver Ponds and Meadows GIS Methods

The shapefile for the NatureServe map (Faber-Langendoen et al. 2007a) was intersected with my shapefile of 1997 Kabetogama beaver impoundments (Johnston and Windels 2015) in ArcGIS 10.1 (ESRI, Redlands, CA), and data were summarized for each NatureServe plant association to determine the percent of association area mapped as beaver ponds and the percent of each plant association occurring in beaver ponds (as opposed to their occurrence outside of beaver impoundments).

Table 6.5 Major plant associations occurring in beaver impoundments of the Kabetogama Peninsula

Ecol. system	Map code (Assoc.)	% of total area mapped as beaver ponds	% of plant association occurring in beaver ponds	Description
Bog	LBC (A2)	2.1	23.9	**Black spruce/Leatherleaf semi-treed bog**: Black spruce (*Picea mariana*) and/or tamarack (*Larix laricina*) are the dominant conifers in this community. Canopy cover is 10–25% and height is usually 2–10 m tall. The ericaceous dwarf-shrub layer usually comprises 90–100% cover, and consists mainly of *Chamaedaphne calyculata* (leatherleaf), occasionally occurring with *Ledum groenlandicum* (labrador tea). The substrate is deep fibric sphagnum peat, which is mineral poor (pH usually <4.3). The water regime is saturated.
Bog	LB (A3)	5.7	69.8	**Leatherleaf bog**: The vegetation is dominated by an open dwarf-shrub/scrub conifer layer with very scattered trees (<10% cover). Ericaceous dwarf-shrubs are dominant, including *Chamaedaphne calyculata* (leatherleaf), *Kalmia polifolia* (bog laurel), *Andromeda glaucophylla* (bog-rosemary), *Ledum groenlandicum* (labrador tea), and the creeping dwarf-shrub *Vaccinium oxycoccos* (small cranberry). Stands have a saturated hydrology with a deep fibric sphagnum peat soil and a pH <4.3.
Poor swamp	BSL (A4)	2.0	9.5	**Black spruce/Labrador tea poor swamp**: Black spruce (*Picea mariana*) trees dominate this type, and are typically 10–20 m tall in the forested stands and 5–10 m tall in the woodland stands. Scattered tamarack (*Larix laricina*) trees are occasional codominants (25% cover). The dwarf-shrubs *Ledum groenlandicum* (labrador tea) and *Chamaedaphne calyculata* (leatherleaf) are nearly always present, but cover is highly variable, ranging from 20 to 90%. The substrate is deep, acidic sphagnum peat that is mineral poor. The water regime is saturated.
Rich swamp	BA (A14)	2.7	27.1	**Black ash**: Mixed Hardwood Swamp—The canopy and subcanopy (if present) of this community most commonly consists solely of black ash (*Fraxinus nigra*). Although soils may occasionally be fairly deep peats reaching depths of >30 cm, soils more commonly consist of 4–10 cm mineral soils or peat over dense clay. Depending on topographic position and substrate, the water regime in these communities can be temporarily to seasonally flooded or saturated.
Rich swamp	TA (A18)	1.3	24.9	**Northern tamarack rich swamp**: The canopy of tamarack (*Larix laricina*) is typically uneven and fairly open, ranging from 20 to 50%. White-cedar and black spruce may also occur in the canopy at low densities (<25% relative cover). A shrub layer of *Alnus incana* (speckled alder) and *Betula glandulifera* (bog-birch) is typically present at 40–90% cover. The substrate is deep peat or shallow peat over clay. The water regime is saturated.
Rich swamp	AS (19)	8.8	39.8	**Speckled alder swamp**: *Alnus incana* (speckled alder) shrubs, around 2–3 m tall, usually form a dense canopy in this community. *Salix* spp. (willow) and *Betula glandulifera* (bog-birch) may also be present at low cover. Occurs in isolated low areas surrounded by upland or as a ring around the edge of less minerotrophic peatlands. Soils may be deep peats, shallow peats, or mineral soils where drainage is impeded by clay or dense glacial till. Depending on substrate and topographic placement, they can be temporarily or seasonally flooded or remain saturated throughout the growing season.

Wet meadow	BJ (A23)	7.4[a]	78.3	**Canada bluejoint eastern meadow**: This community is characterized by a continuous cover of *Calamagrostis canadensis* (bluejoint). *Alnus incana* (speckled alder), *Betula glandulifera* (bog-birch), or *Salix* spp. (willow) infrequently colonize these sites at <25% cover. Other graminoid species present with low cover include *Carex lacustris* (lake-sedge) and *Scirpus cyperinus* (woolgrass). The A23 type occurs predominately in old beaver meadows or along slow moving streams. In beaver meadows, this community is found on relatively dry sites and often occurs on the upland edge of more recent beaver floodings or completely colonizing older, drier beaver meadows. Soils usually contain deep, dense clay preventing or slowing drainage. Water channels and standing or fallen dead trees are frequently present. The water regime is temporarily to seasonally flooded.
Freshwater marsh	PW (A29)	5.2[a]	74.9	**Midwest pondweed submerged aquatic wetland**: This community consists primarily of submerged aquatic plants but can contain <10% cover of floating aquatics. The most common species are *Ceratophyllum demersum* (coontail), *Najas flexilis* (flexuous naiad), *Potamogeton gramineus* (grass-leaf pondweed), and *Sparganium fluctuans* (floating bur-reed). Stands typically occur in fairly sheltered bays of the large lakes, in interior lakes or, rarely, in recent beaver floodings. Water depth is typically 0.5–2 m. The substrate is most commonly clay, though occasionally sand or, in less exposed sites, muck over clay.
Freshwater marsh	WL (A30)	7.4[a]	82.2	**Northern water lily aquatic wetland**: This community is dominated by floating aquatics. In beaver impoundments, *Brasenia schreberi* (water shield) is the most abundant floating species and *Utricularia vulgaris* (greater bladderwort) is the primary submerged aquatic. Water depth is 0.25–2 m and substrate is typically muck, clay, or muck over clay. In beaver impoundments, standing dead trees and patches of emergent vegetation are also common. The water regime is permanently flooded.
Complex	BBX (many)	0.7	90.7	**Beaver basin break-up complex**: May contain A4 Black Spruce/Labrador Tea Poor Swamp, A6 Leatherleaf Poor Fen, A9 Leatherleaf/Sweet Gale Fen, A29 Midwest Pondweed Submerged Aquatic Wetland, A31 Poor Swamp, A35 Poor Fen, A41 Rich Fen, A79 Freshwater Marsh
Complex	DMX (many)	12.7	63.3	**Deep marsh mosaic complex**: May contain A24 Midwest Cattail Deep Marsh, A25 Eastern Reed Marsh, A26 Freshwater Bulrush Marsh, A27 Water Horsetail-Spikerush Marsh, A28 Wild Rice Marsh, A29 Midwest Pondweed Submerged Aquatic Wetland, A30 Northern Water Lily Aquatic Wetland
Complex	SMX (many)	35.0	71.1	**Wet meadow/Fen mosaic complex**: May contain A11 Wiregrass Sedge Shore Fen, A22 Northern Sedge Wet Meadow, A23 Canada Bluejoint Eastern Meadow, A24 Midwest Cattail Deep Marsh, A25 Eastern Reed Marsh
Water	WBP	0.5	81.1	**Water–beaver pond** (<10% vegetated)
Other		6.9		

“Percent of total area mapped as beaver ponds” is the proportion of 1997 Kabetogama Peninsula beaver impoundments (total area = 3397 ha) covered by each association.

[a]Additional area mapped as part of the BBX, DMX, or SMX complex

Table 6.6 Intersection areas (ha) between beaver impoundment cover types (Table 2.5) and USVNP plant communities (Table 6.5)

	PEMFm	PEMF	PEME	PUBF PABF PUBH	PSS1B PSS1E PSS1F	PSS3B PSS3E	PSS7B PSS7E	PFO1A PFO1E	PFO4B PFO4E	PSS1Fm PSS3Fm	PSS5E PSS5F	PFO5E	PFO5F PFO5Fm
AS	3.6	15.1	52.5	3.2	**118.9**	19.2	20.6	7.3	10.3	5.9	13.7	23.2	4.2
BA	0.0	0.1	1.0	0.1	7.5	0.0	0.0	**36.7**	0.5	0.0	1.5	**35.0**	8.5
BBX	4.9	1.4	0.4	4.9	0.0	0.0	0.0	0.0	0.0	3.3	0.0	0.0	**8.2**
BJ	3.8	27.5	**172.3**	6.4	13.4	11.3	3.6	1.7	0.5	1.3	2.1	7.5	0.7
BSL	0.0	0.1	0.2	0.0	2.0	1.1	20.9	2.1	**37.7**	0.4	0.8	4.2	0.0
DMX	47.2	**109.6**	27.6	**185.3**	1.7	0.3	0.2	0.2	0.0	8.4	6.2	10.8	33.0
LB	11.2	6.5	7.8	2.5	5.4	38.4	13.9	0.0	0.4	**80.3**	9.7	8.0	10.2
LBC	0.3	0.1	0.4	0.4	0.8	12.8	**28.5**	0.0	3.4	4.8	6.2	12.0	1.6
PW	2.2	18.8	16.2	**109.0**	0.3	0.0	0.0	0.7	0.0	0.8	2.5	4.1	22.4
SMX	**136.6**	**324.0**	**389.1**	77.0	32.8	28.9	12.7	5.1	2.0	40.9	22.2	70.3	47.9
TA	0.9	0.0	0.4	0.0	1.5	0.5	**17.6**	0.6	**13.5**	0.2	2.0	6.4	0.0
WBP	0.0	0.5	0.2	**12.4**	0.1	0.0	0.0	0.0	0.0	0.0	0.0	0.0	4.8
WL	3.1	63.1	19.0	**149.5**	0.5	0.6	0.0	0.1	0.0	1.1	2.4	2.9	8.4
Total[a]	221.2	588.0	737.4	586.5	225.2	120.2	131.0	93.6	89.9	150.8	76.6	213.0	163.9

Bold values indicate beaver impoundment codes that frequently correspond to the USVNP code in that row

[a]Includes area from minor cover types not listed

The latter two classes corresponded primarily to the open water and aquatic bed beaver ponds on the impoundment maps (PUBF, PUBH, PABF), whereas Canada bluejoint eastern meadows (USNVC code BJ) were usually beaver meadows classified as PEME on the impoundment maps.

Two NatureServe "complex" map units encompassed nearly half of the beaver impoundments: Wet Meadow/Fen Mosaic Complex (code SMX) and Deep Marsh Mosaic Complex (code DMX) (Table 6.5). The beaver impoundment codes (Table 2.5) that were most often used for SMX areas were the three emergent classes: PEME, PEMF, and PEMFm (Table 6.6). The NatureServe DMX class covered a substantial amount of PEMF as well as open water beaver ponds (PUBF, PUBH, PABF) on the impoundment maps.

A NatureServe class that had unexpectedly fidelity to beaver ponds was Leatherleaf Bog (code LB, 70% occurring in beaver ponds) (Table 6.5). Floating peat mats, many of which contain leatherleaf, are frequently associated with beaver ponds (Fig. 2.9). The LB areas corresponded primarily with floating shrub mats (PSS1Fm, PSS3Fm), and secondarily with grounded broad-leaved evergreen shrubs (PSS3) (Table 6.6). The Black Spruce/Leatherleaf Semi-treed Bog (code LBC) occurred less frequently in beaver ponds, and may be a precursor to the LB class in beaver-flooded areas. Another peatland conifer community, Northern Tamarack Rich Swamp (code TA), occurred with similar frequency (25% occurring in beaver ponds) and covered the same impoundment types (PFO4, PSS7) as the LBC map unit. Speckled alder swamps (code AS) on the NatureServe maps corresponded primary to deciduous shrubs (PSS1B, PSS1E, PSS1F), and 40% occurred in beaver impoundments.

The USNVC focuses on live plant species, whereas the National Wetlands Inventory (NWI) classification focuses on vegetation structure. Dead standing tree stands distinguished by the beaver impoundment maps (NWI codes PFO5E, PFO5F, and PFO5Fm) had no specific equivalent in the NatureServe maps, and were spread across a variety of USNVC classes (Table 6.6). However, areas mapped as black ash (USNVC code BA) by the NatureServe maps were about evenly divided between dead and live tree classes on the impoundment maps.

6.4 Nutrient Availability in Grass and Sedge Beaver Meadows

We investigated nutrient availability in grass (*Calamagrostis*), sedge (*Carex lacustris*), and mixed meadows at the Bluefin and Found study sites (Fig. 3.2). Relative nutrient availability was determined using resin bags (Box 6.3), a method which integrates inputs from mineralization, weathering, and effects of soil water regime, mimicking the amount of nutrients that would be made available to plant roots over the incubation period (Gibson et al. 1985; Binkley 1984; Binkley and Matson 1983).

Box 6.3 Nutrient Availability Methods

Field plots were chosen such that grass-dominant (CALADOM) sites had at least 95% cover by *Calamagrostis*, sedge-dominant (CAREDOM) sites had at least 90% cover by *Carex lacustris*, and MIXED sites had at least 30% cover by both *Calamagrostis* and *Carex lacustris*. Approximately 10 g of mixed bed anion/cation exchange resins (Baker AGMI-615 ionic form H/OH) were sewn into nylon stocking bags that were placed ~5 cm deep into the soil surface at each plot and left in place from May to October (Fig. 6.5). After retrieval from the field sites, the resins were air dried and extracted with two rinses of 50 ml 1 N KCl. Resin extracts were analyzed for nitrate, ammonia, and phosphorus using a Technicon Autoanalyzer II. Mean depth to water table was measured by averaging the measurements made during the sampling period at the PVC pipe well associated with each plot.

Fig. 6.5 A retrieved resin bag tethered to a wire flag

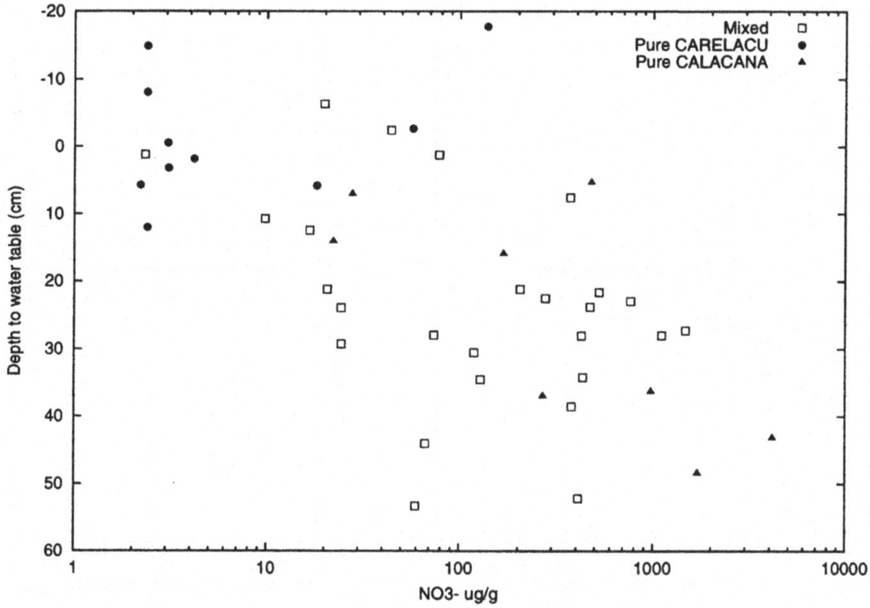

Fig. 6.6 Resin bag concentrations of nitrate (μg NO_3 per gram of resin) relative to average water table depth (cm), by community type. CARELACU = *Carex lacustris*, CALACANA = *Calamagrostis canadensis*

Nitrate was the most variable of the nutrients measured, ranging from 2 to 4000 μg per gram of resin (Fig. 6.6). Nitrate availability was inversely related to depth to water table, inferring that denitrification was occurring in the wettest sites. The eight plots with resin concentrations <10 ug NO_3/g all had water tables near or above the soil surface, and all but one were dominated by *Carex lacustris*. The *Carex lacustris* dominated plots were wetter than the other two plot types (Fig. 6.7). The range of ammonia and phosphorus availability values was much lower than that of nitrate. The three community types were distributed along a nitrate (and to a lesser degree an ammonia) availability gradient (Fig. 6.7).

6.5 Vertical Biomass Allocation, Light Extinction, and Nitrogen in Plant Canopies

Under the dense canopies characteristic of many beaver meadows, competition for light may be very intense because of the rapid downward extinction of light through a short canopy. Canopy structure, the height and arrangement of photosynthetic surfaces within the canopy, is a critical determinant of a plant species' ability to capture light. Nitrogen within the plant canopy is also important because it is a major

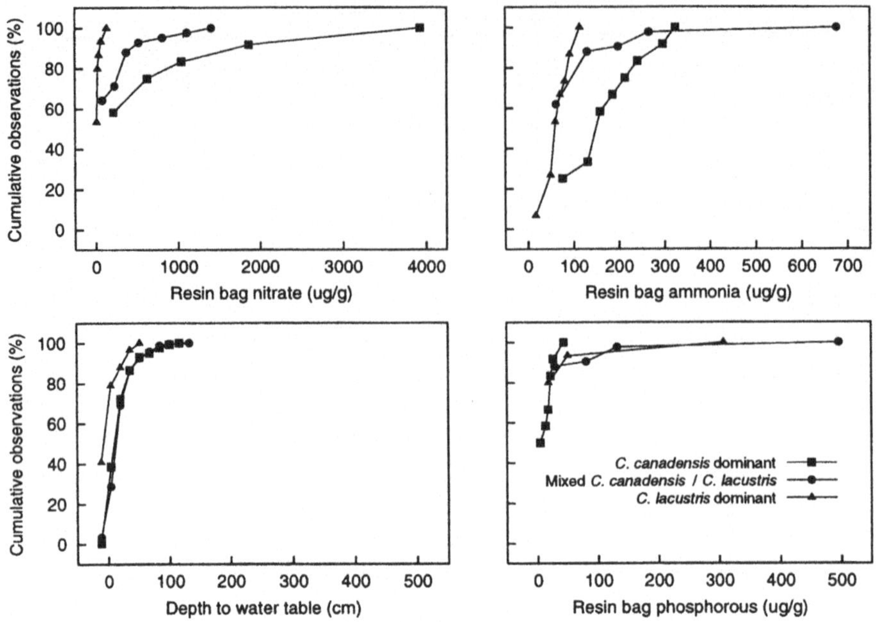

Fig. 6.7 Cumulative observation curves showing resource availability for the three meadow community types

component of Rubisco and other photosynthetic enzymes (Chapin et al. 2011). The way a plant allocates leaf biomass and its N content along this gradient may determine its competitive outcome with its neighbors. Species differences in this vertical distribution of photosynthetic machinery may be the mechanism responsible for successional pathways and diversity of wetland communities. We conducted several studies to examine beaver meadow canopy structure and its effect on light attenuation.

6.5.1 Greenhouse Study: Vertical Distribution of Canopy Nitrogen Concentration

Several studies have shown that the vertical distribution of both biomass and leaf nitrogen determine carbon gain (Aerts and deCaluwe 1994; Werger and Hirose 1991), and that vertical nitrogen distribution reflects the exponential decline in light supply as described in the following model:

$$\ln N_L = \ln N_0 - k\left(B_L / B_T\right)$$

where N_L is nitrogen concentration at depth L, N_0 is nitrogen concentration at the top of the canopy, k is the coefficient of nitrogen distribution, and B_L/B_T is the cumulative fraction of biomass at depth L (Hirose and Werger 1987). We hypothesized that nitrogen supply would affect the carbon gain of beaver meadow plant species through its effects on canopy structure and vertical nitrogen distribution as determined by N_0 and k as defined above. We also expected that different degrees of soil saturation would affect vertical distribution of leaf biomass and nitrogen. To test these ideas, we grew four common wetland graminoids—*Calamagrostis canadensis, Carex lacustris, C. rostrata,* and *Scirpus cyperinus*—in a greenhouse under a factorial combination of four levels of N and P supply nested within two water regimes (field moist capacity and saturated), and measured the vertical distribution of biomass, total nitrogen, and nitrogen concentration (methods described in Box 6.4).

Vertical distribution of biomass and amounts of nitrogen were affected by soil water and nitrogen supplies, but the response was species specific (Figs. 6.9, 6.10 and 6.11, Table 6.7).

Canopy Structure The two *Carex* species exhibited the greatest phenological plasticity in response to nutrient supply rates, nearly doubling in height at the highest fertilization rates. *Calamagrostis* had mid-canopy bulges in leaf biomass and amount of nitrogen, which were accentuated by nitrogen additions (Fig. 6.9). There was no mid-canopy bulge in either biomass or amount of nitrogen for any of the three sedges; biomass and amount of N in each layer increased monotonically with depth in the canopy. *Scirpus* showed almost no plasticity in canopy structure (Fig. 6.11).

Box 6.4 Greenhouse Study Methods (Dewey et al. 1995)

Plants were grown from seed and individuals transplanted into 5 L plastic pails containing a nutrient poor mixture of screened reed-sedge peat, vermiculite, and perlite in equal proportions by volume. Nutrient treatments included a full factorial design of nitrogen = 1 or 15 g m^{-2} × phosphorus = 0.5 g m^{-2} plus controls with no added N or P, nested within two soil moisture levels: 60% water holding capacity (field moist, FM) and saturated (SAT). Each treatment was replicated by 3. The plants were grouped by moisture but randomly distributed by treatments on greenhouse benches and grown in full sunlight (Fig. 6.8). Fertilizer was applied in solution in equal proportions at weeks 2, 4, 6, 8, and 10 after transplant. Soil moisture was adjusted twice weekly.

After the plants reached full height and before senescence they were sectioned into 10 cm increments along planes bisecting the natural leaf inclination. Sections were dried at 60°C and weighed for total biomass, then ground to pass a 40 mesh screen and analyzed for total nitrogen using a CHN analyzer (LECO Corp., St. Joseph, MI). Nitrogen distribution regressions (Table 6.7) were calculated using biomass values and nitrogen concentrations from all sections of a given treatment.

Fig. 6.8 Growing meadow grasses and sedges in the greenhouse

Effect of Soil Moisture *Calamagrostis* had more biomass and nitrogen in each layer and a more accentuated mid-canopy bulge under field moisture capacity than under saturated conditions (Fig. 6.9). *Carex lacustris* had more biomass in each layer under saturated conditions when N supply was high, but more biomass in each layer under field moisture capacity when N supply was low (Fig. 6.10). Soil moisture did not affect *Scirpus* biomass, amount of nitrogen, nor canopy structure (Fig. 6.11).

Effect of Nitrogen Fertilization Additions of nitrogen had a greater effect on canopy N concentrations in *Carex* than in *Calamagrostis* or *Scirpus*; increased N supply increased N concentrations vertically throughout the canopy for *Carex* (Fig. 6.10). Sedge nitrogen concentrations increased more when the soil was at field moisture capacity than when it was saturated. For *Calamagrostis*, increased N supply increased total biomass and consequently nitrogen in upper canopy layers (Fig. 6.9), but not N concentrations.

Vertical Distribution of Nitrogen Concentrations The vertical distribution of nitrogen concentrations conformed best to the exponential decay model at low levels of nitrogen fertilization, with R^2 values ranging from 0.82 to 0.96 (Table 6.7). Model fits were significant but poorer (R^2 range 0.41–0.86) for most of the high fertilization treatments because N concentrations were more evenly distributed throughout the canopy and N in mid-canopy layers was greater than predicted by the model. Model fits were not significant for the control plants (except for *Carex lacustris*) because their short height meant there were too few points to fit the exponential decay model adequately.

Table 6.7 Coefficients of vertical nitrogen distribution from the linear regression, ln N_L = ln N_0-$k(B_L/B_T)$, where N_L is nitrogen concentration at depth L, N_0 is nitrogen concentration at the top of the canopy, k is the coefficient of nitrogen distribution, and B_L/B_T is the cumulative fraction of biomass at depth L

Species	Moisture	*N* rate	N_0	*k*	R^2	*P*
Calamagrostis canadensis	SAT	High	1.41	0.45	0.60	0.001
	SAT	Low	1.48	0.62	0.91	<0.001
	SAT	Control	1.26	0.44	0.52	NS
	FM	High	1.64	0.40	0.74	<0.001
	FM	Low	1.61	0.76	0.96	<0.001
	FM	Control	1.38	0.44	0.53	NS
Scirpus cyperinus	SAT	High	1.89	0.22	0.41	0.006
	SAT	Low	1.52	0.74	0.94	<0.001
	SAT	Control	1.53	0.37	0.58	NS
	FM	High	1.67	−0.13	0.18	NS
	FM	Low	1.55	0.80	0.94	<0.001
	FM	Control	1.29	0.52	0.71	0.001
Carex rostrata	SAT	High	1.87	0.50	0.64	<0.001
	SAT	Low	1.38	0.62	0.93	<0.001
	SAT	Control	1.17	0.26	0.56	NS
	FM	High	2.35	0.49	0.63	<0.001
	FM	Low	1.58	0.72	0.82	<0.001
	FM	Control	1.33	0.26	0.15	NS
Carex lacustris	SAT	High	1.91	0.67	0.72	<0.001
	SAT	Low	1.69	1.03	0.94	<0.001
	SAT	Control	1.61	0.56	0.79	<0.001
	FM	High	2.57	0.78	0.86	<0.001
	FM	Low	1.50	1.02	0.92	<0.001
	FM	Control	1.64	1.09	0.92	<0.001

Model fit is described by regression coefficients (R^2) and P values

This experiment shows that the vertical distribution of leaf biomass and nitrogen depends on the supply of nitrogen, as conditioned by the moisture status of the soil. Increased supply of N increases plant height, plant biomass, leaf nitrogen contents and concentrations, and the vertical distribution of photosynthetic machinery. These effects of increased N supply are constrained by saturated soil conditions and accentuated when the soil is at field moisture capacity. Furthermore, there are large species differences in growth and canopy structure under different moisture and nitrogen supply rates. Fluctuations of water and nutrient levels in wetlands may determine the competitive outcome and diversity of the plant community as well as the net primary productivity of the ecosystem through the vertical distribution of leaf biomass and nitrogen along the light gradient.

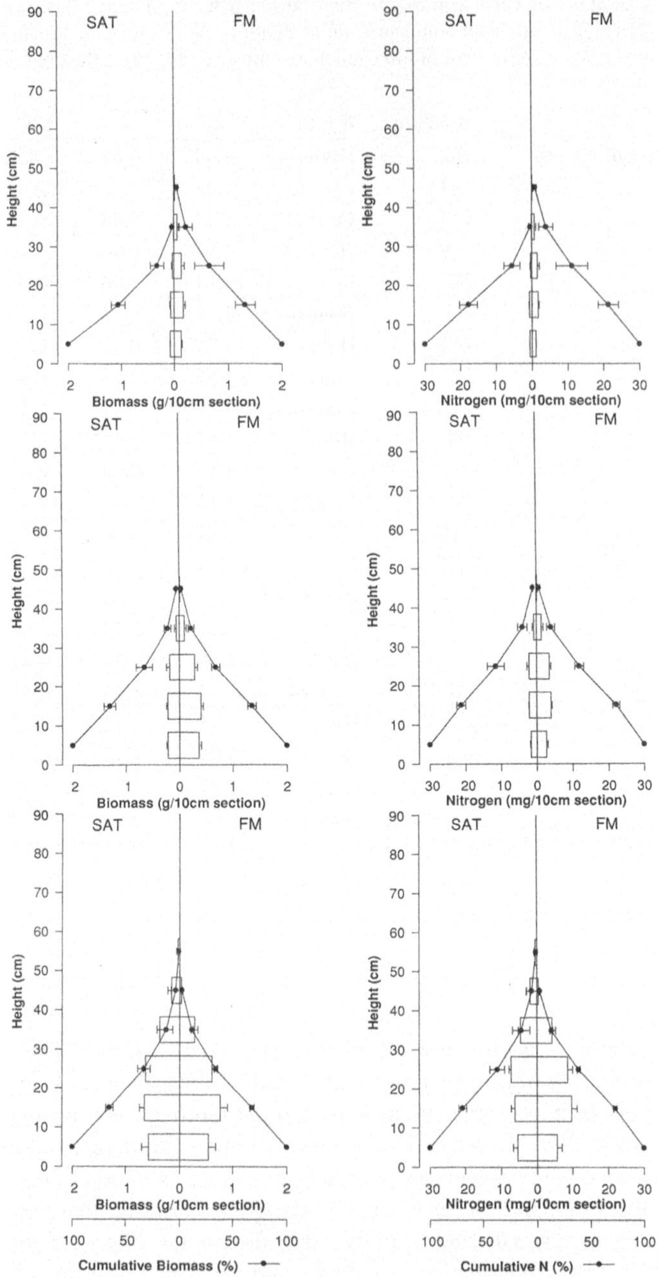

Fig. 6.9 Vertical distribution of canopy biomass and nitrogen under fertilization and moisture treatments, *Calamagrostis canadensis*. *SAT* saturated, *FM* field moisture. Nutrient treatments are: (*top*) control, (*middle*) $N = 1$ g/m^2, $P = 0.5$ g/m^2, (*bottom*) $N = 15$ g/m^2, $P = 0.5$ g/m^2

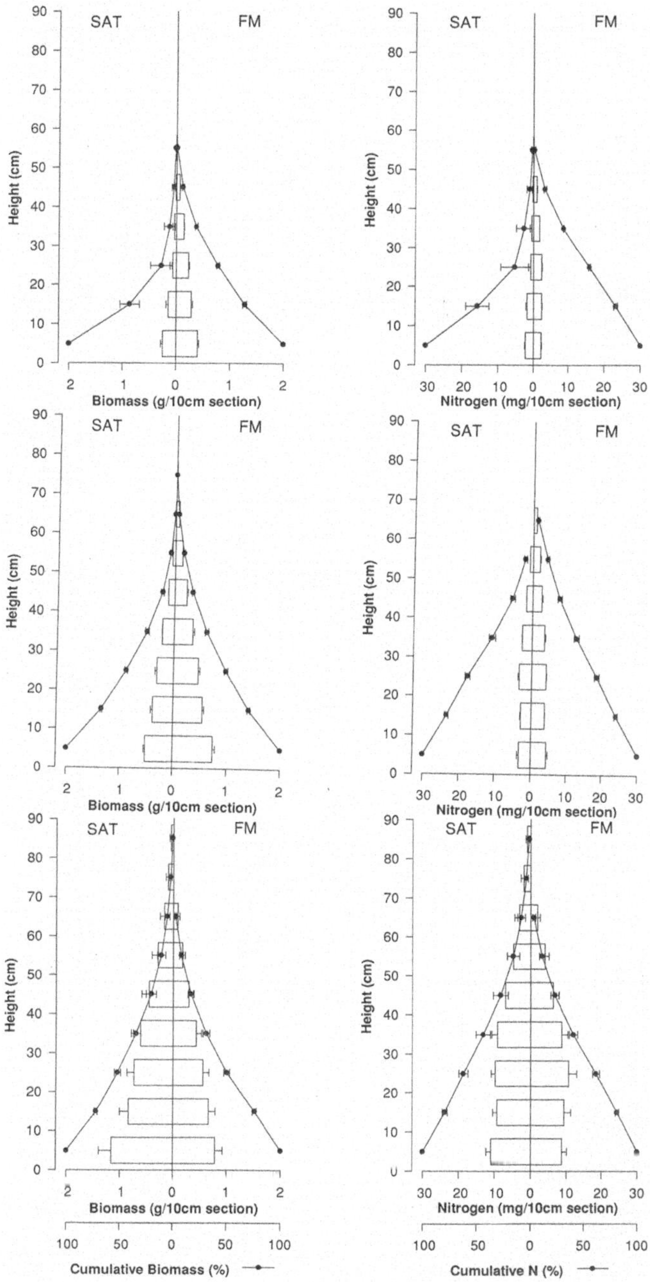

Fig. 6.10 Vertical distribution of canopy biomass and nitrogen under fertilization and moisture treatments, *Carex lacustris*. *SAT* saturated, *FM* field moisture. Nutrient treatments are: (*top*) control, (*middle*) $N = 1$ g/m^2, $P = 0.5$ g/m^2, (*bottom*) $N = 15$ g/m^2, $P = 0.5$ g/m^2

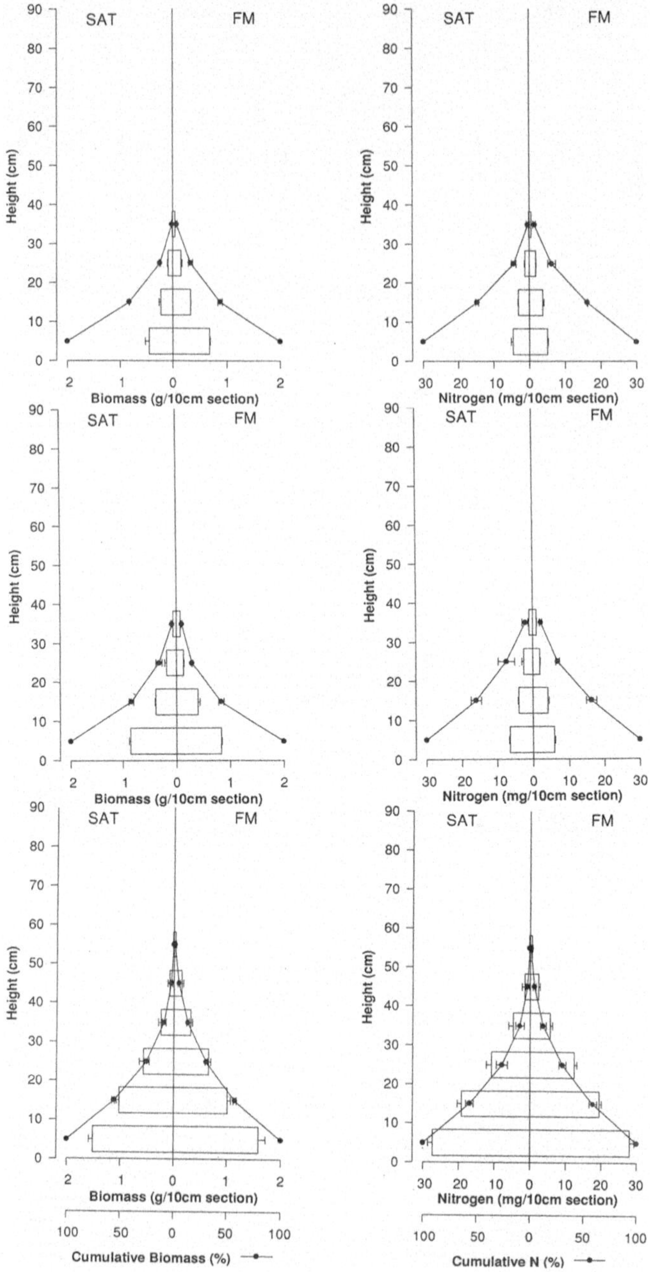

Fig. 6.11 Vertical distribution of canopy biomass and nitrogen under fertilization and moisture treatments, *Scirpus cyperinus*. *SAT* saturated, *FM* field moisture. Nutrient treatments are: (*top*) control, (*middle*) $N = 1$ g/m^2, $P = 0.5$ g/m^2, (*bottom*) $N = 15$ g/m^2, $P = 0.5$ g/m^2

6.5.2 *Field Study: Light Interception*

Undergraduate researcher Michael Bourdaghs and others studied the vertical interception of light, measuring photosynthetically active radiation (PAR) in various beaver meadow plant canopies (Box 6.5, Fig. 6.12): (1) *Calamagrostis* at dry sites, (2) *Calamagrostis* at wet sites (3) *Carex lacustris*, (4) mixtures of *Calamagrostis* and *C. lacustris*, and (5) forbs (giant goldenrod, *Solidago gigantea*) mixed with *Calamagrostis* (Bourdaghs 1998).

The vertical distribution of foliage in these canopies was relatively even, and PAR values adequately fit the Lambert-Beers Law (Fig. 6.13, Table 6.8). The degree of light interception (k) generally increased with increasing wetness (see mean depth to groundwater, Fig. 6.2). The *C. lacustris*, wet *Calamagrostis*, and mixed *Calamagrostis*/*C. lacustris* canopies were very efficient at intercepting the incoming sunlight, such that almost no light reached ground level (Fig. 6.13a, b, c). The forb canopy intercepted 80% of incoming sunlight (Fig. 6.13d). The dry *Calamagrostis* stands, with canopy heights and stem densities that were half those of wet *Calamagrostis* (Table 6.1), intercepted only 41% of full sunlight (Fig. 6.13e).

For the canopies studied, light attenuation was primarily a function of canopy height (Fig. 6.14). The two tallest canopies, *Carex lacustris* and wet *Calamagrostis* (Table 6.1), had highest extinction coefficients, whereas the short dry *Calamagrostis* had the lowest.

Box 6.5 Light Interception Methods

Light intensity was measured three times over the summer in three replicate 1 m^2 plots per type as photosynthetically active radiation (400–700 nm, PAR) in 10 cm increments from the ground level to the top of the canopy using a Sunfleck Ceptometer with a 90 cm wand (Decagon Devices, Pullman WA) (Fig. 6.12). The wand was oriented toward the south (so the operator did not shadow it), and measurements were made at three parallel locations within each 1 m^2 plot and averaged. Relative light intensity (L_D/L_0) was calculated for each depth, where L_D is the intensity of light at depth D below the canopy surface, and L_0 is the ambient light intensity above the canopy. To facilitate comparison among the types, canopy heights were normalized to range from 0 (lowest height) to 1 (top of canopy). The observations were then fitted to the Lambert-Beers Law of exponential decay to calculate extinction coefficients (k) as: $L_D = L_0 \times e^{(-kD)}$.

Fig. 6.12 Terry Brown measures grass canopy light intensity at 10 cm increments using a Sunfleck Ceptometer

6.5.3 Combining the Light Interception and Greenhouse Studies

The results of the light interception and greenhouse studies allowed us to model the competition between *Calamagrostis* and *Carex lacustris*, the two main graminoids in the Kabetogama beaver meadows (Brown et al. 2000). The inputs were a light availability curve based on the field measurements, and height-normalized vertical biomass distribution for *Calamagrostis* and *Carex lacustris* based on the greenhouse studies (Fig. 6.15). The product of biomass and PAR was calculated for each species-nitrogen-moisture combination (Fig. 6.16). The resultant curves represent the plant's *light capture potential*, in other words its ability to position biomass within the canopy so as to capture the available light at that height.

As previously noted, *Calamagrostis* exhibited a mid-canopy bulge in biomass (0.74 g) at a normalized height of 34% when grown under field moist, high N conditions (upper left, Fig. 6.15). However, the available light reaching that depth in the canopy was only about 500 $\mu E\ s^{-1}$. The bulge in *Calamagrostis*' light capture potential actually occurred higher in the canopy (~50% of normalized height), where more light was available and biomass was still high (upper left, Fig. 6.16).

Computing the area between the light capture potential curve and the Y axis allowed us to compare the light capture potential between species. Since the value for *Calamagrostis canadensis* (431) is so much larger than the value for *Carex lacustris* (367), *Calamagrostis* appears to have a competitive advantage over *C. lacustris* under high N, field moist conditions (Fig. 6.16, upper left), which is consistent with our field observation that *Calamagrostis* was dominant under high levels of nitrate

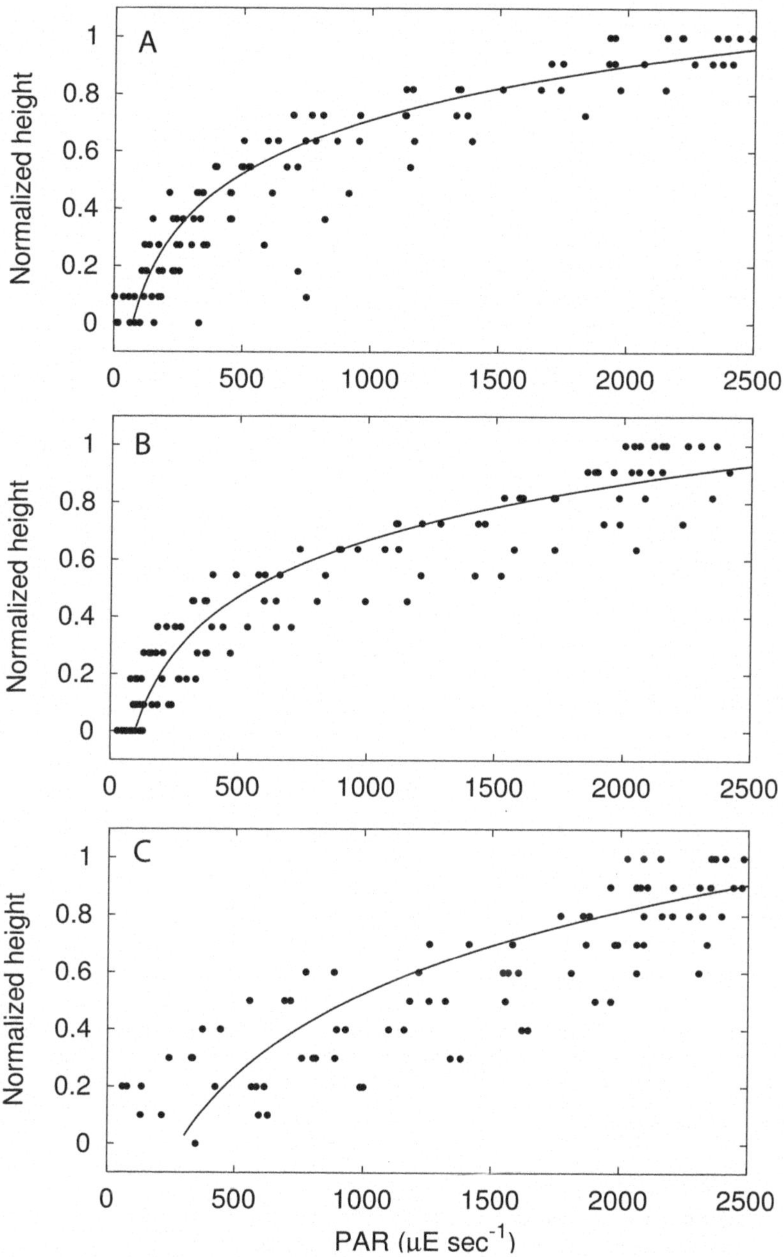

Fig. 6.13 Light interception relative to normalized height. (**a**) *Carex lacustris*. (**b**) wet *Calamagrostis*. (**c**) mixed *Calamagrostis*/*C. lacustris*. (**d**) forbs. (**e**) dry *Calamagrostis*. See Table 6.8 for model coefficients and fit

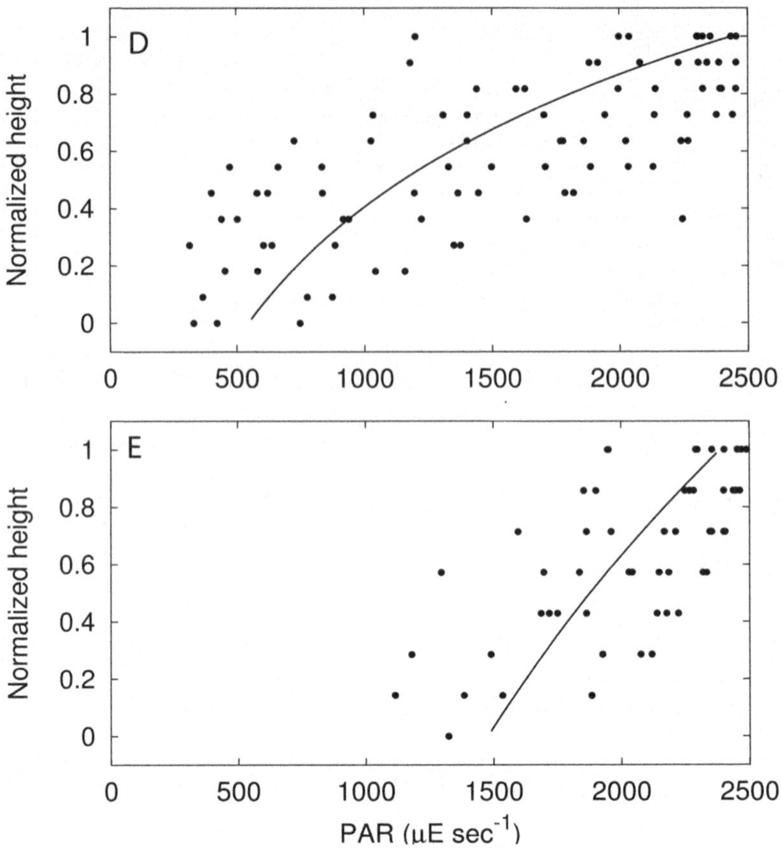

Fig. 6.13 (continued)

Table 6.8 Coefficients of vertical light extinction from the linear regression, ln L_D = ln L_0-kD, where L_D is light (PAR, μE s^{-1}) at normalized depth D, L_0 is light at the top of the canopy, and k is the coefficient of light extinction

	Canopy type	L_0	k	R^2
A	*Carex lacustris*	2892	3.68	0.73
B	Wet *Calamagrostis*	3191	3.50	0.87
C	Mixed *Calamagrostis/C. lacustris*	3123	2.40	0.62
D	Forbs	2438	1.50	0.53
E	Dry *Calamagrostis*	2388	0.48	0.44

Model fit is described by regression coefficients (R^2). Models shown in Fig. 6.13

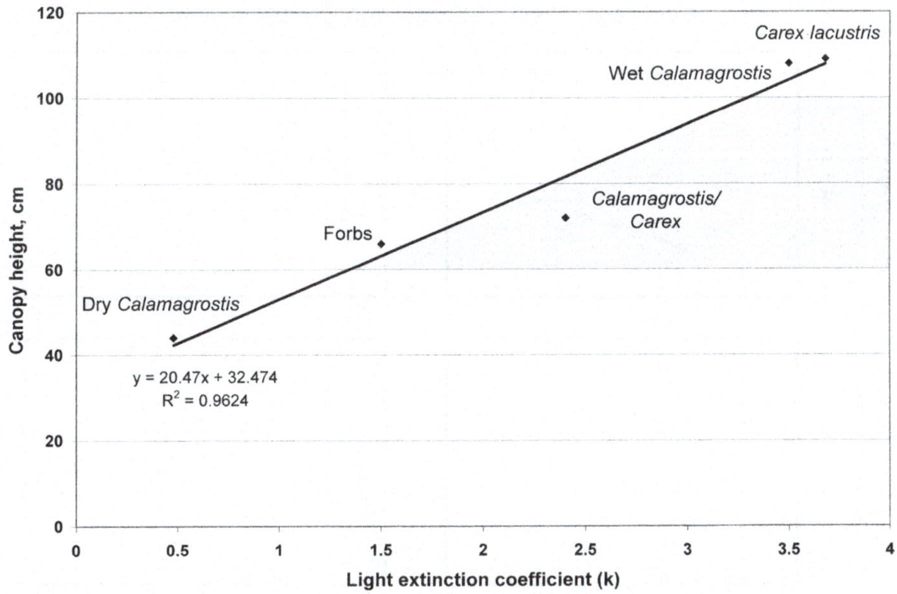

Fig. 6.14 Light extinction coefficients versus average canopy heights for the five canopy types

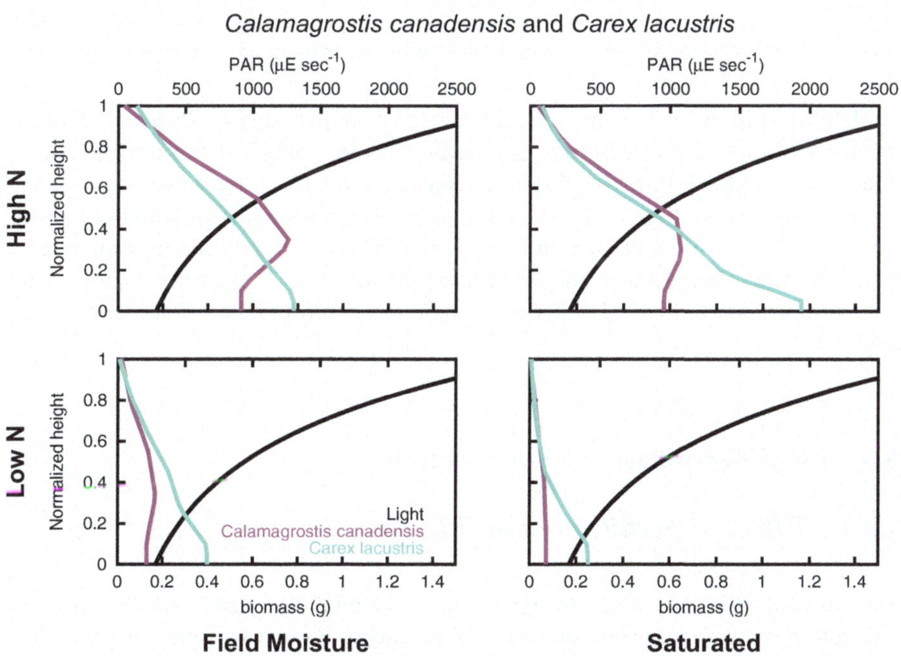

Fig. 6.15 Inputs used to model the competition between *Calamagrostis* and *Carex lacustris*

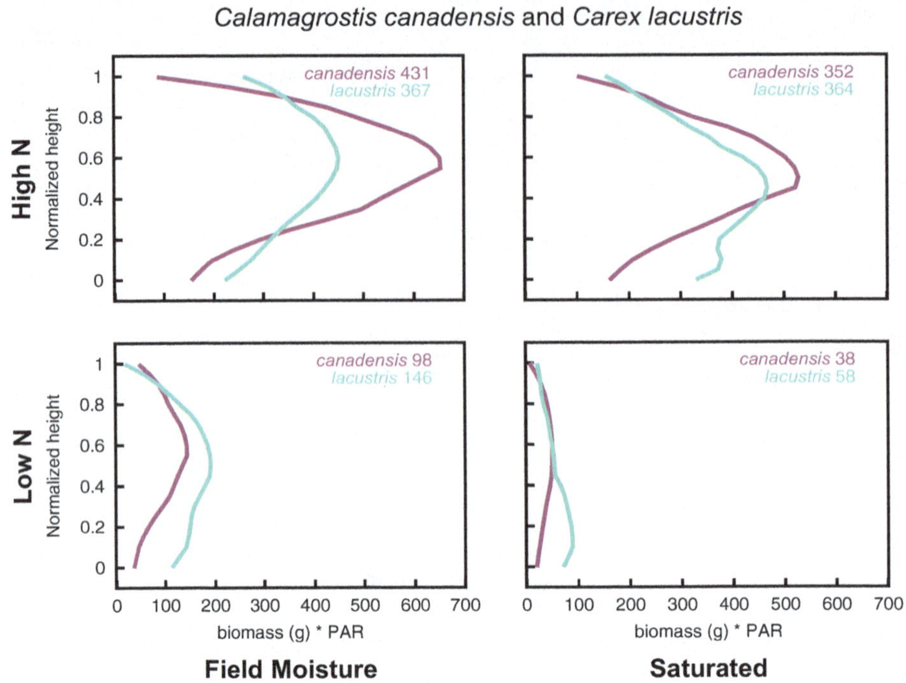

Fig. 6.16 Results of the competition model between *Calamagrostis* and *Carex lacustris*. Numbers in the *upper right* corner represent the area between the light capture curve potential and the Y axis

availability (Fig. 6.7). This model also correctly predicts that *C. lacustris* (58) has greater light capture potential than *Calamagrostis* (38) under low N, saturated conditions (lower right, Fig. 6.16), due to its greater biomass in the lower 40% of the canopy (lower right, Fig. 6.15). The only model that is inconsistent with field observations is the low N, field moisture scenario, which incorrectly predicts that *Carex lacustris* (146) should out-compete *Calamagrostis* (98). Competitive factors other than nitrogen and light availability must therefore control species dominance under this scenario.

6.6 Field Fertilization Experiments

6.6.1 Effect of Fertilization on Biomass

The nutrients nitrogen and phosphorus are essential for plant growth. Biomass production by wetland plant communities is usually limited by nitrogen (Powelson and Lieffers 1992; Olde Venterink et al. 2001; van der Hoek et al. 2004; Morris 1991; Vermeer 1986; Verhoeven and Schmitz 1991), but can be limited or co-limited by

Table 6.9 Nutrient (nitrogen and phosphorus) concentrations and ratios applied to fertilized plots (Erickson 1994)

Treatment	N:P ratio	1991		1992–2000	
		N, g/m^2	P, g/m^2	N, g/m^2	P, g/m^2
1	3:1	27	9	18	6
2	1:1	3	3	2	2
3	9:1	9	1	6	0.667
4	3:1	9	3	6	2
5	1:1	9	9	6	6
6	9:1	27	3	18	2
Control	–	0	0	0	0

Box 6.6 Fertilization Methods

Each 8.55 m^2 plot was encircled by a vertical ring of 45 mil roofing rubber inserted ~10 cm into the soil surface to ensure a relatively impervious plot boundary. Six treatments plus a control were applied (Table 6.9), with two replicates per meadow. N was applied as NH_4Cl. Half of the P was applied as NaH_2PO_4 and half applied as $NaHPO_4$ to maintain a pH similar to that of the surface horizon. The rates of fertilization application applied during the first year (1991) were reduced in subsequent years (1992–1997) to prevent fertilizer burn (Table 6.9). Plant species cover was monitored in each fertilizer plot in 1990–2000.

phosphorus (Shaver et al. 1998; Verhoeven and Schmitz 1991; Olde Venterink et al. 2001). We conducted field fertilization experiments to determine how the supply of nutrients affects the growth of *S. cyperinus* and *Calamagrostis* dominated beaver meadows (Box 6.6, Table 6.9).

Fertilization was initiated in 1990–1991 by Heather Erickson (Erickson 1994) at 14 field plots in each of five Kabetogama beaver meadows, a total of 70 plots (Box 6.6, Table 6.9). Biomass was sampled at the end of the 1991 growing season by clipping 30 × 60 cm subplots, oven-drying the samples at 70°C, and weighing. A year's worth of fertilization significantly increased biomass in the group of *Calamagrostis* meadows (ANOVA, $F_{20,21} = 3.26$, $P = 0.005$), but not in the *S. cyperinus* meadows (ANOVA, $F_{13,14} = 0.35$, $P = 0.967$) (Tables 6.2 and 6.10). Of the *Calamagrostis* meadows, only Upper Shoepack and Found meadows exhibited a significant response to fertilization (ANOVA, $F_{6,7} = 3.86$, $P = 0.05$ and $F_{6,7} = 6.92$, $P = 0.011$, respectively). Nitrogen was the primary limiting nutrient at Found meadow (ANOVA, $F_{1,7} = 5.65$, $P = 0.049$), but phosphorus had the primary influence at Upper Shoepack meadow (ANOVA, $F_{1,7} = 4.81$, $P = 0.064$). There was no effect of nutrient ratios (Erickson 1994).

Table 6.10 Biomass, g/m^2 (Mean ± SE) in plots treated with fertilizer

		Treatment					
Site	Type	1	2	3	4	5	6
Lower Bluefin	S	413 ± 98	439 ± 132	541 ± 32	423 ± 29	555 ± 194	346 ± 38
Lower Shoepack	S	429 ± 24	475 ± 25	633 ± 224	596 ± 64	570 ± 226	610 ± 122
Upper Bluefin	C	234 ± 132	282 ± 25	388 ± 120	384 ± 43	327 ± 40	349 ± 12
Upper Shoepack	C	248 ± 24	475 ± 24	412 ± 36	498 ± 4	386 ± 18	360 ± 95
Found	C	226 ± 39	319 ± 42	377 ± 64	546 ± 42	494 ± 69	387 ± 54

S = *S. cyperinus* dominated, C = *Calamagrostis* dominated. Compare with 1991 biomass data for untreated plots in Table 6.2

6.6.2 *Effect of Fertilization on Species Composition*

Nutrient additions may also shift the composition of plant communities as species that grow faster under nutrient-rich conditions out-compete less responsive species (Johnston and Brown 2013; Woo and Zedler 2002; van der Hoek et al. 2004; Vermeer 1986). In order to determine if the fertilization that occurred during 1991 through 1997 had any long-term effects on plant community composition, I used all the plant cover data recorded during the 11 years of the study (1990–2000) for the 70 fertilizer plots (Box 6.6), dividing the dataset by *S. cyperinus*-dominated meadows (Lower Shoepack, Lower Bluefin), and *Calamagrostis*-dominated meadows (Upper Shoepack, Upper Bluefin, Found). I grouped the fertilizer treatments into high (treatments 1 and 6), medium (treatments 3–5), and low (treatment 2 and control). I analyzed the plant cover data using ANOSIM (analysis of similarities) which analyzes groups similar to a standard univariate ANOVA. I also performed non-metric multidimensional scaling (MDS) and a SIMPER analysis on the data. All species composition analyses were done in PRIMER (Clarke and Gorley 2006).

The ANOSIM analysis showed that there was no significant effect of fertilization on species composition of the *Calamagrostis* meadows (global $R = 0.012$, $P = 0.247$), but there was an effect on the *S. cyperinus* meadows (global $R = 0.094$, $P < 0.001$). Inspection of the pairwise tests comparing the treatments showed that the species composition of the low treatment differed significantly from the medium and high treatments, but that the medium and high treatments were not different from each other.

The SIMPER analysis of the treatment groups showed that *Calamagrostis* contributed much less to the similarity of the low treatment plots than it did to the medium or high treatment plots (Table 6.11). The low treatment plots were also more diverse: nine species contributed 1.2% or more to low treatment plot similarity, whereas only four species contributed that much to high treatment plot similarity. The high treatment plots also had less *S. cyperinus* than did the low treatment plots.

This pattern was mirrored in the non-metric multidimensional analysis, where the first MDS axis was positively correlated ($r = 0.861$) with *Calamagrostis* and nega-

Table 6.11 Species average similarity contribution percentage to beaver meadow fertilization treatment groups

	Low	Medium	High
Calamagrostis canadensis	22.90	39.41	53.63
Carex lacustris	13.80	10.15	11.44
Impatiens capensis	5.05	2.24	
Lysimachia thyrsiflora	3.20	5.13	2.8
Polygonum punctatum	1.90	1.24	
Polygonum sagittatum	1.38		
Scirpus cyperinus	36.70	35.99	27.67
Scutellaria galericulata	2.09		
Triadenum fraseri	5.43		
Cumulative contribution to similarity	92.45	94.16	95.54
Average similarity	52.20	53.21	56.68

Species shown contributed >1.2% to average similarity among sites in treatment group

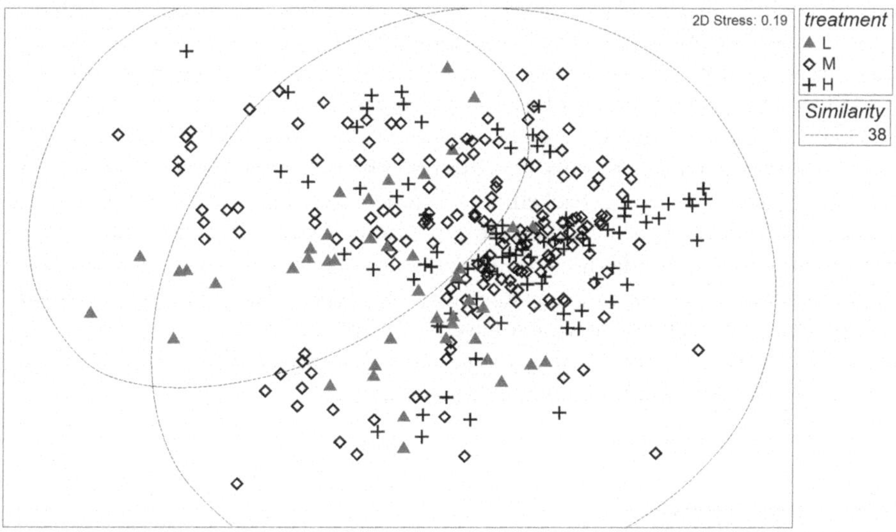

Fig. 6.17 MDS analysis of plant species cover in fertilizer treatment plots. *L* = low (treatments 2 and control), *M* = medium (treatments 3–5), *H* = high (treatments 1 and 6). Dashed lines encompass plots with a hierarchical clustering similarity of 38%

tively correlated ($r = -0.806$) with *S. cyperinus*. The low treatment plots tend to be located on the left (*S. cyperinus* dominated) side of the MDS plot, and the high treatment plots tend to be located on the right side (*Calamagrostis* dominated) (Fig. 6.17). The finding that nitrogen fertilization increased the proportion of grass cover is consistent with findings of Vermeer (Vermeer 1986) in a Netherlands fen.

6.7 Mouse Wars: How Voles Keep Trees Out of Beaver Meadows

Coniferous trees surround beaver meadows, but the rate of woody plant succession into Kabetogama beaver meadows is slow (Sect. 6.2.3). Terwilliger and Pastor (Terwilliger and Pastor 1999) describe the challenges to woody plant succession:

> Re-establishment of conifers in beaver meadows would involve dispersal of seeds away from the parent plant, seed survival, and germination. Seedlings then must survive resource limitations, frost heaving, physical damage, herbivores and successfully compete with pre-existing vegetation. In addition, survival and growth of conifers require symbiotic association with ectomycorrhizal fungi.

Conifer trees rely on symbiotic root-associated fungi, called ectomycorrhizae, to aid in water and nutrient uptake, but beaver meadow soils lack the fungi necessary to form ectomycorrhizae. One way that ectomycorrhizae could become reestablished on conifer seedlings is from the droppings of fungi-eating rodents, such as the red-backed vole (*Clethrionomys gapperi*). Ectomycorrhizae grew on 22–33% of black spruce seedlings experimentally grown in beaver meadow soils inoculated with red-backed vole droppings, and the black spruce seedlings had significantly higher shoot, root, and total dry weights after 46 weeks of growth in inoculated soils compared to uninoculated controls (Terwilliger and Pastor 1999).

However, trapping studies at eight beaver meadows showed that red-back voles didn't occupy the meadows, only the adjacent forest (Terwilliger and Pastor 1999). Therefore, although the red-backed vole is a potential vector for ectomycorrhizal fungal spores into beaver meadows, the expression of this potential is limited by its habitat use patterns. Furthermore, red-back voles may be excluded from meadows by the aggressive behavior of meadow voles (*Microtus pennsylvanicus*), which *do* occupy beaver meadows but which have fewer fungal spores in their feces. This complex interaction among plants, fungi, and voles may explain why most woody plant encroachment occurred along meadow edges.

6.8 Conclusions

- The four most abundant plant species in Kabetogama Peninsula beaver meadows are *Calamagrostis canadensis, Carex lacustris, Scirpus cyperinus*, and *Rubus idaeus* spp. *strigosus*.
- Bluejoint grass (*Calamagrostis*) exhibits phenotypic plasticity that allows it to occupy a range of moisture conditions within beaver meadows
- Beaver meadows contain three general plant groups that are geographically as well as floristically separated: wet, upland edge, and core. Plant diversity is greatest in the upland edge communities and least in the wet communities.
- Beaver impoundments flood a range of wetland ecological systems: bog, poor swamp, rich swamp, wet meadow, and freshwater marsh.

- Beaver meadow sites dominated by the sedge *Carex lacustris* were wettest and contained the least nitrate in comparison to sites dominated by *Calamagrostis* or mixed *Calamagrostis/C. lacustris*.
- Experimental fertilization in the greenhouse approximately doubled the height of *Carex* species, and caused a mid-canopy bulge in the vertical leaf distribution of *Calamagrostis*.
- Field measurements of light attenuation showed that it was primarily a function of meadow canopy height.
- Experimental fertilization in the field increased biomass in woolgrass (*Scirpus cyperinus*) meadows but not *Calamagrostis* meadows. Long-term (11 years) fertilization increased the proportion of *Calamagrostis* relative to other species, decreasing plant diversity.
- The feces and behavior of voles (along with ectomycorrhizae) help explain the slow re-establishment of woody species in beaver meadows.

References

Aerts R, deCaluwe H (1994) Effects of nitrogen supply on canopy structure and leaf nitrogen distribution in *Carex* species. Ecology 75(5):1482–1490

Arrhenius O (1921) Species and area. J Ecol 9:95–99

Bartel RA, Haddad NM, Wright JP (2010) Ecosystem engineers maintain a rare species of butterfly and increase plant diversity. Oikos 119(5):883–890

Binkley D (1984) Ion exchange resin bags: factors affecting estimates of nitrogen availability. Soil Sci Soc Am J 48(5):1181–1184

Binkley D, Matson PA (1983) Ion exchange resin bag method for assessing forest soil nutrient availability. Soil Sci Soc Am J 47:1050–1052

Bonner J, Anderson J, Rentch J, Grafton W (2009) Vegetative composition and community structure associated with beaver ponds in Canaan valley, West Virginia, USA. Wetl Ecol Manag 17(5):543–554

Bourdaghs M (1998) Light interception in northern beaver meadow canopies (REU research report). Natural Resources Research Institute, Duluth

Bray JR, Curtis JT (1957) An ordination of the upland forest communities in southern Wisconsin. Ecol Monogr 27:325–349

Brown TN, Bourdaghs M, Dewey BW, Johnston CA, Pastor J Vertical N distribution and light interception as a mechanism for direct competition between sedges and grasses. In: Abstracts, 85th annual meeting of the ecological society of America, snowbird, UT, 6–10 August 2000. p. 63

Chapin FS, Matson PA, Vitousek PM (2011) Principles of terrestrial ecosystem ecology, 2nd edn. Springer Science+Business Media, New York

Clarke KR, Gorley RN (2006) PRIMER v6: user manual/tutorial. PRIMER-E, Plymouth

Conn JS (1990) Seed viability and dormancy of 17 weed species after burial for 4.7 years in Alaska. Weed Sci 38:134–138

Coté J-F, Thibault J-R (1988) Allelopathic potential of raspberry foliar leachates on growth of Ectomycorrhizal fungi associated with black spruce. Am J Bot 75(7):966–970

Darris D (2005) Plant fact sheet for bluejoint (*Calamagrostis canadensis*). USDA, Natural Resources Conservation Service, Plant Materials Center, Corvallis

Dewey B, Pastor J, Johnston CA, Erickson H (1995) Biomass canopy allocation under different N, P, and moisture levels for selected wetland graminoid species. Bulletin of the ecological society of America, program and abstracts 80th annual ESA meeting 76(2):63

Erickson HE (1994) Nitrogen and phosphorus availability, ecosystem processes, and plant community dynamics in boreal wetland meadows. Ph.D. thesis, University of Washington, Seattle

Faber-Langendoen D, Aaseng N, Hop K, Lew-Smith M (2007a) Field guide to the plant community types of Voyageurs National Park. U.S. Geological Survey, Reston

Faber-Langendoen D, Aaseng N, Hop K, Lew-Smith M, Drake J (2007b) Vegetation classification, mapping, and monitoring at voyageurs National Park, Minnesota: an application of the U.S. National Vegetation Classification. Appl Veg Sci 10(3):361–374

Flora of North America Editorial Committee (2002) Flora of North America north of Mexico, vol 23. Oxford University Press, New York

Gibson DJ, Colquhoun IA, Greig-Smith P (1985) New method for measuring nutrient supply rates in soils using ion-exchange resins. In: Fitter AH (ed) Ecological interactions in soils: plants, animals, and microbes. British Ecological Society, London

Hirose T, Werger MJA (1987) Maximizing daily canopy photosynthesis with respect to the leaf nitrogen allocation pattern in the canopy. Oecologia 72:520–526

Hogg EH, Lieffers VJ (1991a) The impact of *Calamagrostis canadensis* on soil thermal regimes after logging in northern Alberta. Can J For Res 21(3):387–394

Hogg EH, Lieffers VJ (1991b) Seasonal changes in shoot regrowth potential in Calamagrostis canadensis. Oecologia 85(4):596–602

Hood GA, Bayley SE (2009) A comparison of riparian plant community response to herbivory by beavers (Castor canadensis) and ungulates in Canada's boreal mixed-wood forest. For Ecol Manag 258(9):1979–1989

Hughes JW, Fahey TJ (1991) Colonization dynamics of herbs and shrubs in a disturbed northern hardwood forest. J Ecol 79(3):605–616

Johnston CA (1988) Productivity of wet soils: biomass of cultivated and natural vegetation. Oak Ridge National Laboratory, Oak Ridge

Johnston CA, Brown T (2013) Water chemistry distinguishes wetland plant communities of the Great Lakes coast. Aquat Bot 104:111–120

Johnston CA, Ghioca DM, Tulbure M, Bedford BL, Bourdaghs M, Frieswyk CB, Vaccaro L, Zedler JB (2008) Partitioning vegetation response to anthropogenic stress to develop multi-taxa wetland indicators. Ecol Appl 18(4):983–1001

Johnston CA, Windels SK (2015) Using beaver works to estimate colony activity in boreal landscapes (online spatial data). CUAHSI Spatial Data Service, Consortium of Universities for the Advancement of Hydrologic Science Inc., Medford. http://spatial.cuahsi.org/johnstonC01/

Johnston CA, Zedler JB (2012) Identifying preferential associates to initiate restoration plantings. Restor Ecol 20:764–772

Kadlec JA (1958) An analysis of a woolgrass (*Scirpus cyperinus*) community in Wisconsin. Ecology 39(2):327–332

Kadlec JA (1961) A further comment on the ecology of woolgrass (Scirpus cyperinus). Ecology 42(3):591–592

Keddy PA, Reznicek AA (1982) The role of seed banks in the persistence of Ontario's coastal plain flora. Am J Bot 69(1):13–22

Landhäusser SM, Mulak TL, Lieffers VJ (2007) The effect of roots and litter of Calamagrostis canadensis on root sucker regeneration of *Populus tremuloides*. Forestry 80(4):481–488

Landhäusser SM, Wein RW (1993) Postfire vegetation recovery and tree establishment at the arctic treeline – climate-change vegetation-response hypotheses. J Ecol 81:665–672

Larson JL (1999) Woolgrass: a plant profile. Ecol Restor 17(4):210

Larson JL, Stearns F (1987) Seed bank of a woolgrass (*Scirpus cyperinus*) sedge meadow six years after strip mining. Bull Bot Club Wisconsin 21(1):16–17

Lichvar RW, Butterwick M, Melvin NC, Kirchner WN (2014) The National wetland plant list: 2014 update of wetland ratings. Phytoneuron 41:1–42

Lieffers VJ, Macdonald SE, Hogg EH (1993) Ecology of and control strategies for *Calamagrostis canadensis* in boreal forest sites. Can J For Res 23(10):2070–2077

Lieffers VJ, Stadt KJ (1994) Growth of understory *Picea glauca*, *Calamagrostis canadensis*, and *Epilobium angustifolium* in relation to overstory light transmission. Can J For Res 24(6):1193–1198

Longcore JR, McAuley DG, Pendelton GW, Bennatti CR, Mingo TM, Stromborg KL (2006) Macroinvertebrate abundance, water chemistry, and wetland characteristics affect use of wetlands by avian species in Maine. Hydrobiologia 567(1):143–167

MacArthur RH, MacArthur JW (1961) On bird species diversity. Ecology 42:594–598

MacDonald SE, Lieffers VJ (1991) Population variation, outcrossing, and colonization of disturbed areas by *Calamagrostis canadensis*: evidence from allozyme analysis. Am J Bot 78(8):1123–1129

MacDonald SE, Lieffers VJ (1993) Rhizome plasticity and clonal foraging of *Calamagrostis canadensis* in response to habitat heterogeneity. J Ecol 81(4):769–776

McKendrick JD, Brundage AL, Burton VL (1977) Quality of bluejoint hay is influenced by time of harvest. Agroborealis 9:26–29

McMaster RT, McMaster ND (2000) Vascular flora of beaver wetlands in western Massachusetts. Rhodora 102(910):175–197

Monson PH (1986) An analysis of the effects of fluctuating water levels on littoral zone macrophytes in the Namakan reservoir/rainy Lake system, voyageurs national park. Voyageurs National Park, International Falls

Morris JT (1991) Effects of nitrogen loading on wetland ecosystems with particular reference to atmospheric deposition. Annu Rev Ecol Syst 22:257–279

Mou P, Fahey TJ, Hughes JW (1993) Effects of soil disturbance on vegetation recovery and nutrient accumulation following whole-tree harvest of a northern hardwood ecosystem. J Appl Ecol 30:661–675

Norby RJ, Kozlowski TT (1980) Allelopathic potential of ground cover species on *Pinus resinosa* seedlings. Plant Soil 57(2–3):363–374

Olde Venterink H, van der Vliet RE, Wassen MJ (2001) Nutrient limitation along a productivity gradient in wet meadows. Plant Soil 234(2):171–179

Pastor J, Downing A, Erickson HE (1996) Species-area curves and diversity-productivity relationships in beaver meadows of voyageurs National Park, Minnesota, USA. Oikos 77(3):399–406

Peach M, Zedler J (2006) How tussocks structure sedge meadow vegetation. Wetlands 26(2):322–335

Pielou EC (1975) Ecological diversity. Wiley, New York

Pollock MM, Naiman RJ, Hanley TA (1998) Plant species richness in riparian wetlands – a test of biodiversity theory. Ecology 79(1):94–105

Powelson RA, Lieffers VJ (1992) Effect of light and nutrients on biomass allocation in *Calamagrostis canadensis*. Ecography 15(1):31–36

Quinlan SE, Cuccarese S (2004) Native Alaskan and exotic plants used by wildlife. Alaska Department of Fish and Game, Anchorage

Remillard MM, Gruendling GK, Bogucki DJ (1987) Disturbance by beaver (*Castor canadensis* Kuhl) and increased landscape heterogeneity. In: Turner MG (ed) Landscape heterogeneity and disturbance. Ecological studies, vol. 64. Springer-Verlag, New York, pp 103–122

Ricard J-P, Messier C (1996) Abundance, growth and allometry of red raspberry (*Rubus idaeus* L.) along a natural light gradient in a northern hardwood forest. For Ecol Manag 81(1):153–160

Rivard PG, Woodard PM, Rothwell RL (1990) The effect of water table depth on white spruce (Piceaglauca) seedling growth in association with the marsh reed grass (Calamagrostiscanadensis) on wet mineral soil. Can J For Res 20(10):1553–1558

Ruel J-C (1992) Impact de la compétition exercée par le framboisier (*Rubus idaeus* L.) et les feuillus de lumière sur la croissance du sapin (*Abies balsamea* Mill.) en régéněracion. Can J For Res 22:1408–1416

Schlichting CD (1986) The evolution of phenotypic plasticity in plants. Annu Rev Ecol Syst 17:667–693

Shaver GR, Johnson LC, Cades DH, Murray G, Laundre JA, Rastetter EB, Nadelhoffer KJ, Giblin AE (1998) Biomass and CO_2 flux in wet sedge tundras: responses to nutrients, temperature, and light. Ecol Monogr 68(1):75–97

Shaver GR, Melillo JM (1984) Nutrient budgets of marsh plants: efficiency concepts and relation to availability. Ecology 65(5):1491–1510

Simonavičiūtė L, Ulevičius A (2007) Structure of phytocenoses in beaver meadows in Lithuania. Ekologija 53(3):34–44

Stout AB (1912) A biological and statistical analysis of a typical wild hay meadow. Trans Wis Acad Sci. Arts Lett 17:405–469

Sugihara G (1981) $S = cA^z$; $z \approx 0.25$: a reply to Connor and McCoy. Am Nat 117:790–793

Terwilliger J, Pastor J (1999) Small mammals, ectomycorrhizae, and conifer succession in beaver meadows. Oikos 85(1):83–94

Townsend JE (1953) Beaver ecology in western Montana with special reference to movements. J Mammal 34(4):459–479

Travis SE, Marburger JE, Windels S, Kubátová B (2010) Hybridization dynamics of invasive cattail (Typhaceae) stands in the Western Great Lakes region of North America: a molecular analysis. J Ecol 98(1):7–16

van der Hoek D, van Mierlo AJEM, van Groenendael JM, Lepš J (2004) Nutrient limitation and nutrient-driven shifts in plant species composition in a species-rich fen meadow. J Veg Sci 15(3):389–396

Verhoeven JTA, Schmitz MB (1991) Control of plant growth by nitrogen and phosphorus in mesotrophic fens. Biogeochemistry 12(2):135–148

Vermeer J (1986) The effect of nutrients on shoot biomass and species composition of wetland and hayfield communities. Acta Oecol, Oecol Plant 7(1):31–41

Werger MJA, Hirose T (1991) Leaf nitrogen distribution and whole canopy photosynthetic carbon gain in herbaceous stands. Vegetatio 97(1):11–20

Westbrook CJ, Cooper DJ, Baker BW (2011) Beaver assisted river valley formation. RiverRes Appl 27(2):247–256

Wetzel PR, van der Valk AG (2005) The biomass and nutrient levels of Calamagrostis canadensis and Carex stricta under different hydrologic and fungicide regimes. Can J Bot 83(1):124–130

Whitney GG (1982) The productivity and carbohydrate economy of a developing stand of *Rubus idaeus*. Can J Bot 60:2697–2703

Whitney GG (1986) A demographic analysis of Rubus idaeus and Rubus pubescens. Can J Bot 64(12):2916–2921

Wilcox DA, Pavlovic NB, Mueggler ML (1985) Selected ecological characteristics of *Scirpus cyperinus* and its role as an invader of disturbed wetlands. Wetlands 5(1):87–97

Woo I, Zedler JB (2002) Can nutrients alone shift a sedge meadow towards dominance by the invasive *Typha* x *glauca*? Wetlands 22(3):509–521

Wynia RL (2006) Plant guide for bluejoint reedgrass (*Calamagrostis canadensis*), USDA, Natural Resources Conservation Service, Plant Materials Center, Manhattan

Yetka LA, Galatowitsch SM (1999) Factors affecting revegetation of *Carex lacustris* and *Carex stricta* from rhizomes. Restor Ecol 7(2):162–171

Chapter 7
The Biogeochemistry of Boreal Beaver Ponds

Abstract The flooding of soils by beaver dams causes the depletion of oxygen and development of anaerobic conditions that alter biogeochemical processes. Field studies showed that, as anaerobic conditions increased, porewater concentrations of Fe^{2+} increased and SO_4^{2-} decreased, consistent with the predictions of redox reactions. Porewater PO_4^{3-} concentrations also increased under anaerobic conditions, presumably due to dissolution of iron phosphates following iron reduction. Ammonium concentrations increased with increasing saturation; under aerobic conditions ammonium would normally be converted to nitrate. Porewater Ca^{2+} and Mg^{2+} concentrations also increased with increasing saturation, but the mechanism was probably carbonate dissolution due to increased CO_2 pressure in the soil during waterlogging, rather than changes in redox potential. Soil type influenced beaver meadow chemistry: glaciolacustrine soils had higher total N, oxalate-extractable Fe, cation exchange capacity, and base cations (K, Ca, Mg) than did glaciofluvial soils. A split moving window was used along a beaver meadow transect to identify biogeochemical ecotones, the locations where sharp breaks occurred in anaerobiosis and nitrate concentrations. A distributed parameter biogeochemical model was applied to the Finlander watershed of the Kabetogama Peninsula, and indicated that beaver impoundments there would decrease nitrogen concentration, chemical oxygen demand, and sediment yield measured at the watershed outlet cell.

The chemistry of beaver ponds is a function of inputs from water, air, plants, and soil, as well as chemical and biological transformations that occur under the specialized environmental conditions created when a beaver dam floods an area. This chapter describes some of the biogeochemical processes that occur in boreal beaver ponds and meadows, focusing on results from the Kabetogama Peninsula.

7.1 Atmospheric Inputs and Outputs

At the landscape scale, biogeochemical inputs to the Kabetogama Peninsula come primarily from precipitation (Table 7.1). Due to the peninsula's long distance from agricultural and industrial air pollutants, concentrations of NH_4^+ and SO_4^{2-} in

C.A. Johnston, *Beavers: Boreal Ecosystem Engineers*,
DOI 10.1007/978-3-319-61533-2_7

Table 7.1 Concentrations (mean ± SE) of selected ions in precipitation at Voyageurs National Park, 2001–2014

Parameter	Concentration (mg/L)
NO_3^-	0.80 ± 0.053
NH_4^+	0.35 ± 0.016
K^+	0.02 ± 0.001
Ca^{2+}	0.16 ± 0.009
Mg^{2+}	0.02 ± 0.001
SO_4^{2-}	0.56 ± 0.040
Cl^-	0.03 ± 0.001

Data summarized from the National Atmospheric Deposition Program (NADP 2017)

precipitation are low relative to other locations in the U.S. The average annual pH of precipitation at VNP (2001–2014) was 5.4, which is normal for unpolluted rainwater. In contrast, the pH of precipitation in the Adirondack Mountains of New York, which has acid rain, was only 4.1 in 1980 (Driscoll et al. 2003).

Nitrogen can also be supplied by microbial fixation of atmospheric N_2, either by free-living microbes (asymbiotic) or by microbes living in root nodules (symbiotic). Nitrogen-fixing microbes have the capacity to convert N_2 into ammonium (NH_4^+) that supports their own growth and then enters the soil organic N pool. Nitrogen fixation by asymbiotic sediment microbes in a beaver-dammed second-order Quebec stream contributed 68% of the annual N budget, compared with only 4.2% in an undammed stream, the difference being due to the greater sediment area available for microbial colonization in the beaver pond as compared with riffles (Naiman and Melillo 1984; Francis et al. 1985).

Microbial activity can release N_2 back into the atmosphere by the reduction process of denitrification, which converts aqueous nitrate (NO_3^-) into gaseous form. Denitrification requires anaerobic conditions, such as the inundated or saturated soils of beaver ponds and beaver meadows. Two other microbial processes that affect the forms of inorganic N in the soil are nitrification, which converts ammonium into nitrate in aerobic soils, and dissimilatory nitrate reduction, which converts nitrate into ammonium in anaerobic soils (Tiedje 1988).

Researchers in Rhode Island used ^{15}N–nitrate mass balance techniques on soil core mesocosm incubations from three beaver ponds to examine rates of nitrate transformation among these various pathways (Lazar et al. 2015). Of the added NO_3^--N, 22–39% was transformed during the course of the incubation. Denitrification had the highest rates of transformation (97–236 mg N m^{-2} day^{-1}), followed by assimilation into the organic soil N pool (41–93 mg N m^{-2} day^{-1}) and ammonium generation (11–14 mg N m^{-2} day^{-1}), presumably from dissimilatory nitrate reduction. Their denitrification rates were higher than those reported in the literature from other shallow ponds and wetlands, but their nitrate input rates were also high due to residential and agricultural development in the watersheds of their beaver ponds. The authors estimated that in nitrate enriched watersheds, beaver pond denitrification can remove approximately 50–450 kg NO_3^--N per km^2 of catchment area, an important sink for watershed nitrate.

Table 7.2 Average (mean ± SE) redox potential, pH, N fixation (measured as acetylene fixation), and denitrification rates (measured as N_2O production) of moisture gradient zones

	Hydrologic zone			
Parameter	Forest	Moist meadow	Wet meadow	Beaver pond
Redox potential (mV)	536 ± 15	338 ± 22	−18 ± 25	−64 ± 45
Soil/sediment pH	5.06 ± 0.26	5.13 ± 0.15	5.78 ± 0.07	6.06 ± 0.06
Nitrogen fixation (mol m^{-2} h^{-1})	2.9 ± 0.9	2.9 ± 1.4	2.3 ± 0.6	1.5 ± 0.4
Denitrification (mol m^{-2} h^{-1})	0.14 ± 0.02	0.33 ± 0.06	0.39 ± 0.05	0.30 ± 0.04

From Naiman RJ, Pinay G, Johnston CA, Pastor J (1994) Beaver influences on the long-term biogeochemical characteristics of boreal forest drainage networks. Ecology 75 (4): 905–921, with permission of John Wiley & Sons

My colleagues measured nitrogen fixation and denitrification rates on Kabetogama beaver pond soil cores using the acetylene reduction (Knowles 1982) and acetylene block techniques (Yoshinari and Knowles 1976), respectively. Sediment N fixation ranged from 1.5 to 2.0 mol m^{-2} h^{-1} across hydrologic zones representing beaver pond, wet meadow, moist meadow, and forest (Table 7.2), but was not statistically different across the zones (Naiman et al. 1994). However, denitrification rates were two times higher in the beaver pond, wet meadow, and moist meadow soils than in the never-impounded forest soils. Measurement of soil redox potential using platinum electrodes confirmed that the forest soils were aerated (highly positive redox potentials), whereas the moist meadow, wet meadow, and pond soils were progressively more anaerobic (Table 7.2).

Two woody plant species associated with some VNP beaver ponds, alder (*Alnus incana*) and sweet gale (*Myrica gale*), have root nodules containing *Frankia* actinomycete bacteria that support symbiotic N fixation (Ekblad and Huss-Danell 1995; Schwintzer 1979), but the magnitude of N inputs from these sources is unknown.

7.2 Chemical Changes Along Moisture Gradients

Soils contain about 50% pore space, and in upland soils that pore space contains about half air and half water. When soils are flooded, the air they contain is displaced by water, and any remaining soil oxygen is quickly depleted by belowground respiration. A number of soil chemical reactions require oxygen, so soil flooding fundamentally changes the concentrations and molecular forms of key elements. Submerging a soil brings about the following electrochemical changes (Ponnamperuma 1972): (a) a decrease in redox potential, (b) an increase in pH of acid soils, (c) changes in specific conductance and ionic strength, (d) drastic shifts in mineral equilibria, (e) cation and anion exchange reactions, and (f) sorption and desorption of ions.

We studied the effect of flooding on nutrients in Kabetogama beaver ponds and meadows by sampling soil porewater along moisture gradients at 94 locations in the Clyde Creek, Lost Ponds, and Shoepack Ponds study sites (Fig. 3.2). Samples were

Fig. 7.1 Porous ceramic cup sampling device with tubes attached for pumping out water samples. Located at the edge of the residual Upper Shoepack Pond. The abandoned beaver lodge is visible in the background

collected using vacuum lysimeters with porous ceramic cups that were installed 25 cm below the soil or sediment surface (Fig. 7.1). Porewater from unimpounded stream sediments was also collected to compare with beaver-altered sites. Observed trends in porewater chemical concentrations are discussed below in conjunction with the processes that appear to have affected them.

7.3 Processes Behind the Observed Chemical Trends

7.3.1 Oxidation-Reduction Reactions

Nitrate, ammonium, iron, orthophosphate, and sulfate concentrations are potentially affected by changes in redox potential, either directly as part of oxidation-reduction (redox) reactions or indirectly because of redox compounds with which they interact. Oxidation-reduction (redox) reactions involve the transfer of electrons from one reactant to another, and occur in a predictable sequence (Table 7.3). Microorganisms harvest the most energy by transferring electrons to oxygen, so oxygen reduction predominates when it is present, but the reactions on the list occur as the soil becomes increasingly anaerobic. Reduction reactions consume H^+ ions (Table 7.3), which is why soil pH increases with decreasing redox potential (Table 7.2).

Nitrate is the first compound to be reduced after oxygen becomes depleted, so we expected porewater nitrate concentrations to decrease with increasing wetness. However, there was no significant difference in NO_3^- concentrations across the

Table 7.3 Redox reactions affecting inorganic substances in wetland soils (van Breeman and Brinkman 1976; Ponnamperuma 1972)

Reduction of NO_3^- (denitrification): $NO_3^- + 2H^+ + 2e^- \rightarrow NO_2^- + H_2O$
Reduction of Mn^{4+} to Mn^{2+}: $MnO_2 + 4H^+ + 2e^- \rightarrow Mn^{2+} + 2H_2O$
Reduction of Fe^{3+} to Fe^{2+}: $Fe(OH)_3 + 3H^+ + e^- \rightarrow Fe^{2+} + 3H_2O$
Reduction of SO_4^{2-} to H_2S: $SO_4^{2-} + 10H^+ + 8e^- \rightarrow H_2S + 4H_2O$
Reduction of CO_2 to CH_4: $CO_2 + 8H^+ + 8e^- \rightarrow CH_4 + 2H_2O$

moisture zones (Table 7.4). All zones contained very little NO_3^-, and the concentrations of NO_3^- in wet meadow porewater were an order of magnitude less than those in precipitation (Table 7.1). This finding implies that denitrification, biotic uptake, and/or dissimilatory nitrate reduction may have depleted porewater NO_3^- in all the zones.

Iron is also a redox element, changing oxidation state from Fe^{3+} to Fe^{2+} as it is reduced. Ferrous iron (Fe^{2+}), the reduced form of iron, is much more soluble than its oxidized counterpart, ferric iron (Fe^{3+}). Therefore, we expected to see an increase in aqueous Fe^{2+} concentrations with increasing wetness, and that was the result (Table 7.4). These changes in iron oxidation state also explain the changes in soil color observed in the beaver meadow soils (Fig. 5.12): compounds containing oxidized ferric iron are rust colored, whereas ferrous iron compounds impart the greenish gray color that is associated with hydric soils (Vasilas et al. 2010).

We also expected to see a decrease in SO_4^{2-} as it was reduced to produce H_2S gas (Table 5.3), and that was observed (Table 7.4). A comprehensive solute budget of a beaver pond in the Adirondack Mountains of New York State reported similar findings regarding the fate of redox elements: the pond exported Fe^{2+} and was a sink for NO_3^- and SO_4^{2-}, attributed to microbially mediated redox reactions (Cirmo and Driscoll 1993; Smith et al. 1991). The authors found that ferrous iron concentrations were particularly high under ice cover in the winter.

Redox potential has a strong effect on P dynamics through its interaction with Fe. Ferric iron (Fe^{3+}) phosphates are insoluble, but the transformation of ferric to ferrous (Fe^{2+}) forms releases PO_4^{3-} into solution (Patrick et al. 1973). This process explains our finding that porewater PO_4^{3-} concentrations were significantly greater in pond sediments than in the soils of beaver meadows or forest (Table 7.4). Anaerobic conditions during extended winter ice cover promotes the release of soluble reactive P from beaver ponds (Klotz 1998).

The reduction of CO_2 to CH_4 is important to the carbon cycle, and is discussed more in Chap. 8.

7.3.2 Accumulation of Ammonium

Ammonium is not a redox reactant, but its presence is affected by moisture as a result of its microbially mediated transformation into nitrate under aerobic conditions, a process called nitrification. Ammonium (NH_4^+) is continually produced during

Table 7.4 Concentrations (mg/L, mean ± SE) of nutrients and ions associated with pore water during 1986–1987 for the five hydrologic zones

	Hydrologic zone					
Parameter	Forest (A)	Moist meadow (B)	Wet meadow (C)	Beaver pond (D)	Stream (E)	Tukey comparison
Sample size (*n*)	108–127	140–215	178–306	135–232	35–59	
Total nitrogen (mg/L)	0.92 ± 0.07	1.23 ± 0.17	1.88 ± 0.15	4.35 ± 0.52	1.99 ± 0.41	DECBA
Nitrate-N (mg/L)	0.03 ± 0.02	0.08 ± 0.07	0.06 ± 0.04	0.00 ± 0.00	0.01 ± 0.00	BCAED
Ammonium-N (mg/L)	0.09 ± 0.02	0.05 ± 0.01	0.38 ± 0.08	2.94 ± 0.51	1.01 ± 0.41	DECAB
Total phosphorus (mg/L)	0.04 ± 0.01	0.02 ± 0.00	0.03 ± 0.00	0.11 ± 0.02	0.10 ± 0.03	DEACB
Orthophosphate-P (mg/L)	0.03 ± 0.01	0.02 ± 0.01	0.03 ± 0.01	0.14 ± 0.03	0.11 ± 0.07	DECAB
Potassium (mg/L)	1.19 ± 0.18	0.71 ± 0.09	1.24 ± 0.20	2.31 ± 0.33	1.56 ± 0.19	DECAB
Calcium (mg/L)	4.46 ± 0.49	6.76 ± 0.94	17.44 ± 2.02	23.86 ± 3.01	8.54 ± 2.40	DECBA
Magnesium (mg/L)	1.52 ± 0.23	2.79 ± 0.44	7.94 ± 1.12	11.15 ± 1.50	3.68 ± 1.01	DECBA
Iron (mg/L)	0.41 ± 0.10	2.03 ± 0.44	10.80 ± 1.68	12.61 ± 2.85	7.37 ± 2.88	DCEBA
Sulfate (mg/L)	5.97 ± 0.38	6.72 ± 0.57	2.45 ± 0.76	0.72 ± 0.18	0.58 ± 0.12	BACDE
Chloride (mg/L)	3.87 ± 0.61	3.09 ± 0.25	3.35 ± 0.23	4.53 ± 0.36	2.19 ± 0.21	DACBE

Values connected by a line in the Tukey comparison are not significantly different ($P > 0.05$). From Naiman RJ, Pinay G, Johnston CA, Pastor J (1994) Beaver influences on the long-term biogeochemical characteristics of boreal forest drainage networks. Ecology 75 (4): 905–921, with permission of John Wiley & Sons decomposition as microbes use organic matter to support their own growth and respiration, secreting NH_4^+ as a waste product that can subsequently be nitrified. Because nitrification relies on oxygen, however, soil flooding can block it, allowing NH_4^+ to accumulate under anaerobic conditions. Dissimilatory nitrate reduction, whereby NO_3^- is converted to NH_4^+, could also contribute to this NH_4^+ accumulation (Tiedje 1988).

As a result of these processes, we hypothesized that NH_4^+ would accumulate in the pore waters of beaver pond sediments and soils at VNP. Our hypothesis was correct: NH_4^+ and total N concentrations *were* significantly greater in beaver ponds (the most reduced sites) than in the other four moisture types (Table 7.4). Input concentrations of NH_4^+ in precipitation (0.35 mg/L) were much lower than those in pond porewater (2.94 mg/L), implying that soil NH_4^+ is being generated by microbial activity. Other boreal region studies have also found beaver ponds to be net exporters of NH_4^+ (Cirmo and Driscoll 1993; Devito et al. 1989). Ammonium accumulation rates are particularly high under winter ice cover in beaver ponds, when pond waters become anoxic (Devito and Dillon 1993; Driscoll et al. 1987; Cirmo and Driscoll 1993).

7.3.3 *Carbonate Dissolution*

An unexpected finding at VNP was the highly significant effect of moisture on Ca and Mg concentrations in soil pore water sampled from Kabetogama beaver ponds and meadows (Table 7.4). Neither element is involved in a redox reaction, but both occur as soil carbonates, which accumulate in fine-textured soils around the edges of beaver ponds and other wetlands (Johnston 2000; Arndt and Richardson 1992). These carbonates can be solubilized when flooded due to increased CO_2 pressure in the soil during waterlogging, which drives these formulas to the right, releasing calcium and magnesium ions into solution (Novozamsky and Beek 1976):

$$CaCO_3 + H_2O + CO_2 \leftrightarrow Ca^{2+} + 2HCO^{3-}$$

$$MgCO_3 + H_2O + CO_2 \leftrightarrow Mg^{2+} + 2HCO^{3-}$$

7.3.4 *Sedimentation*

By reducing stream velocity, beaver ponds promote sedimentation of nutrient-containing mineral particles. This benefits western U.S. mountain streams where upstream erosion delivers sediment particles to beaver ponds (Maret et al. 1987; Butler and Malanson 1995), but is probably less important in boreal beaver ponds, particularly those in bedrock-dominated Precambrian Shield catchments where erosion rates are low (but see Section 7.7 on "Watershed Modeling of Nutrient Fluxes"). The beaver meadow soil profiles described for the Kabetogama Peninsula showed little pedogenic evidence of sedimentation, and there were few soils mapped as having alluvial parent materials (see Chap. 5). Accumulation rates by this mechanism were estimated at 0.15–0.55 g P m^{-2} $year^{-1}$ and 2.5–6.6 g N m^{-2} $year^{-1}$ in a Canadian beaver pond (Devito and Dillon 1993).

7.3.5 Biotic Uptake and Mineralization

Biotic uptake by microbes, algae, and macrophytes in beaver ponds transforms inorganic N and P forms into organic forms. A 5-year study of a Canadian beaver pond showed that the annual retention of inorganic N and P was roughly balanced by net export of organic N and P, resulting in low annual retention rates, and that total organic N outputs were greater than inputs in every year of the study (Devito and Dillon 1993). This transformation from inorganic to organic is an ecosystem service that benefits downstream waters by converting nutrients to less available forms.

Mineralization is the conversion of organic forms of nutrients back into inorganic forms by decomposition processes. Mineralization rates are very important in natural ecosystems because nutrient inputs from the atmosphere or rock weathering are so low—organisms rely upon the cycling of nutrients for continued growth. My colleagues and I conducted an 80-week laboratory experiment to measure soil N mineralization rates, comparing a beaver meadow sedge peat with peat collected at the surface and at 1 m depth from a *Sphagnum* bog (Updegraff et al. 1995). Beaver meadow soil had the highest N mineralization rates under anaerobic conditions, but surface peat had the highest N mineralization under aerobic conditions (Table 7.5). Carbon mineralization was also measured, and is discussed in Chap. 8.

7.4 The Importance of Geologic Setting

Recall from Chap. 2 that beaver ponds require three ingredients: beavers, water, and *land suitable for flooding*. The land that is flooded plays an important role in the biogeochemistry of beaver ponds. Our study sites had soils derived from two different types of glacial deposits (Chap. 5): glaciofluvial soils deposited by meltwater flowing from continental glaciers, and glaciolacustrine soils deposited in Glacial Lake Agassiz that covered the region about 11,000 years ago. The soils that we sampled had comparable redox gradients, but the two soil types had very different particle size distributions: the glaciofluvial soils were sandy, and the glaciolacustrine soils were clayey (Johnston et al. 1995). Sand particles have little ability to sorb cations, so the glaciofluvial soils had very low cation exchange capacity, which is the number of positive charges that a soil can hold available for exchange with the

Table 7.5 Cumulative N mineralization for parallel 80-week incubations of soils from three locations

Incubation conditions	Source	N mineralized (mg N/g N)
Aerobic		
15 °C	Bog surface	48.9 ± 5.1[a]
	Bog 1 m depth	21.9 ± 1.9[b]
	Beaver meadow	24.1 ± 1.9[b]
30 °C	Bog surface	72.7 ± 6.2[a]
	Bog 1 m depth	39.1 ± 5.1[b]
	Beaver meadow	65.5 ± 4.9[a]
Anaerobic		
15 °C	Bog surface	11.9 ± 1.4[b]
	Bog 1 m depth	8.3 ± 0.9[b]
	Beaver meadow	22.3 ± 2.2[a]
30 °C	Bog surface	36.8 ± 3.6[b]
	Bog 1 m depth	13.6 ± 0.7[c]
	Beaver meadow	53.5 ± 6.1[a]

Data are expressed as the sum of leachate NO_3-N plus NH_4-N as a proportion of total N in the samples. Mean ± SE, $n = 10$ for bog soils, $n = 9$ for beaver meadow soil. Means within column and temperature followed by the same letter are not significantly different. From Updegraff K, Pastor J, Bridgham SD, Johnston CA (1995) Environmental and Substrate Controls over Carbon and Nitrogen Mineralization in Northern Wetlands. Ecological Applications 5 (1): 151–163, with permission from John Wiley & Sons

soil water solution (Table 7.6). Clays have very high cation exchange capacity, so the CEC values for the glaciolacustrine soils were high. The concentrations of base cations (K, Ca, and Mg) extracted from soil samples were much higher in glaciolacustrine soils than in glaciofluvial soils, due at least in part to the higher CEC of the clayey soils. Total N and oxalate-extractable iron were also higher in the glaciolacustrine soils (Table 7.6).

A two-way analysis of variance considered the effect of moisture, soil type (glaciolacustrine vs. glaciofluvial), and their interaction on porewater concentrations in beaver meadow soils (Table 7.7). Consistent with the earlier results (Table 7.4), porewater SO_4-S, Fe, Ca, and Mg concentrations were significantly affected by moisture. However, porewater concentrations of NH_4-N and PO_4-P were affected only by the interaction between moisture and soil. Concentrations of K, Ca, and Mg were significantly greater in glaciolacustrine soils, consistent with their greater concentration in soil extracts, but Fe concentrations were paradoxically *lower* in the porewater of the glaciolacustrine soils, despite the fact that oxylate-extractable Fe concentrations were higher in glaciolacustrine soils than in glaciofluvial soils (Table 7.6). These results illustrate that: (1) concentrations in soil extracts may not mirror concentrations in porewater, and (2) the geologic setting contributes to biogeochemical inputs and retention.

Table 7.6 Chemical characteristics of B and C horizons sampled from glaciolacustrine and glaciofluvial soils

	No. samples	Total N ($g\ kg^{-1}$)	Oxalate Fe ($mg\ kg^{-1}$)	CEC ($cmol\ kg^{-1}$)	K ($mg\ kg^{-1}$)	Ca ($mg\ kg^{-1}$)	Mg ($mg\ kg^{-1}$)	Ca:Mg
Glaciolacustrine	6	1.02	3.77	72.78	2.90	43.98	17.35	2.70
Glaciofluvial	7	0.64	1.49	3.21	0.31	2.50	0.39	6.61

Analysis by atomic absorption spectrophotometer following extraction with 1 *M* NH_4OAc

Table 7.7 Effects of moisture, soil texture, and their interaction on porewater concentrations of selected elements

Element	Moisture	Soil type	Interaction
NO_3-N	ns	ns	ns
NH_4-N	ns	ns	*
Total N	ns	ns	ns
PO_4-P	ns	ns	*
SO_4-S	↓	ns	ns
Fe	↑	Sand > Clay	*
K	ns	Clay > Sand	*
Ca	↑	Clay > Sand	*
Mg	↑	Clay > Sand	*

↓ significant decreasing concentration with increasing moisture, ↑ significant increasing concentration with increasing moisture, *ns* not significant
*$P < 0.05$, interaction between moisture and soil type (Johnston et al. 1995)

7.5 Soil Nutrient Ecotones

An ecotone is a zone of transition between adjacent ecological systems, having a set of characteristics uniquely defined by space and time scales and by the strength of the interactions between adjacent ecological systems (Holland 1988). In the beaver meadow soils that we studied, some nutrient concentrations changed gradually across the meadow area, whereas others changed abruptly. A study done at lower Shoepack Pond (Box 7.1) examined fine-scale spatial variation along a transect that crossed several ecotones.

This moving split-window analysis showed that ecotones related to anaerobiosis were stable and coincided well with visible physical and vegetative changes (Fig. 7.2a). There was a large, persistent peak that coincided with the edge between pond and meadow, a secondary peak at the opposite end of the transect that occurred at the high water mark between meadow and upland, and a central peak that occurred where the soil changed from an Argiaquoll (high carbon content) to an Ochraqualf (low carbon content). The locations and magnitude of these peaks remained relatively constant through the three sample periods.

Box 7.1: Split Moving Window Analysis of Ecotones

Porewater nutrient concentrations were sampled at a 25 cm depth, taking measurements every 3 m along a 51 m transect on three different dates in late summer/early autumn (Johnston et al. 1992). The seven original soil water attributes were reduced to two principal components, which were analyzed by a moving split window approach that computed the squared Euclidean distance (SED) within a 6 sample window (3 samples per window split) for each of the 51 positions along the transect (Fig. 7.2). The first principal component represented anaerobiosis and the second principal component represented nitrate concentrations.

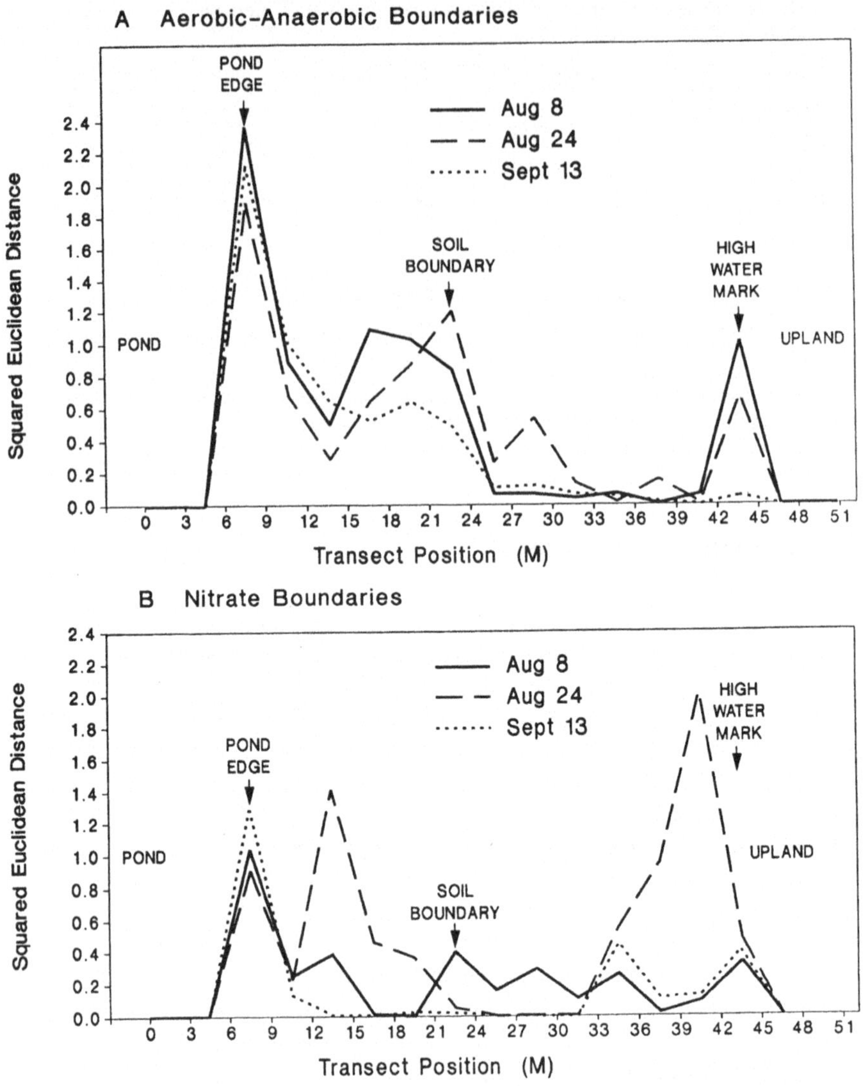

Fig. 7.2 (**a**) Ecotones determined by moving split-window analysis (window width = 6) of principal components data related to soil-water anaerobiosis. (**b**) Ecotones determined by moving split-window analysis (window width = 6) of principal components data related to soil nitrate availability. From Johnston CA, Pastor J, Pinay G (1992) Quantitative methods for studying landscape boundaries. In: Castri Fd, Hansen A (eds) Landscape Boundaries: Consequences for Biotic Diversity and Ecological Flows. Springer-Verlag, New York, pp. 107–125, with permission

The nitrate ecotones, in contrast, were much more mobile. There was a persistent ecotone at the pond/meadow edge, similar to that observed for the first principal component (Fig. 7.2b). However, there was never a pronounced nitrate ecotone at the high water mark, and peaks that occurred in August 24 were ephemeral and unexplained. This analysis illustrates that the soil and microbial processes that affect nitrate are very dynamic in time and space (Johnston et al. 1992).

7.6 Net Nutrient Retention

The discussion so far has been about in situ processes, ignoring the fact that nutrient-laden water flows into beaver ponds from sideslopes and inlet streams, and out of beaver ponds as outlet streams. At the beaver pond scale, this flowing water supplies additional biogeochemical inputs. Comprehensive nutrient budgets for two Precambrian Shield beaver ponds showed that inlet streams and runoff supplied 65.2 and 81.4% of total P and 50.0 and 41.4% of total inorganic N inputs (Devito et al. 1989).

There have been mixed results regarding net retention of nutrients by boreal beaver ponds, which can be positive (i.e., inputs > outputs) or negative (i.e., inputs < outputs) (Klotz 1998; Devito et al. 1989). There can also be interannual variation in retention, where the same site retains nutrients in 1 year and releases them the next. In the 5-year study of a Canadian beaver pond, annual retention rates ranged from +0.09 to −0.8 kg P ha^{-1} $year^{-1}$ and +1.22 to −12.69 kg N ha^{-1} $year^{-1}$ (Devito and Dillon 1993). The magnitude of these values is small, however, in comparison to that reported for a Maryland Coastal Plain beaver pond, which retained 35 kg P ha^{-1} $year^{-1}$ and 109 kg N ha^{-1} $year^{-1}$ (Correll et al. 2000). Net retention (and export) is small in most boreal beaver ponds because inputs are usually small.

Box 7.2: AGNPS Model Methods

The AGNPS model was applied to the 534 ha Finlander catchment, which has an elevation range of 339–414 m and flows north into Finlander Bay of Rainy Lake (catchment #6, Fig. 2.2). GIS routines were used before the modeling to extract input variables and after the modeling to display the model results. Raster GIS hydrologic tools (Environmental Systems Research Institute Inc. 1992) were essential for data extraction, and the same capabilities are provided by current versions of ArcGIS.

Watershed boundaries were manually delineated and digitized from the Kempton Bay 1:24,000 U.S. Geological Survey (USGS) topographic map (Fig. 7.3). This polygon was used to clip elevation data from a USGS digital elevation model (DEM) with 30 × 30 m cells. In order to route water through the watershed, the DEM was preprocessed to fill "sinks" that would artificially block water flow; this required 3 iterations of the multistep fill process.

The filled DEM was then resampled to aggregate the 30 × 30 m cells into 90 × 90 m (0.81 ha) grid cells to facilitate modeling. Each of these 659 grid cells was uniquely numbered according to AGNPS conventions, beginning at the cell in the northwest corner and sweeping from west to east and north to south. The GIS hydrologic tools were then used to derive a stream network for the entire watershed (Fig. 7.3), using the sequence of commands FLOWDIRECTION, FLOWACCUMULATION, and STREAMLINE. The output of FLOWDIRECTION was used with a custom DOCELL routine in Arc Macro Language to calculate receiving cell identification numbers, and FLOWDIRECTION codes (1, 2, 4, 8, 16, 32, 64, 128) were renumbered to the 1–8 aspect codes required by AGNPS (Nawrocki et al. 1994). Slope percent was calculated directly in the GIS, and "slope shape factor" was determined using the CURVATURE command, positive values indicating convex slopes and negative values indicating concave slopes (water and wetlands were assigned 0 values). A complete list of AGNPS input variables and their derivation is in Table 7.8.

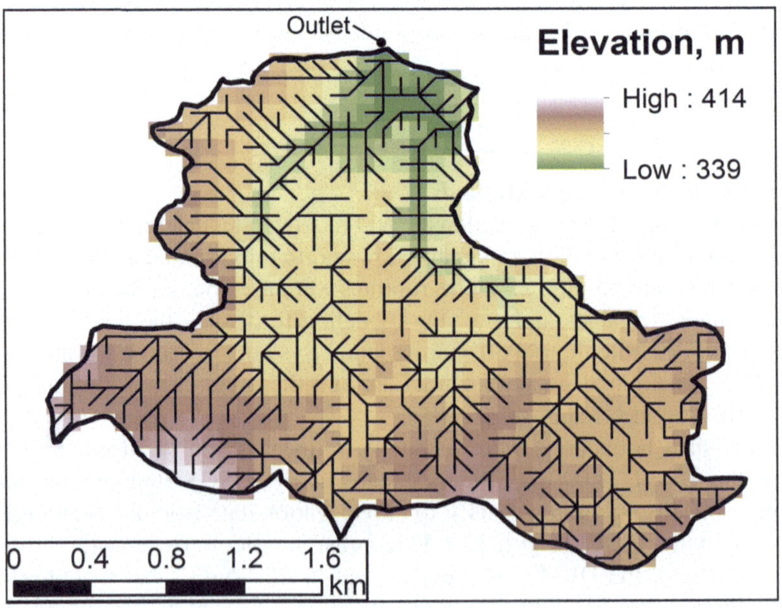

Fig. 7.3 Topographic relief of the Finlander Bay catchment, with stream network generated from digital elevation data

Table 7.8 AGNPS input variables, their derivation, and the range of input values used

Input parameter	Derivation	Range of input values
Cell number	Integer values assigned to grid cells	1–659
Receiving cell number	GIS FLOWDIRECTION and DOCELL program	2–660
Aspect	GIS ASPECT & lookup table	0–8
SCS curve number	Land cover lookup	77, 78, 85, 100
Land slope (%)	GIS SLOPE	0–7.2
Slope shape factor	GIS CURVATURE & lookup table	1 = uniform, 2 = convex, 3 = concave
Slope length (ft)	GIS SLOPE & lookup table	0, 100, 125
Channel slope (%)	Manually computed from topo map	0–4.4
Manning's coefficient	Land cover lookup	0.03, 0.08, 0.13, 0.99
Soil erodability factor (K)	Land cover lookup	0, 0.28
Cover/management factor (C)	Land cover lookup	0, 0.01
Practice factor (P)	Land cover lookup	0, 1.0
Surface condition constant	Land cover lookup	0, 0.01, 0.59
Soil texture	Land cover lookup	0 = water, 1 = sand, 3 = clay, 4 = organic
Chemical oxygen demand (COD)	Land cover lookup	0, 25, 60, 65
Impoundment factor	Beaver impoundment maps	1 = impoundment, 0 = no impoundment
Area draining to impoundment (acres)	Beaver impoundment maps	4–132
Outlet pipe diameter (in.)	–	10

7.7 Watershed Modeling of Nutrient Fluxes

Beaver ponds rarely occur in isolation in boreal landscapes, and the cumulative effect of multiple beaver ponds on downstream water quality may differ from their individual effect (Johnston 1994). Tatiana Nawrocki and I investigated this by applying an established non-point source pollution model to evaluate the effect of beaver dam building on water quality (Box 7.2, Fig. 7.3, Table 7.8) (Nawrocki et al. 1994). The model was AGNPS, a distributed parameter model developed to simulate non-point source pollution from agriculture (Young et al. 1989), but which has been applied to watersheds with a broad range of land cover types (Mankin et al. 2003; Corbett et al. 1997; Bhuyan et al. 2002; Grunwald and Frede 1999). AGNPS is an event-based model and operates on a cell basis so that the parameters of each cell in the watershed can be accounted for in the analysis. The model has explicit procedures for evaluating the effects of impoundments on watershed runoff, a benefit for this application.

Two land cover scenarios were tested, one "with ponds" that represented conditions in 1988 (Allen et al. 1993), and one "without ponds" that represented 1940 conditions, before beavers had recolonized the area (Fig. 7.4). The only differences between the two scenarios were: (1) the conversion of some forests and wetlands to beaver ponds and meadows and (2) the insertion of impoundments at 17 cell locations representing active beaver dams. The land cover was converted to raster format, and used to derive several input variables using a lookup table (Table 7.9), based on guidance provided in the model's user manual (Young et al. 1986). The land cover scenarios were tested with seven different rainfall intensities ranging in reoccurrence probability from a 1-year storm (2 in. of rainfall in 24 h, Energy Intensity = 10 foot-tons/acre-inch) to a 100-year storm (5 in. of rainfall in 24 h, Energy Intensity = 71 foot-tons/acre-inch).

The model output indicated that the land cover and impoundment changes caused by beaver would decrease nitrogen (N) concentration, COD, and sediment yield measured at the catchment outlet cell (Table 7.10). The magnitude of the reduction varied with rainfall intensity: as rainfall increased from 2 to 5 in., the reduction of sediment yield increased from 7 to 12% but the reduction of COD concentration decreased from 16 to 9%. There was a simulated reduction of nitrogen concentration for 2, 2.3, and 3 in. rain events, but no effect or an *increase* in nitrogen concentration at higher rainfalls, suggesting that there may be a flushing of nitrogen during high-intensity storms. The modeled change in phosphorus concentration was so small as to be negligible.

Cell-specific model results were exported from AGNPS for display in ArcGIS, which illustrated the spatial distribution of the simulated effects. The reduction in N concentration occurred mainly in the east branch of the stream network; N concentrations actually *increased* in headwater cells of the west branch, but these effects were not transmitted downstream (Fig. 7.5). This underscores the idea that geologic setting can modify the effects of beaver engineering.

The "without ponds" spatial distribution map for COD (Fig. 7.6) basically shows the location of wetlands that existed prior to beaver colonization of the catchment, which had COD concentrations equivalent to the COD input parameter used, 25 ppm (see Table 7.9). These wetlands appear as disconnected patches, and the majority of the catchment (including the outlet) has the value of 65 ppm that was assigned as an input parameter to forest. In contrast, the stream network is clearly visible in the "with ponds" spatial distribution, connecting beaver ponds, beaver meadows, and pre-existing wetlands into a system that has cumulative benefits to downstream COD reduction.

The difference between cell-specific model runs can also be used to show biogeochemical "hot spots," patches that show disproportionally high reaction rates relative to the surrounding matrix (McClain et al. 2003). Figure 7.7 shows cell-specific differences in sediment yields for the "without ponds" minus the "with ponds" values, with positive values indicating more sediment retention as a result of the land cover changes cause by beavers. This type of analysis could be used to promote beaver relocation to sites that would provide the greatest water quality benefit.

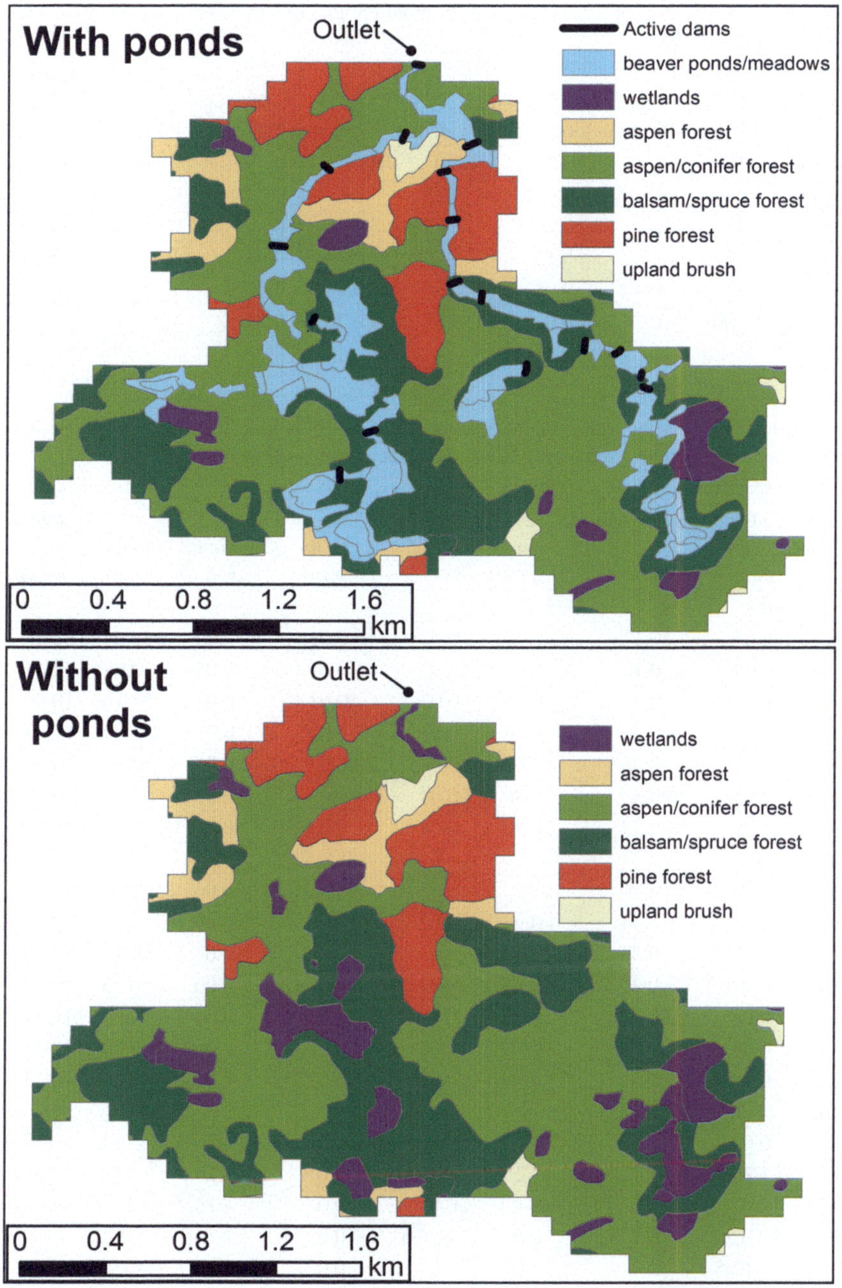

Fig. 7.4 Two land use scenarios used in AGNPS application for Finlander catchment

Table 7.9 Lookup table for assigning AGNPS input parameter values based on land use classes (Nawrocki et al. 1994)

Land cover	Curve #	Soil texture	Manning's coefficient	COD factor	K factor	C factor	P factor	Surface condition constant
Water	100	0	0.99	0	0	0	0	0
Wetland	85	3	0.99	25	0	0	0	0
Peat	85	4	0.99	25	0	0	0	0
Meadow	78	3	0.13	60	0.28	0.01	1	0.59
Forest	77	1	0.08	65	0.28	0.01	1	0.59
Bedrock	100	1	0.03	65	0.28	0.01	1	0.01

Table 7.10 Output results from AGNPS simulation of the Finlander catchment "with ponds" and "without ponds" scenarios, and their percent difference (Nawrocki et al. 1994)

24-h rainfall (in.)	Sediment adsorbed N (lb/acre)	Dissolved N (lb/acre)	N conc. (ppm)	Sediment adsorbed P (lb/acre)	Dissolved P (lb/acre)	P conc. (ppm)	COD (ppm)	Sediment yield (tons/acre)
With ponds								
2.0	0.04	0.11	0.92	0.02	0.01	0.05	49.44	6.56
2.3	0.04	0.15	0.91	0.02	0.01	0.05	50.98	7.83
3.0	0.06	0.23	0.89	0.03	0.01	0.05	53.10	10.69
3.5	0.07	0.30	0.88	0.03	0.02	0.05	53.99	12.65
4.0	0.07	0.38	0.87	0.04	0.02	0.05	54.63	14.54
4.5	0.08	0.45	0.86	0.04	0.03	0.05	55.11	16.37
5.0	0.09	0.53	0.86	0.04	0.03	0.05	55.48	18.17
Without ponds								
2.0	0.04	0.10	0.96	0.02	0.01	0.06	59.44	7.05
2.3	0.05	0.14	0.93	0.02	0.01	0.06	59.77	8.50
3.0	0.06	0.23	0.90	0.03	0.01	0.06	60.24	11.75
3.5	0.07	0.29	0.88	0.04	0.02	0.06	60.45	13.97
4.0	0.08	0.37	0.87	0.04	0.02	0.06	60.60	16.11
4.5	0.09	0.44	0.86	0.04	0.03	0.06	60.72	18.33
5.0	0.10	0.52	0.85	0.05	0.03	0.05	60.81	20.61
Difference (%)								
2.0	0.0	10.0	−4.2	0.0	0.0	−16.7	−16.8	−7.0
2.3	−20.0	7.1	−2.2	0.0	0.0	−16.7	−14.7	−7.9
3.0	0.0	0.0	−1.1	0.0	0.0	−16.7	−11.9	−9.0
3.5	0.0	3.4	0.0	−25.0	0.0	−16.7	−10.7	−9.4
4.0	−12.5	2.7	0.0	0.0	0.0	−16.7	−9.9	−9.7
4.5	−11.1	2.3	0.0	0.0	0.0	−16.7	−9.2	−10.7
5.0	−10.0	1.9	1.2	−20.0	0.0	0.0	−8.8	−11.8

Negative difference values indicate greater retention by the "with ponds" scenario

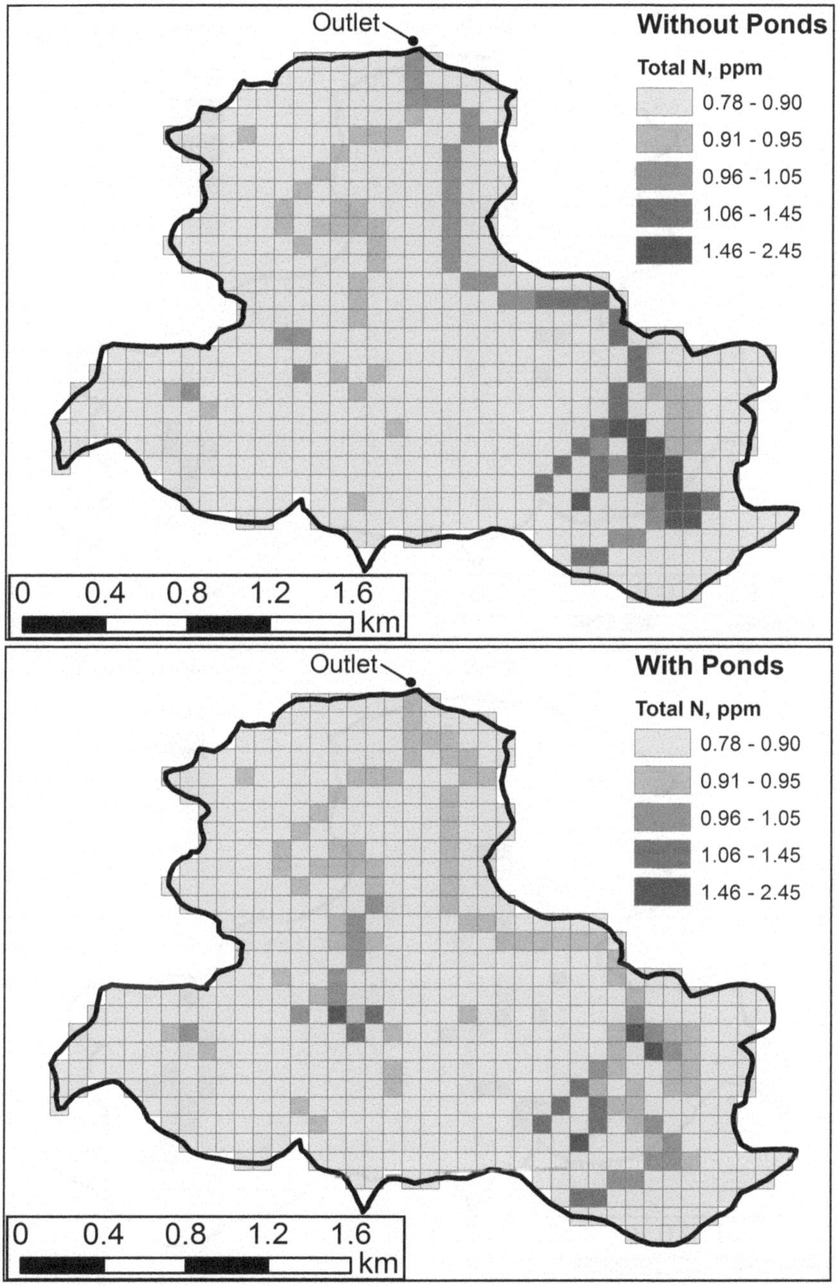

Fig. 7.5 Simulated nitrogen concentration of runoff from cells in Finlander Bay catchment, with and without beaver ponds, for a 2 in. rainfall. The darkest cells have the highest N concentrations

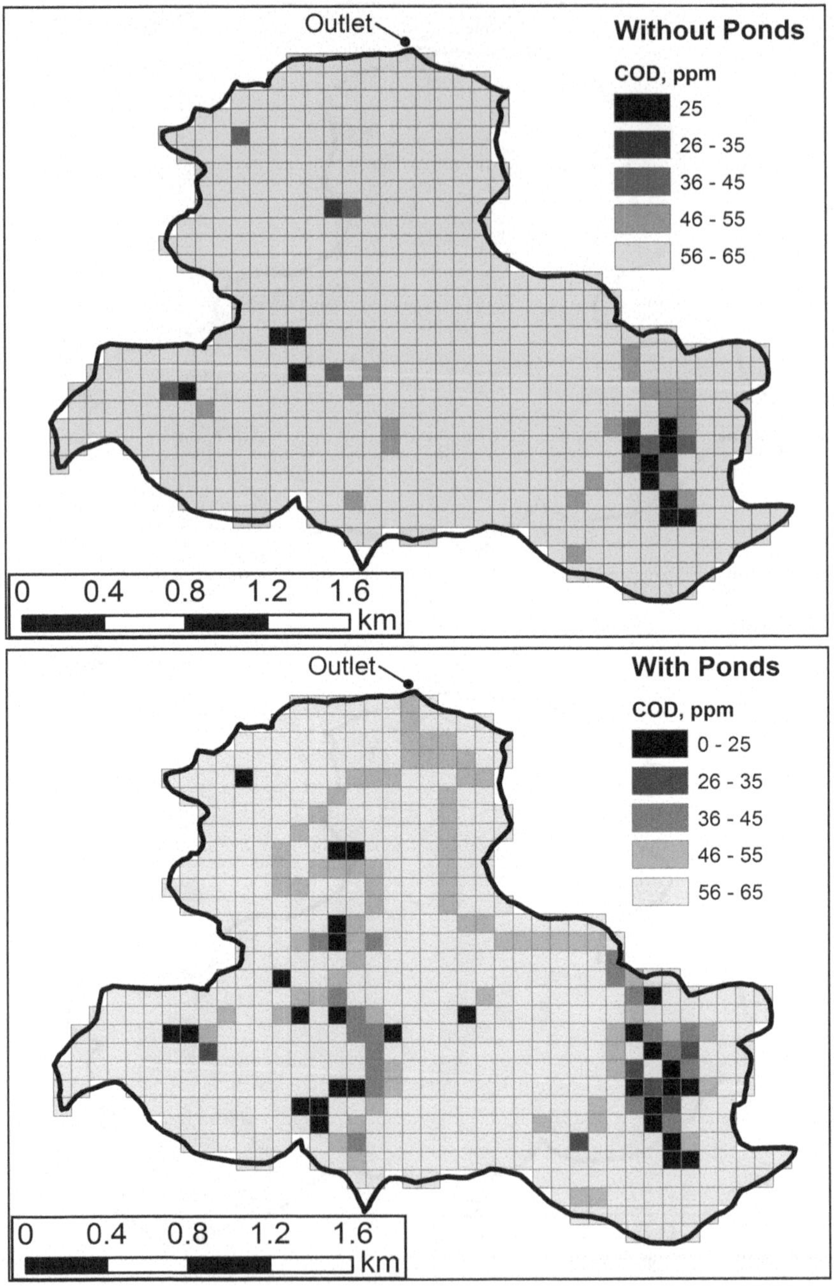

Fig. 7.6 Simulated chemical oxygen demand (COD) concentration of runoff from cells in Finlander Bay catchment, with and without beaver ponds, for a 2 in. rainfall. The darkest cells have the *lowest* COD concentrations

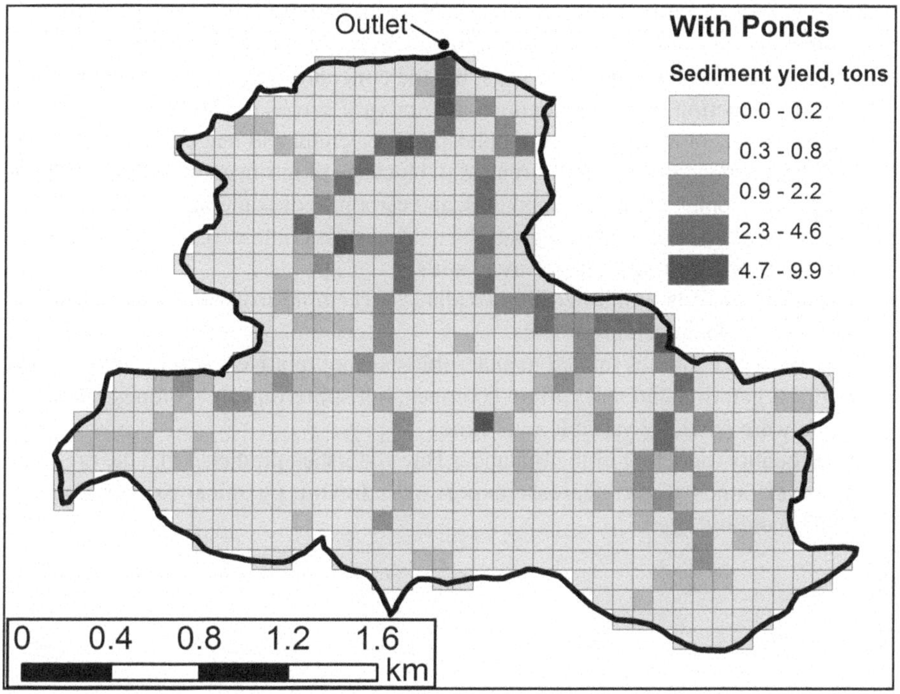

Fig. 7.7 Decrease in sediment yield as a result of beaver pond creation. Sediment yield from "with ponds" scenario was subtracted from "no ponds" scenario

There are no field measurements with which to validate the AGNPS simulation results, but they make sense in terms of the biogeochemical processes that typically occur in a catchment. Given that the model was developed for much more polluted watersheds and there was no human disturbance in either scenario, we were somewhat surprised to see any differences at all. These results illustrate that beaver engineering can potentially affect areas far downstream, and that their cumulative effects may be greater than the effects of individual ponds.

7.8 Conclusions

- The Kabetogama Peninsula receives few plant nutrients (N, P) from precipitation
- We studied the biogeochemistry of moisture gradients extending from residual beaver ponds up into never-impounded forest.
- Redox potentials and porewater sulfate concentrations increase, and porewater iron concentrations decrease along this gradient. Both iron and sulfate are involved in redox reactions.

- Microbial nitrogen fixation contributed N about equally along this gradient.
- Denitrification rates were two times higher in moist meadow, wet meadow, and beaver pond soils than in forest soils, but the concentration of nitrate (involved in the redox reaction of denitrification) did not vary across moisture zones.
- Concentrations of ammonium, orthophosphate, calcium, and magnesium were highest in beaver pond porewaters, attributed to blockage of nitrification, solubilization of phosphorus-containing iron compounds, and carbonate dissolution.
- Soil samples and porewater of glaciolacustrine soils contained significantly more K, Ca, and Mg than did glaciofluvial soils.
- Glaciofluvial soils contained less oxalate-extractable iron but more porewater Fe than did glaciolacustrine soils.
- Subterranean ecotones related to anaerobiosis were relatively stable and matched visible vegetation and soil carbon ecotones, but subterranean ecotones related to nitrate were spatially variable over time.
- Water quality modeling of the Finlander Bay watershed predicted that the presence of beaver ponds would decrease export of sediment, chemical oxygen demand (COD), and nitrogen.

References

Allen B, Salés J, Bonde J, Meysembourg P, Johnston CA (1993) Effects of past disturbance on present-day forests at Voyageurs National Park. In: ARC/INFO maps 1992. Environmental Systems Research Institute, Redlands, p 63

Arndt JL, Richardson JL (1992) Carbonate and gypsum chemistry in saturated, neutral pH soil environments. In: Robarts RD, Bothwell ML (eds) Aquatic ecosystems in semi-arid regions: implications for resource management. vol N.H.R.I. Symposium Series 7. Environment Canada, Saskatoon, Saskatchewan

Bhuyan SJ, Marzen LJ, Koelliker JK, Harrington JA, Barnes PL (2002) Assessment of runoff and sediment yield using remote sensing, GIS, and AGNPS. J Soil Water Conserv 57(6):351–363

Butler DR, Malanson GP (1995) Sedimentation rates and patterns in beaver ponds in a mountain environment. Geomorphology 13(1–4):255–269

Cirmo CP, Driscoll CT (1993) Beaver pond biogeochemistry – acid neutralizing capacity generation in a headwater wetland. Wetlands 13(4):277–292

Corbett CW, Wahl M, Porter DE, Edwards D, Moise C (1997) Nonpoint source runoff modeling: a comparison of a forested watershed and an urban watershed on the South Carolina coast. J Exp Mar Biol Ecol 213(1):133–149

Correll DL, Jordan TE, Weller DE (2000) Beaver pond biogeochemical effects in the Maryland coastal plain. Biogeochemistry 49(3):217–239

Devito KJ, Dillon PJ (1993) Importance of runoff and winter anoxia to the P and N dynamics of a beaver pond. Can J Fish Aquat Sci 50(10):2222–2234

Devito KJ, Dillon PJ, Lazerte BD (1989) Phosphorus and nitrogen retention in five Precambrian shield wetlands. Biogeochemistry 8(3):185–204

Driscoll C, Wyskowski B, Cosentini C, Smith M (1987) Processes regulating temporal and longitudinal variations in the chemistry of a low-order woodland stream in the Adirondack region of New York. Biogeochemistry 3(1):225–241

Driscoll CT, Driscoll KM, Roy KM, Mitchell MJ (2003) Chemical response of lakes in the Adirondack region of New York to declines in acidic deposition. Environ Sci Technol 37(10):2036–2042

Ekblad ALF, Huss-Danell K (1995) Nitrogen fixation by *Alnus incana* and nitrogen transfer from *A. incana* to *Pinus sylvestris* influenced by macronutrients and ectomycorrhiza. New Phytol 131(4):453–459

Environmental Systems Research Institute Inc. (1992) Cell-based modeling with GRID 6.1, supplement – hydrologic and distance modeling tools. Redlands

Francis M, Naiman R, Melillo J (1985) Nitrogen fixation in subarctic streams influenced by beaver (*Castor canadensis*). Hydrobiologia 121(3):193–202

Grunwald S, Frede HG (1999) Using the modified agricultural non-point source pollution model in German watersheds. Catena 37(3–4):319–328

Holland MM (1988) SCOPE/MAB technical consultations on landscape boundaries. In: diCastri F, Hansen AJ, Holland MM (eds) A new look at ecotones: merging international projects on landscape boundaries. Biology international, vol. special issue 17. International Union of Biological Sciences, Paris, pp 47–104

Johnston CA (1994) Cumulative impacts to wetlands. Wetlands 14:49–55

Johnston CA (2000) Wetland soil and landscape alteration by beavers. In: Richardson JL, Vepraskas MJ (eds) Wetland soils: their genesis, morphology, hydrology, landscapes, and classification. CRC Press, Boca Raton, pp 391–408

Johnston CA, Pastor J, Pinay G (1992) Quantitative methods for studying landscape boundaries. In: Fd C, Hansen A (eds) Landscape boundaries: consequences for biotic diversity and ecological flows. Springer, New York, pp 107–125

Johnston CA, Pinay G, Arens C, Naiman RJ (1995) Influence of soil properties on the biogeochemistry of a beaver meadow hydrosequence. Soil Sci Soc Am J 59(6):1789–1799

Klotz RL (1998) Influence of beaver ponds on the phosphorus concentration of stream water. Can J Fish Aquat Sci 55(5):1228–1235

Knowles R (1982) Free-living dinitrogen-fixing bacteria. In: Page AL, Miller RH, Keeney DR (eds) Methods of soil analysis, part 2: chemical and microbiological properties, vol agronomy monograph no. 9, 2nd edn. ASA-SSSA, Madison, pp 1071–1092

Lazar JG, Addy K, Gold AJ, Groffman PM, McKinney RA, Kellogg DQ (2015) Beaver ponds: resurgent nitrogen sinks for rural watersheds in the northeastern United States. J Environ Qual 44(5):1684–1693

Mankin KR, Wang SH, Koelliker JK, Huggins DG, de Noyelles F (2003) Watershed-lake water quality modeling: verification and application. J Soil Water Conserv 58(4):188–197

Maret TJ, Parker M, Fannin TE (1987) The effect of beaver ponds on the nonpoint source water quality of a stream in southwestern Wyoming. Water Res 21(3):263–268

McClain EM, Boyer WE, Dent LC, Gergel ES, Grimm BN, Groffman MP, Hart CS, Harvey WJ, Johnston AC, Mayorga E, McDowell HW, Pinay G (2003) Biogeochemical hot spots and hot moments at the interface of terrestrial and aquatic ecosystems. Ecosystems 6(4):301–312

NADP. National Trends Network (2017) National Atmospheric Deposition Program. http://nadp.sws.uiuc.edu/NADP. Accessed 2017

Naiman RJ, Melillo JM (1984) Nitrogen budget of a subarctic stream altered by beaver (Castor canadensis). Oecologia 62(2):150–155

Naiman RJ, Pinay G, Johnston CA, Pastor J (1994) Beaver influences on the long-term biogeochemical characteristics of boreal forest drainage networks. Ecology 75(4):905–921

Nawrocki T, Johnston CA, Salés J (1994) GIS and modeling in ecological studies: analysis of beaver pond impacts on runoff and its quality. Natural Resources Research Institute, Duluth

Novozamsky I, Beek J (1976) Common solubility equilibria in soils. In: Bolt GH, Bruggenwert MGM (eds) Soil chemistry: a basic elements. Elsevier, Amsterdam, pp 96–125

Patrick WH, Gotoh S, Williams BG (1973) Strengite dissolution in flooded soils and sediments. Science 179:564–565

Ponnamperuma FN (1972) The chemistry of submerged soils. Adv Agron 24:29–96

Schwintzer CR (1979) Nitrogen fixation by Myrica Gale root nodules Massachusetts wetland. Oecologia 43(3):283–294

Smith ME, Driscoll CT, Wyskowski BJ, Brooks CM, Cosentini CC (1991) Modification of stream ecosystem structure and function by beaver (*Castor canadensis*) in the Adirondack Mountains, New York. Can J Zool 69:55–61

Tiedje JM (1988) Ecology of denitrification and dissimilatory nitrate reduction to ammonium. In: Zehnder AJB (ed) Environmental microbiology of anaerobes. Wiley, New York, pp 179–244

Updegraff K, Pastor J, Bridgham SD, Johnston CA (1995) Environmental and substrate controls over carbon and nitrogen mineralization in northern wetlands. Ecol Appl 5(1):151–163

van Breeman N, Brinkman R (1976) Chemical equilibria and soil formation. In: Bolt GH, Bruggenwert MGM (eds) Soil chemistry: a basic elements. Elsevier, Amsterdam, pp 141–170

Vasilas LM, Hurt GW, Noble CV (eds) (2010) Field indicators of hydric soils in the United States, 7.0 edn. USDA Natural Resources Conservation Service in cooperation with the National Technical Committee for Hydric Soils, Washington

Yoshinari T, Knowles R (1976) Acetylene inhibition of nitrous oxide reduction by denitrifying bacteria. Biochem Biophys Res Commun 69:705–710

Young RA, Onstad CA, Bosch DD, Anderson WP (1986) AGNPS agricultural nonpoint source pollution model, a watershed analysis tool: part 1 guide to model users. USDA Agricultural Research Service, Morris

Young RA, Onstad CA, Bosch DD, Anderson WP (1989) AGNPS: a nonpoint-source pollution model for evaluating agricultural watersheds. J Soil Water Conserv 44(2):168–173

Chapter 8
Beaver Ponds and the Carbon Cycle

Abstract Carbon takes a variety of forms in beaver ponds and beaver meadows, including live plant biomass, standing dead biomass, soil organic matter, soil carbonates, dissolved organic carbon, and trace gases. The pools of carbon in these storage compartments and the fluxes between them are discussed. The slow decomposition rate of beaver meadow plant litter under anaerobic conditions promotes the accumulation of organic (O) soil horizons, which have a calculated mean residence time of 69 years. The carbon per unit area in soils that were formerly impounded by beavers (15.1 ± 6.8 km C m^{-2}) was nearly twice that of adjacent never-impounded forest soils (8.2 ± 2.9 km C m^{-2}). Beaver meadow sedge peat mineralization was compared with that of bog peat in long-term (80-week) laboratory incubations. Beaver meadow sedge peat had significantly higher carbon mineralization rates under all incubation conditions except aerobic incubation at 15 °C (other treatments were anaerobic incubation at 15 and 30 °C and aerobic incubation at 30 °C). Field measurement of trace gas fluxes showed that beaver ponds and beaver meadows with water table at or above the soil surface emitted methane, but beaver meadows with water tables below the soil surface did not.

8.1 Components of the Carbon Cycle

Beaver ponds and meadows alter the growing environment for plants and the decomposition environment for microbes, both of which interact to affect carbon storage in the environment. Carbon is of interest to scientists because of its role in climate change—both carbon dioxide (CO_2) and methane (CH_4) are greenhouse gases, and their concentrations are increasing in the atmosphere. Ecosystems that retain carbon provide a global benefit by reducing atmospheric carbon. Ecosystem engineering by beavers has mixed effects on the carbon cycle.

Carbon occurs in the atmosphere, in biota, and in the soil (Fig. 8.1). Plants photosynthesize, capturing atmospheric carbon and storing it within their biomass. Their dead plant parts fall to the earth's surface as litter, some of which is decomposed by microbes, and some of which is incorporated into soil organic matter (OM). Microbial respiration during decomposition releases CO_2 and CH_4 under

C.A. Johnston, *Beavers: Boreal Ecosystem Engineers*,
DOI 10.1007/978-3-319-61533-2_8

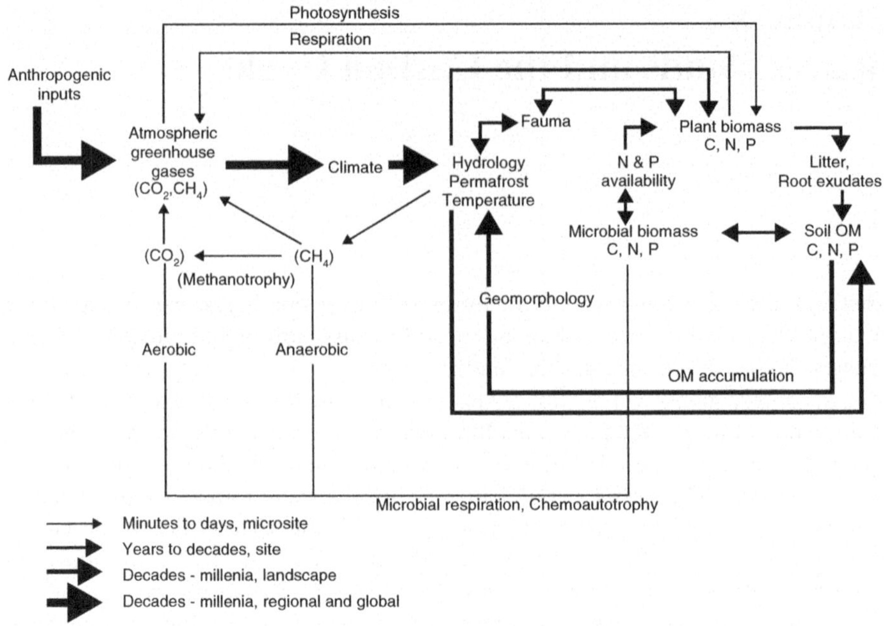

Fig. 8.1 A conceptual model of the interactions between greenhouse gases, climate, and terrestrial ecosystem processes in northern wetlands. The thickness of the *arrows* indicates the spatial and temporal scale at which each process occurs. *OM* organic matter. From Bridgham SD, Johnston CA, Pastor J, Updegraff K (1995) Potential feedbacks of northern wetlands on climate change. BioScience 45:262–274, with permission of Oxford University Press

aerobic and anaerobic conditions, respectively. Carbon sequestration, the removal and storage of carbon from the atmosphere, occurs when carbon is stored in relatively stable non-gaseous forms, such as tree boles or soil organic matter (Johnston et al. 2004). It is important to consider all of these carbon storage compartments, as well as their rates of carbon uptake and release, in order to understand the balance between atmospheric carbon gases and organic carbon storage in the environment.

8.2 Carbon in Ponds

Attached and free-floating aquatic algae conduct photosynthesis, and beaver ponds increase algal biomass in comparison with unimpounded streams (Hanson and Campbell 1963; Coleman and Dahm 1990). Carbon in this form is important as a source of food to beaver pond fauna (Hall and Pritchard 1975), but the amount of carbon taken up by algal photosynthesis is small and rapidly returned to the atmosphere, so it is a negligible storage pool of carbon (Roulet et al. 1997). An energy budget showed that photosynthesis provided only 21.5% of the energy flux into a beaver pond, whereas particulate organic matter conveyed by stream water ("drift") provided 70.4% of energy inputs (Hodkinson 1975b). Of the total yearly energy input, 26% was respired, 18% was exported downstream, and 56% accumulated in the sediments. Carbon stored in beaver pond sediments can be flushed, however, by extreme events; a beaver dam failure on a Canadian stream caused carbon export to spike from 11 to 851 kg DOC day^{-1} (Hillman et al. 2004).

Dissolved organic carbon (DOC) occurs in beaver ponds and other boreal waters, and is important to the acid–base chemistry of acid-sensitive freshwater systems (Schiff et al. 1990). Some studies have reported beaver ponds to be a net annual source of DOC to downstream waters (Cirmo and Driscoll 1993; Margolis et al. 2001), whereas others found DOC fluxes into and out of the beaver ponds to be about equal (Devito et al. 1989; Moore 2003). A long-term study in Finland reported that DOC concentration increased during the first three beaver-impoundment years, but declined back to initial levels after the impoundment had lasted for 4–6 years (Vehkaoja et al. 2015). As with photosynthesis, water column DOC is a negligible storage mechanism for carbon.

8.3 Carbon Uptake and Storage by Vascular Plants Associated with Beaver Ponds

Vascular plants associated with beaver ponds are a significant carbon storage pool. These rooted plants are associated with all life stages of beavers ponds: pre-impoundment forests, floating-leaved and submersed aquatic plants growing in ponds, and beaver meadows. The carbon content of the different life-forms associated with those stages varies with their structure (Table 8.1).

If we assume that carbon constitutes 47% of live biomass, then the forests surrounding Kabetogama beaver ponds (Table 4.9) contained an average of 4900 g C m^{-2} in their aboveground biomass (Table 8.1). In contrast, the beaver meadows we studied contained only a small fraction of that amount because grasses and sedges are much shorter than trees. Converting forests to ponds would decrease that carbon pool even more because the biomass of phytoplankton and submerged macrophytes is so small.

Table 8.1 Quantity of carbon in live aboveground biomass (standing stock) on the Kabetogama Peninsula

Vegetation type	Carbon standing stock (g C m^{-2})	Reference
Aquatic vegetation	27	Morris (2014)
Beaver meadows	106–298	Erickson (1994)
Riparian forests	1993–9206	Chapter 4

Assumed biomass carbon content of 47%

With any carbon storage pool, the duration of storage is important. Aquatic vegetation typically has an annual life cycle in boreal regions, so the 27 g C m^{-2} that is taken up in the spring would subsequently be released as the vegetation decomposes in the fall, resulting in no net carbon sequestration. Trees live longer: a typical trembling aspen stand begins to deteriorate after 55–60 years, and would therefore retain carbon in woody tissues for about six decades (Wisconsin Department of Natural Resources 2003).

If we consider the annual rate of carbon dioxide removal from the atmosphere, however, forests and meadows are more comparable. Net primary production (NPP), the quantity of new plant matter produced annually, determines the amount of carbon that is taken up by plant communities. Even though forests contain substantial biomass because of their vertical growth and longevity of woody tissues, their aboveground NPP is only slightly greater than that of beaver meadows. The aboveground NPP of boreal forests in 12 North American locations averaged 296 g C m^{-2} year^{-1} (Gower et al. 2001), whereas the aboveground NPP of Kabetogama beaver meadows would be about 106–298 g C m^{-2} year^{-1} (the same as the carbon standing stock). No data exist for total NPP (aboveground + belowground) within the Kabetogama beaver meadows, but total NPP data from other sedge meadow studies averaged 560 g C m^{-2} year^{-1} (Johnston 1988). Thus, the changes in plant communities caused by converting forests to meadows would probably have little effect on the rate of carbon uptake.

8.4 Plant Litter and Decomposition

Senescent tissues that are shed by plants fall to the soil surface, where they become part of the litter layer. Deciduous trees and shrubs shed all of their leaves in the autumn, evergreen trees shed their older needles, and herbaceous plants die back and fall to the ground. This plant litter decomposes at varying rates, depending on the plant species, the plant parts (e.g., leaves, twigs), litter chemical composition (especially lignin content), and the moisture, temperature, and aeration of the decomposition environment (Polunin 1984).

Decomposition rates are often quantified using litter bags, in which mass loss is recorded over time for litter of known initial mass enclosed in a mesh bag that is anchored on the soil surface. Exponential decay constants are computed by fitting a negative exponential curve to a plot of mass remaining over time, or by estimating k as the natural log of the fraction of mass remaining divided by time t (Table 8.2).

Litterbag incubations showed that the bog shrub leatherleaf (*Chamaedaphne calyculata*) decomposed very slowly ($k = 0.18$ year^{-1}), whereas the leaves of deciduous trees and shrubs had faster decomposition rates (k = 0.30–0.98 year^{-1}). The decomposition rate for sedges and grasses varied across different studies, but was generally slower than that of deciduous leaves. Based on these values, the time required for 95% decomposition of these leaves would range from 3 to 21 years.

Table 8.2 Exponential decay constants (*k*) of boreal wetland plants and shrub leaves, as determined by litter bag field incubations of aboveground tissues

Species	*k* (year^{-1})	Source
Graminoids		
Calamagrostis canadensis	0.26	Reader and Stewart (1972)
Carex sp.	0.45	Chamie and Richardson (1978)
C. acutiformis	0.30	Aerts and de Caluwe (1997)
C. aquatilis	0.17–0.31	Bartsch and Moore (1985)
C. aquatilis	0.48	Thormann et al. (2001)
C. chordorrhiza	0.22	Bartsch and Moore (1985)
C. diandra	0.80	Aerts and de Caluwe (1997)
C. lasiocarpa	0.34	Aerts and de Caluwe (1997)
C. lasiocarpa	0.37–0.60	Szumigalski and Bayley (1996)
C. limosa	0.18–0.28	Bartsch and Moore (1985)
C. rostrata	0.25	Reader and Stewart (1972)
C. rostrata	0.14–0.26	Bartsch and Moore (1985)
C. rostrata	0.65	Aerts and de Caluwe (1997)
Deschampsia cespitosa	0.67	Hodkinson (1975a)
Juncus tracyi	0.41	Hodkinson (1975a)
Shrub and tree leaves		
Betula pumila	0.47	Chamie and Richardson (1978)
Chamaedaphne calyculata	0.17	Chamie and Richardson (1978)
Chamaedaphne calyculata	0.19	Reader and Stewart (1972)
Pinus contorta	0.22	Hodkinson (1975a)
Populus tremuloides	0.58	McClaugherty et al. (1985)
Salix sp.	0.46	Chamie and Richardson (1978)
Salix sp.	0.98	Hodkinson (1975a)
S. bebbiana	0.34	Reader and Stewart (1972)
S. planifolia	0.57	Thormann et al. (2001)
S. serissima	0.30	Reader and Stewart (1972)

Values based on 1-year incubations. Larger values decompose more quickly than smaller values

8.5 Deadwood

Beavers cut and kill trees through their foraging and flooding activities, leaving behind a substantial amount of deadwood in various forms: snags (i.e., standing dead trees), stumps, downed logs, coarse woody debris, and fine wood debris (Vehkaoja 2016). Snags provide important habitat for roosting bats (Menzel et al. 2001; Nummi et al. 2011) and cavity nesting birds (Hartke and Hepp 2004; Merendino et al. 1995; Lockmiller 1979), and fallen wood provides substrate for the estimated 400,000–1 million wood-inhabiting species (including invertebrates, lichens, and fungi) (Stokland et al. 2012). Snags of black ash trees can remain vertical for at least two decades (Fig. 2.24), and submerged fallen trees decompose slowly (Fig. 8.2). Beavers also gather wood into their structures; an average beaver dam contains about 3.3 Mg of organic carbon (Johnston and Naiman 1990).

Fig. 8.2 Fallen trees and stumps litter a drained pond bottom exposed by the collapse of a beaver dam

Woody debris contains about 10% of the carbon stored in forests within the conterminous U.S. (Turner et al. 1995), but the proportion is probably greater in beaver-altered forests. Average deadwood volumes surrounding five beaver-dammed Finnish lakes (31 m^3 ha^{-1}) were two to four times greater than non-beaver sites, and a sixth site had an exceptionally large deadwood volume (277 m^3 ha^{-1}) due to flooding mortality from a beaver-dammed stream (Thompson et al. 2016).

Woody tissues decompose more slowly than do herbaceous tissues (Chamie and Richardson 1978), particularly under anaerobic conditions. Hodkinson (Hodkinson 1975a) estimated the k value for a submerged *Populus balsamifera* log as 0.011 $year^{-1}$, requiring 246 years for 95% decomposition. Beaver-gnawed wood chunks have been found buried in alluvium and peat, some as old as 12,000 years (James and Lanman 2012; Rains 1987; Robinson et al. 2007; Garrison 1967; Tedford and Harington 2003; Ferguson and Osborn 1981).

8.6 Transfer of Plant Biomass to the Soil

In forests, the main mechanism of carbon transfer to the soil is by seasonal litterfall as leaves dehisce from branches. Some of the leaf litter that falls to the soil surface decomposes, returning carbon to the atmosphere as CO_2, and some of it remains as recalcitrant organic matter that accumulates in a litter layer. The Insula soil that is very common on the uplands of Voyageurs National Park is described as having a 4 cm thick surface litter layer (O horizon), and this is consistent with my field observations (Fig. 5.11).

In herbaceous wetlands (including beaver meadows), there is no seasonal leaf dehiscence, and litterfall occurs mostly by fragmentation during winter and spring storms after the first killing frost (Davis and van der Valk 1978). Dead plant materials can remain standing well into the next growing season, forming a thatch above the soil surface (Fig. 10.8). The grasses and sedges in beaver meadows can also be killed by rapid inundation as they are reflooded by repaired beaver dams (Fig. 8.3).

Floating mats in beaver impoundments consist of sedges such as *Carex lacustris* and *C. lasiocarpa* which are rooted in a peat substrate that lifts up and down with water level changes (Faber-Langendoen et al. 2007). Occasionally, sections of mat break off and migrate to another location within a pond (Fig. 2.9). When a beaver pond drains, floating mats become draped over the pond bottom, forming an organic

Fig. 8.3 Sedge tussocks killed by flooding of a beaver meadow are re-exposed after the pond drains. Soil probe (~1 m) shown for scale

Fig. 8.4 When the beaver pond drained, this formerly floating sedge mat settled on top of a boulder

surface layer over the mineral soil and rocks beneath (Fig. 8.4). In this manner, photosynthesizing vegetation can cover bedrock and boulders that would not otherwise support primary production.

Transfer of belowground biomass to the soil occurs as roots die and are sloughed off. In the case of tussock forming sedges such as *Carex stricta* and *Scirpus cyperinus*, the accumulated root biomass builds mounds that are 15–30 cm tall (Johnston et al. 2007; Peach and Zedler 2006). These mounds increase wetland surface area by 40% (Peach and Zedler 2006), and also increase the volume of organic matter stored on the surface of the mineral soil.

8.7 Carbon Storage Within Beaver Meadow Soils

Several factors combine to promote the accumulation of partially decomposed organic (O) layers on the surface of beaver meadow soils (Fig. 8.5): high net primary production, plant litter that is moderately recalcitrant, cool soil temperatures and anaerobic conditions that deter microbial activity, and peat forming structures such as tussocks and floating mats (Johnston 2014). In the Rocky Mountains, relict beaver meadows represent ~8% of total carbon storage within the landscape, but the value was closer to 23% when beaver actively maintained wet meadows (Wohl 2013).

Fig. 8.5 Surface organic horizon (Oi, Oe) derived from beaver meadow detritus. Large black numbers on ruler are inches. Reprinted from Johnston CA (2014) Beaver pond effects on carbon storage in soils. Geoderma 213:371–378, with permission from Elsevier

I studied soil carbon storage in conjunction with the field study of beaver meadow soils previously described in Chap. 5. I sampled all horizons volumetrically, so as to be able to calculate carbon mass per unit volume. Total carbon was determined from two replicates per layer using a CHN analyzer (LECO Corporation, St. Joseph, MI) (Johnston 2014).

The O horizons contained the most carbon of any soil horizon, ten times the amount by volume found in B horizons and 20 times the amount found in C soil horizons (Fig. 8.6). The O layer in most beaver meadow soils was perched on the mineral soil surface (see photo in Fig. 5.13), such that the majority of the soil carbon mass was at the top of the profile (Fig. 8.7, Aqualf). A variation of this arrangement was the burial of an O horizon by colluvium washed in from the pond edge (Fig. 5.7).

Some beaver-impounded soils were already peatlands prior to impoundment, and carbon storage at those sites had a much different vertical pattern. This was the case for a Haplosaprist sampled at the Lost Pond site (Fig. 5.10). The entire soil

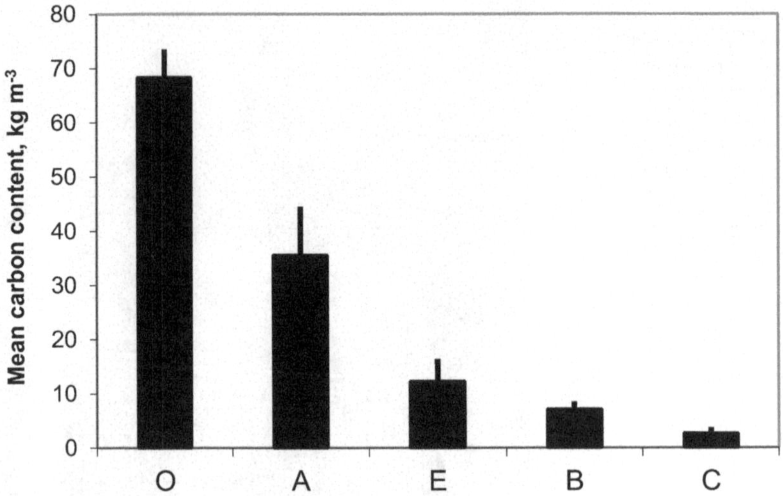

Fig. 8.6 Mean carbon content of soil horizons associated with Kabetogama Peninsula beaver meadows. *Bars* are one standard error

profile to a depth of >65 cm consisted of organic matter (Fig. 8.7), and its carbon content increased from about 38 kg C m^{-3} at the surface to 100 kg C m^{-3} at depth. This was judged to be relict carbon that was present before beaver flooded the area, rather than carbon that accumulated as a result of beaver activity.

Another soil type that contained carbon at depth was an Argiaquoll, described in Chap. 5. This soil profile had 8 cm of surface litter overlying a thick (39 cm) A horizon. It was not an organic soil, but rather incorporated organic carbon within its mineral A horizon. Unlike the Aqualf profile, in which soil carbon content dropped to only 8.9 kg m^{-3} below the O horizon, the Argiaquoll maintained soil carbon at concentrations of 18–34 kg m^{-3} in its upper mineral soil (Fig. 8.7). As with the Haplosaprist, the carbon in this soil profile was judged to be a relict of long-term storage in a wetland that existed prior to beaver activity.

I measured the total carbon per unit area to a depth of 60 cm (TC_{60}) in 12 soils, five of which had been impounded by beaver and seven of which had not (the Haplosaprist and Argiaquoll soils with relict carbon were excluded), and used a two-way ANOVA to determine the effect of beaver impoundment and parent material on TC_{60}. The TC_{60} values for beaver-impounded mineral soils ranged from 7.9 to 26.3 kg C m^{-2}, comparable to values reported in the literature for freshwater mineral soil wetlands (Bridgham et al. 2006; Batjes 1996; Tarnocai et al. 1998). Beaver-impounded soils contained nearly twice the amount of carbon (15.1 ± 6.8 kg C m^{-2}) as the never-impounded soils (8.2 ± 2.9 kg m^{-2}), a significant difference ($F = 5.45$, $P = 0.0478$). There was no significant effect of parent material (i.e., glaciofluvial versus glaciolacustrine) ($F = 0.74$, $P = 0.414$) nor the interaction between beaver and parent material ($F = 0.47$, $P = 0.512$) (Johnston 2014).

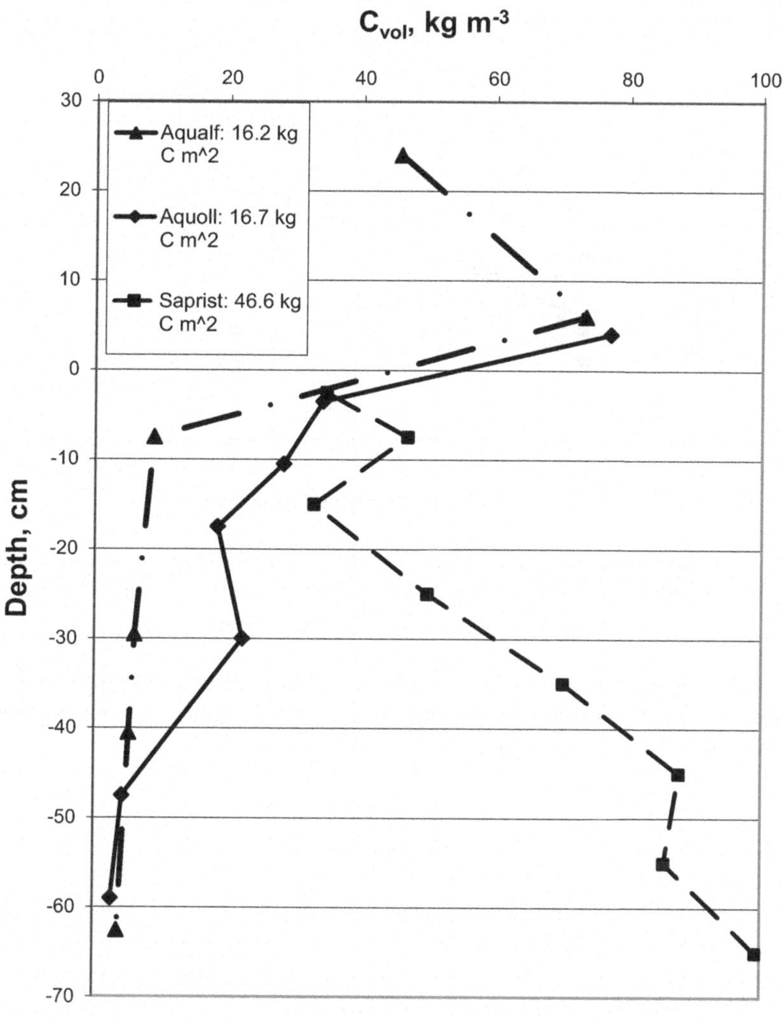

Fig. 8.7 Vertical distribution of C_{vol} within an Aqualf, a Haplosaprist, and an Argiaquoll. Reprinted from Johnston CA (2014) Beaver pond effects on carbon storage in soils. Geoderma 213:371–378, with permission from Elsevier

The TC_{60} of the 12 soils sampled was strongly related to the thickness of the O horizon, which was greatest in aquic soils (i.e., beaver-impounded) (Fig. 8.8). For example, the 24-cm thick O horizon in the Aqualf soil shown in Fig. 8.7 contained 12.6 kg C m^{-2}, which was 78% of the carbon stored in the soil profile. This relationship between O horizon thickness and carbon content is consistent with the finding that O horizons had the greatest volumetric C concentrations (Fig. 8.6). The accumulation of surface organic material is a major mechanism by which beaver activity promotes soil carbon sequestration.

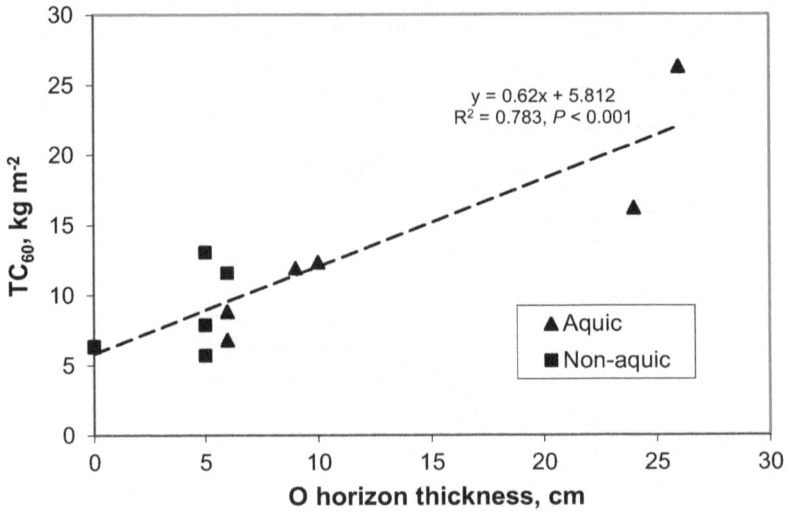

Fig. 8.8 Linear regression between O horizon thickness and total soil carbon to a depth of 60 cm (TC_{60}). Aquic = formerly impounded beaver meadow, non-aquic = never impounded

By knowing the amount of litter added to the soil surface and the mass of the litter layer, we can estimate the decomposition rate (*k* value) for beaver meadow soils. Assuming that the litter layer is at steady state, then:

$$k = \frac{\text{litterfall}}{\text{litterpool}}$$

If we estimate beaver meadow net primary productivity to be 450 g/m^2 based on Erickson's field data (Erickson 1994) and use the litterpool mass from the soil O horizon measurements, the average k is 0.0146 year^{-1} and mean residence time ($1/k$) is 69 years. Thus, the storage duration of carbon in the O horizon is comparable to that stored in mature tree boles.

Carbonate accumulation is a mechanism for storing *inorganic* carbon in beaver meadow soils. As noted in Chap. 7, calcium carbonate accumulates in fine-textured soils around the edges of the Kabetogama beaver ponds. We saw evidence of carbonate accumulation in soil profile morphology and chemistry. A soil at the Shoepack study site had a horizon at depth 64 cm that contained 52 kg C m^{-3}, nearly as much as its surface O horizon (60 kg C m^{-3}). However, it was evident from the horizon color (10 YR 5/3) and position in the soil that the carbon was due to carbonates. That soil horizon also had the highest C:N ratio of any sampled, 122, which was twice the highest value observed in organic soil horizons. The presence of carbon without nitrogen indicates pedogenic rather than biologic origin.

8.8 Laboratory Soil Carbon Mineralization

Although carbon storage in soil is relatively long-term, the decomposition of organic carbon materials continues to release carbon gases to the atmosphere over time. Chapter 7 described the 80-week laboratory experiment that we used to examine nitrogen mineralization of organic soil collected from a Kabetogama beaver meadow and bog (Table 7.5). This same experiment also examined carbon mineralization, the conversion of organic carbon-containing materials into CO_2 or CH_4 gases (Updegraff et al. 1995; Bridgham et al. 1995).

The results of this 80-week incubation show that: (1) mineralization of the beaver meadow organic soil was twice as great under aerobic versus anaerobic conditions, (2) mineralization of the beaver meadow organic soil doubled when the incubation temperature was doubled from 15 to 30 °C, and (3) mineralization was significantly higher for the beaver meadow organic soil than the bog soils under all incubation conditions except aerobic incubation at 15 °C (Fig. 8.9). A 59-week incubation at 30 °C that compared beaver meadow soils with several other types of organic substrate (bogs, poor fens, rich fens, swamp forests) confirmed that methane production under anaerobic incubation was sensitive to substrate type: greatest

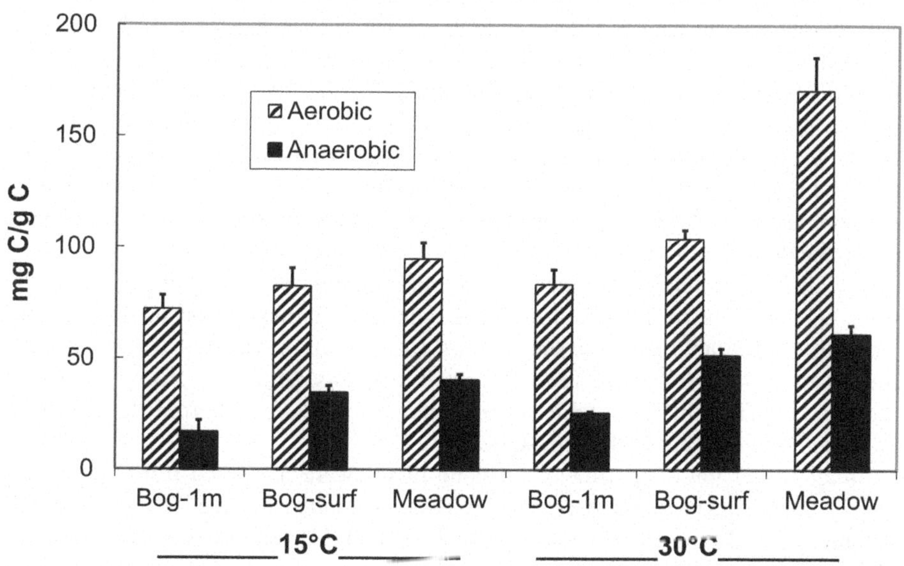

Fig. 8.9 Cumulative carbon mineralization over 80-week incubations at 15 and 30 °C in the surface organic layer of a beaver meadow and in bog peat from the surface (Bog-surf) and from 85 to 100 cm depth (Bog-1 m) at Voyageurs National Park. Anaerobic carbon mineralized included both carbon dioxide and methane production, although carbon dioxide predominated. Data given as milligrams carbon mineralized per gram total soil carbon. *Bars* are one standard error. From Bridgham SD, Johnston CA, Pastor J, Updegraff K (1995) Potential feedbacks of northern wetlands on climate change. BioScience 45:262–274, with permission of Oxford University Press

for the beaver meadow soil and least for the bog (Bridgham et al. 1995). Carbon dioxide production was relatively insensitive to substrate.

8.9 Field Measurements of Trace Gas Fluxes from Beaver Meadows and Ponds

The methane produced in a water-logged, anaerobic soil is often subsequently oxidized to carbon dioxide by methanotrophic bacteria in an overlying aerobic zone, which is of prime importance in reducing atmospheric methane emissions from wetlands. Thus, theoretical methane emission rates under controlled laboratory conditions may not reflect actual emission rates in the field.

My colleagues and I investigated field fluxes of carbon trace gases at Voyageurs National Park in relation to organic matter type (bog versus beaver meadow), soil temperature, and water table depth, which approximates the aerobic–anaerobic soil boundary. We erected mylar tents as gas chambers next to the areas from which the soil incubation samples had been collected (Fig. 8.10). In the beaver meadow, gas fluxes were measured from both control and fertilized (nitrogen plus phosphorus) sites. Measurements were done in the dark, so carbon dioxide fluxes represent net ecosystem respiration (aboveground and belowground) and soil microbial respiration.

As in the case of the laboratory incubations, carbon dioxide emission was greater from the beaver meadows (X's and diamonds) than the bogs (squares) and carbon dioxide emissions increased significantly with increasing temperature (Fig. 8.11a). There was no significant effect of water table depth on carbon dioxide emissions (Fig. 8.11b). There was, however, a nonlinear effect of water table on methane production, such that beaver meadows with water tables at or above the soil surface emitted methane at a rate of 10–45 mg CH_4-C m^{-2} day^{-1} (Fig. 8.11d).

Recall from Table 7.3 that methane production requires highly anaerobic conditions, which occur in inundated soils. Methanogenesis may be occurring at depth when the water table is below the soil surface, but any methane that rises upward must pass through an aerobic layer that likely contains methanotrophs which would consume it. When the water table is at or above the surface, the aerobic layer is thin, allowing methane flux to the atmosphere.

In an earlier field experiment, methane fluxes were measured from a hydrosequence consisting of four different zones associated with the Shoepack beaver pond: upland forest, wet meadow, submergent marsh, and deep open water (Naiman et al. 1991). Within each zone, four 10 L bell jars fitted with rubber gas sampling septa were set into the ground (forest and meadow sites) or mounted on floating wooden frames (submerged marsh and deep water sites) to collect gas emissions. Gas samples were collected by syringe weekly from late May to early October; winter emissions were estimated from the average of flux rates measured before ice formation and after ice-out, and added to the growing season measurements to calculate annual methane flux. Annual methane fluxes from the deep water (8.73 g C m^{-2} $year^{-1}$)

Fig. 8.10 Gas chamber tent used to measure carbon trace gas emissions from bogs and beaver meadows. *Top photo* shows the tent setup in a leatherleaf bog (*Chamaedaphne calyculata*). Sites were accessed by a boardwalk (*right* side of photo). *Bottom photo* shows the gas septum in the tent used to collect gas samples with a syringe

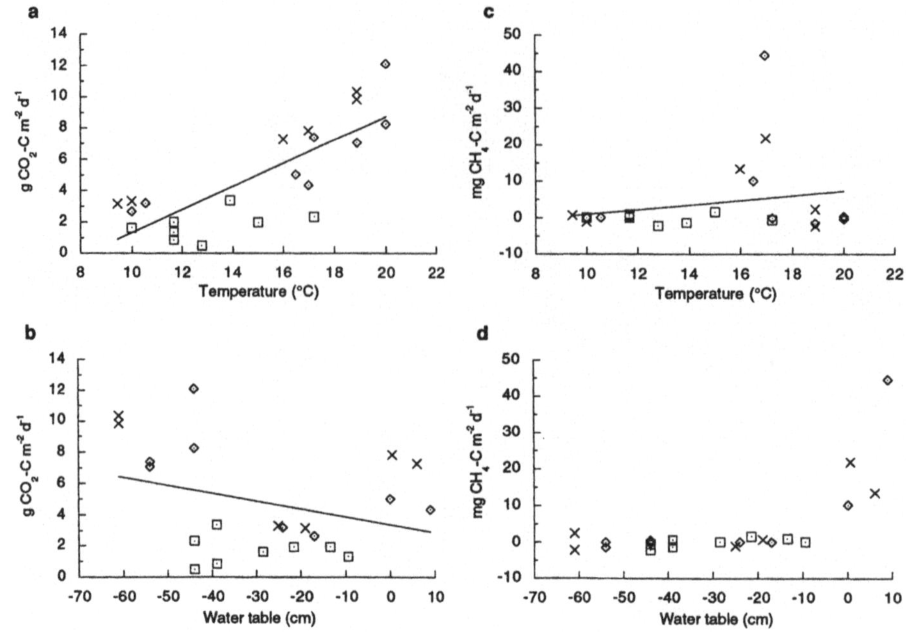

Fig. 8.11 Relationship between net ecosystem respiration as carbon dioxide emissions (**a, b**) and methane emissions (**c, d**) and soil temperature (**a, c**) and water-table level (**b, d**) in a bog (*squares*) and a fertilized (*x's*) and unfertilized (*diamonds*) beaver meadow in Voyageurs National Park. Soil temperatures were measured at a depth of 10 cm. *Regression lines* are given for each panel (**a**: $r^2 = 0.64$, $P < 0.001$, **b**: $r^2 = 0.10$, $P = 0.15$, **c**: $r^2 = 0.05$, $P = 0.33$), except for (**d**) where a nonlinear relationship holds. From Bridgham SD, Johnston CA, Pastor J, Updegraff K (1995) Potential feedbacks of northern wetlands on climate change. BioScience 45:262–274, with permission of Oxford University Press

and submergent zone (10.78 g C m^{-2} year^{-1}) were 24–43 times higher than from the forest (0.25 g C m^{-2} year^{-1}) or wet meadow (0.36 g C m^{-2} year^{-1}), despite the fact that the wet meadow soils contained nearly twice as much carbon (9.6 kg m^{-2}) as the other three hydrosequence zones (Naiman et al. 1991).

These two field campaigns illustrate the influence of surface water ponding on methane fluxes. Even though beaver meadow soils with high carbon content have the *potential* to produce substantial methane, actual fluxes to the atmosphere are negligible when the water table is below the surface. Bubble formation (i.e., ebullition) may be a factor: methane formed deep in pond sediments could bypass methanotrophs by forming bubbles that rise to the water surface. Other researchers have similarly found beaver ponds to be "hot spots" of carbon dioxide and methane emissions based on gas fluxes into field chambers (Bubier et al. 1993; Weyhenmeyer 1999) and eddy covariance towers (Dove et al. 1999; Roulet et al. 1997).

8.10 Extrapolating Methane Fluxes to the Landscape

Knowing the area of beaver ponds and other wetlands in the landscape allowed us to extrapolate these field measurements to landscape-scale fluxes. I used the maps we created of wetland soils that existed as of 1940 (i.e., prior to beaver population expansion) (Fig. 5.8) and beaver ponds created as of 1961 and 1988 (Johnston and Windels 2015), to classify all wetlands into five types that had been demonstrated to have different rates of methane flux (Table 8.3). I multiplied the cumulative area of each wetland type by a methane flux rate that was consistent with published studies to determine how the increase in beaver pond area over time would have affected landscape-scale methane release.

The results illustrate the significance of beaver pond building to methane fluxes (Fig. 8.12). Beaver activity increased the Kabetogama Peninsula's wetland area by 21% from 1940 to 1961 and by 7% from 1961 to 1988 (Fig. 8.12a). However, the proportion of ponds and marshes, which have the greatest source strength for methane flux (Table 8.3), increased from 4 to 30% of all wetlands. Consequently there was a 3.7-fold increase in total methane flux from all wetlands on the Kabetogama Peninsula (Fig. 8.12b). Because wetlands constitute approximately one-fifth of this landscape, this raised the average methane flux per unit area of land from 0.26 to 0.87 g C m^{-2} $year^{-1}$. This example illustrates how landscape changes that are relatively small in extent can greatly influence cumulative gas fluxes (Bridgham et al. 1995). Methane is a greenhouse gas that has 23 times the warming power of CO_2, so this increase is concerning.

Table 8.3 Wetland types and assumed flux rates used in landscape-scale extrapolation of methane fluxes

Wetland type	Water depth	Methane flux (g C m^{-2} $year^{-1}$)	Reference
Ponds and marshes	Persistent water above soil surface	10.0	Naiman et al. (1991) and Roulet et al. (1992)
Shrub swamps	Water at or below surface	3.6	Roulet et al. (1992)
Fens and wet meadows	Water at or below surface	0.8	Bridgham et al. (1995), Roulet et al. (1992), and Naiman et al. (1991)
Open bogs	Water below surface	0.7	Roulet et al. (1992) and Bridgham et al. (1995)
Forested swamps	Water below surface	0.1	Roulet et al. (1992)

From Bridgham SD, Johnston CA, Pastor J, Updegraff K (1995) Potential feedbacks of northern wetlands on climate change. BioScience 45:262–274, with permission from Oxford University Press

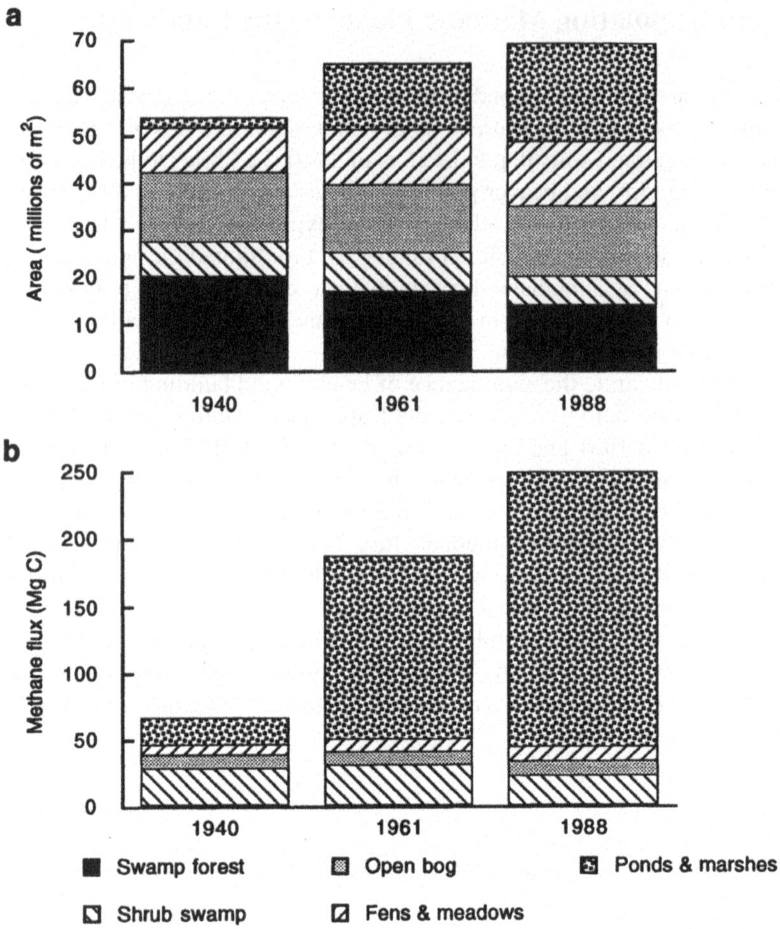

Fig. 8.12 (**a**) Area of various types of wetlands on the Kabetogama Peninsula from 1940 to 1988. Changes in the areas of different wetland categories are largely the result of an increase in beaver populations. (**b**) Changes in methane flux due to changes in wetland area shown in (**a**). Methane fluxes were assigned to each wetland category from Table 8.3. From Bridgham SD, Johnston CA, Pastor J, Updegraff K (1995) Potential feedbacks of northern wetlands on climate change. BioScience 45:262–274, with permission of Oxford University Press

8.11 Conclusions

So what's the balance? Are beavers good guys or bad guys with regard to carbon cycling? Do the beneficial effects of carbon sequestration that occur in beaver meadows outweigh the detrimental effects of methane emissions from beaver ponds? We don't know the answer to that question yet.

- Carbon takes many forms within beaver ecosystems: living and dead, solid and gas, organic and inorganic, aboveground and belowground.

- Soils and vegetation contain the largest stores of carbon in beaver ecosystems.
- The concentration of dissolved organic carbon (DOC) in the water column is important to the acid–base chemistry of beaver ponds, but is a negligible storage mechanism for carbon.
- Vascular plants associated with beaver ponds and meadows take up atmospheric carbon dioxide during photosynthesis and incorporate it into their tissues. The quantity of carbon uptake determines net primary production (NPP).
- Forests contain greater standing stocks of carbon than meadows due to their height, but beaver meadows and boreal forests have similar NPP.
- The decomposition rate of plant litter in beaver meadows is a function of the plant species and the decomposition environment, particularly soil aeration and temperature. Beaver meadow organic soils decomposed half as fast under anaerobic conditions.
- The transfer of plant litter to the soil surface is more complex in beaver meadows than in forests due to the lack of leaf dehiscence, the formation of thatch, and the presence of peat forming structures such as tussocks and floating mats.
- There was twice as much carbon per unit area stored in beaver meadow mineral soils than in forest soils on the Kabetogama Peninsula.
- Beaver meadows and ponds are a source of methane gas when the water table is at or above the surface. Pond creation due to the beaver population expansion at Voyageurs National Park has increased landscape-level methane fluxes.
- It is not known whether the carbon sequestration in beaver pond soils offsets the release of greenhouse gases by beaver ponds.

References

Aerts R, de Caluwe H (1997) Nutritional and plant-mediated controls on leaf litter decomposition of *Carex* species. Ecology 78(1):244–260

Bartsch I, Moore TR (1985) A preliminary investigation of primary production and decomposition in four peatlands near Schefferville, Québec. Can J Bot 63(7):1241–1248

Batjes NH (1996) Total carbon and nitrogen in the soils of the world. Eur J Soil Sci 47(2):151–163

Bridgham SD, Johnston CA, Pastor J, Updegraff K (1995) Potential feedbacks of northern wetlands on climate change. Bioscience 45:262–274

Bridgham SD, Megonigal JP, Keller JK, Bliss NB, Trettin C (2006) The carbon balance of north American wetlands. Wetlands 26(4):889–916

Bubier JL, Moore TR, Roulet NT (1993) Methane emissions from wetlands in the midboreal region of northern Ontario, Canada. Ecology 74(8):2240–2254

Chamie JPM, Richardson CJ (1978) Decomposition in northern wetlands. In: Good RE, Whigham DF, Simpson RL (eds) Freshwater wetlands: ecological processes and management potential. Academic, New York, pp 115–130

Cirmo CP, Driscoll CT (1993) Beaver pond biogeochemistry - acid neutralizing capacity generation in a headwater wetland. Wetlands 13(4):277–292

Coleman RL, Dahm CN (1990) Stream geomorphology: effects on periphyton standing crop and primary production. J N Am Benthol Soc 9(4):293–302

Davis CB, van der Valk AG (1978) Litter decomposition in prairie glacial marshes. In: Good RE, Whigham DF, Simpson RL (eds) Freshwater wetlands: ecological processes and management potential. Academic, New York, pp 99–113

Devito KJ, Dillon PJ, Lazerte BD (1989) Phosphorus and nitrogen retention in five Precambrian shield wetlands. Biogeochemistry 8(3):185–204

Dove A, Roulet NT, Crill PM, Chanton J, Boubonniere RA (1999) Methane dynamics of a northern boreal beaver pond. Ecoscience 6(4):577–586

Erickson HE (1994) Nitrogen and phosphorus availability, ecosystem processes, and plant community dynamics in boreal wetland meadows. PhD thesis. University of Washington, Seattle

Faber-Langendoen D, Aaseng N, Hop K, Lew-Smith M (2007) Field guide to the plant community types of Voyageurs National Park. U.S. Geological Survey, Reston

Ferguson A, Osborn G (1981) Minimum age of deglaciation of upper Elk Valley, British Columbia. Can J Earth Sci 18:1635–1636

Garrison GC (1967) Pollen stratigraphy and age of an early postglacial beaver site near Columbus, Ohio. Ohio J Sci 67(2):96–105

Gower ST, Krankina O, Olson RJ, Apps M, Linder S, Wang C (2001) Net primary production and carbon allocation patterns of boreal forest ecosystems. Ecol Appl 11(5):1395–1411

Hall HA, Pritchard G (1975) The food of larvae of Tipula Sacra Alexander in a series of abandoned beaver ponds (Diptera: Tipulidae). J Anim Ecol 44:55–66

Hanson WD, Campbell RS (1963) The effects of pool size and beaver activity on distribution and abundance of warmwater fishes in a North Missouri stream. Am Midl Nat 69(1):136–149

Hartke KM, Hepp GR (2004) Habitat use and preferences of breeding female wood ducks. J Wildl Manag 68(1):84–93

Hillman GR, Feng JC, Feng CC, Wang YH (2004) Effects of catchment characteristics and disturbances on storage and export of dissolved organic carbon in a boreal headwater stream. Can J Fish Aquat Sci 61(8):1447–1460

Hodkinson ID (1975a) Dry weight loss and chemical changes in vascular plant litter of terrestrial origin, occurring in a beaver pond ecosystem. J Ecol 63(1):131–142

Hodkinson ID (1975b) Energy flow and organic matter decomposition in an abandoned beaver pond ecosystem. Oecologia 21(2):131–139

James CD, Lanman RB (2012) Novel physical evidence that beaver historically were native to the Sierra Nevada. Calif Fish Game 98(2):129–132

Johnston CA (1988) Productivity of wet soils: biomass of cultivated and natural vegetation. Oak Ridge National Laboratory, Oak Ridge

Johnston CA (2014) Beaver pond effects on carbon storage in soils. Geoderma 213:371–378

Johnston CA, Bedford BL, Bourdaghs M, Brown T, Frieswyk C, Tulbure M, Vaccaro L, Zedler JB (2007) Plant species indicators of physical environment in Great Lakes coastal wetlands. J Great Lakes Res 33(sp3):106–124

Johnston CA, Groffman P, Breshears DD, Cardon ZG, Currie W, Emanuel W, Gaudinski J, Jackson RB, Lajtha K, Nadelhoffer K, Nelson D, Post WM, Retallack G, Wielopolski L (2004) Carbon cycling in soil. Front Ecol Environ 2(10):522–528

Johnston CA, Naiman RJ (1990) Browse selection by beaver: effects on riparian forest composition. Can J For Res 20:1036–1043

Johnston CA, Windels SK (2015) Using beaver works to estimate colony activity in boreal landscapes (online spatial data). CUAHSI Spatial Data Service, Consortium of Universities for the Advancement of Hydrologic Science Inc, Medford. http://spatial.cuahsi.org/johnstonC01/

Lockmiller RL (1979) Use of beaver ponds by southeastern woodpeckers in winter. J Wildl Manag 43(1):263–266

Margolis BE, Castro MS, Raesly RL (2001) The impact of beaver impoundments on the water chemistry of two Appalachian streams. Can J Fish Aquat Sci 58(11):2271–2283

McClaugherty CA, Pastor J, Aber JD, Melillo JM (1985) Forest litter decomposition in relation to soil nitrogen dynamics and litter quality. Ecology 66(1):266–275

Menzel MA, Carter TC, Ford WM, Chapman BR (2001) Tree-roost characteristics of subadult and female adult evening bats (Nycticeius humeralis) in the upper coastal plain of South Carolina. Am Midl Nat 145(1):112–119

Merendino MT, McCullough GB, North NR (1995) Wetland availability and use by breeding waterfowl in southern Ontario. J Wildl Manag 59(3):527–532

Moore TR (2003) Dissolved organic carbon in a northern boreal landscape. Glob Biogeochem Cycles 17(4):n/a–n/a

Morris DM (2014) Aquatic habitat use by North American moose (Alces alces) and associated richness and biomass of submersed and floating-leaved aquatic vegetation in north-central Minnesota. Lakehead University, Thunder Bay

Naiman RJ, Manning T, Johnston CA (1991) Beaver population fluctuations and tropospheric methane emissions in boreal wetlands. Biogeochemistry 12:1–15

Nummi P, Kattainen S, Ulander P, Hahtola A (2011) Bats benefit from beavers: a facilitative link between aquatic and terrestrial food webs. Biodivers Conserv 20(4):851–859

Peach M, Zedler JB (2006) How tussocks structure sedge meadow vegetation. Wetlands 26(2):322–335

Polunin NVC (1984) The decomposition of emergent macrophytes in fresh water. Adv Ecol Res 14:115–165

Rains B (1987) Holocene alluvial sediments and a radio-carbon dated relict beaver dam, Whitehead Creek, Edmonton, Alberta. Can Geogr 31:272–277

Reader RJ, Stewart JM (1972) The relationship between net primary production and accumulation for a Peatland in southeastern Manitoba. Ecology 53(6):1024–1037

Robinson S, Beaudoin AB, Froese DG, Doubt J, Clague JJ (2007) Plant macrofossils associated with an early Holocene beaver dam in interior Alaska. Arctic 60(4):430–438

Roulet NT, Ash R, Moore TR (1992) Low boreal wetlands as a source of atmospheric methane. J Geophys Res 97(D4):3739–3749

Roulet NT, Crill PM, Comer NT, Dove A, Boubonniere RA (1997) CO_2 and CH_4 flux between a boreal beaver pond and the atmosphere. J Geophys Res 102:29313–29319

Schiff SL, Aravena R, Trumbore SE, Dillon PJ (1990) Dissolved organic carbon cycling in forested watersheds: a carbon isotope approach. Water Resour Res 26(12):2949–2957

Stokland JN, Siitonen J, Jonsson BG (2012) Biodiversity in dead wood. Cambridge University Press, Cambridge

Szumigalski AR, Bayley SE (1996) Decomposition along a bog to rich fen gradient in central Alberta, Canada. Can J Bot 74(4):573–581

Tarnocai C, Kettles IM, Lacelle B (1998) The amount of organic carbon in various soil orders and ecological provinces in Canada. In: Lal R, Kimble JM, Follett RF, Stewart BA (eds) Soil processes and the carbon cycle. CRC Press, Boca Raton, pp 81–92

Tedford RH, Harington CR (2003) An Arctic mammal fauna from the early Pliocene of North America. Nature 425(6956):388–390

Thompson S, Vehkaoja M, Nummi P (2016) Beaver-created deadwood dynamics in the boreal forest. For Ecol Manag 360:1–8

Thormann MN, Bayley SE, Currah RS (2001) Comparison of decomposition of belowground and aboveground plant litters in peatlands of boreal Alberta, Canada. Can J Bot 79(1):9–22

Turner DP, Koerper GJ, Harmon ME, Lee JJ (1995) A carbon budget for forests of the conterminous United States. Ecol Appl 5(2):421–436

Updegraff K, Pastor J, Bridgham SD, Johnston CA (1995) Environmental and substrate controls over carbon and nitrogen mineralization in northern wetlands. Ecol Appl 5(1):151–163

Vehkaoja M (2016) Beaver in the drainage basin: an ecosystem engineer restores wetlands in the boreal landscape. University of Helsinki, Helsinki

Vehkaoja M, Nummi P, Rask M, Tulonen T, Arvola L (2015) Spatiotemporal dynamics of boreal landscapes with ecosystem engineers: beavers influence the biogeochemistry of small lakes. Biogeochemistry 124(1):405–415

Weyhenmeyer CE (1999) Methane emissions from beaver ponds: rates, patterns, and transport mechanisms. Glob Biogeochem Cycles 13(4):1079–1090

Wisconsin Department of Natural Resources (2003) Silviculture and forest aesthetics handbook. Wisconsin Department of Natural Resources, Madison

Wohl E (2013) Landscape-scale carbon storage associated with beaver dams. Geophys Res Lett 40(14):3631–3636

Chapter 9
Fish Assemblages in a Beaver-Influenced Successional Landscape

Isaac J. Schlosser and Larry W. Kallemeyn

Abstract Beavers are increasingly viewed as "ecological engineers," having broad effects on physical, chemical, and biological attributes of north-temperate landscapes. We examine the influence of both local successional processes associated with beaver activity and regional geomorphic boundaries on spatial variation in fish assemblages along the Kabetogama Peninsula in Voyageurs National Park, northern Minnesota, USA. Based on the results, we present a hierarchical conceptual model suggesting how geomorphic boundaries and beaver pond succession interact to influence fish assemblage attributes. The presence of a productive and diverse fish assemblage in headwater streams of north-temperate areas requires the entire spatial and temporal mosaic of successional habitats associated with beaver activity, including those due to the creation and abandonment of beaver ponds. The ultimate impact of the local successional mosaic on fishes, however, will be strongly influenced by the regional geomorphic context in which the mosaic occurs.

Ecologists increasingly examine spatial and temporal environmental variation to understand how fundamentally different processes determine population dynamics, species interactions, and assemblage structure over hierarchically structured scale-dependent levels of complexity (Allen and Starr 1982; Naveh and Lieberman 1984). As differing environmental levels in the hierarchy are transcended, variation in dispersal and invasion can affect in a geographically specific manner population size, demography, and co-adaptation as well as food web interactions and assemblage structure (Tonn et al. 1990; Endler 1991). Furthermore, hierarchical environmental structure often occurs in the regional context of landscape mosaics characterized by a highly dynamic successional driven patch structure (Pickett and White 1985; Chesson and Huntley 1997; Wiens 1997; Hanski and Simerbloff 1997). Patch quality, connectivity, context, and boundary characteristics have profound influences on dispersal and

I.J. Schlosser (✉)
Department of Biology, University of North Dakota, Grand Forks, ND 58201, USA
e-mail: isaac.schlosser@email.und.edu

L.W. Kallemeyn
USGS Columbia Environmental Research Center, International Falls, MN 56649, USA

C.A. Johnston, *Beavers: Boreal Ecosystem Engineers*,
DOI 10.1007/978-3-319-61533-2_9

subsequent food web interactions (Pickett and White 1985; Wiens 1997). A fundamental objective of ecology is to subsequently understand how hierarchically dependent attributes of environmental variation, especially patch structure and boundary dynamics, interact in the context of landscape succession to influence dispersal and ecological success (Hanski and Simerbloff 1997; Rahel 2002).

Beaver (*Castor canadensis*) are recognized as critical "ecological engineers," having broad effects on physical, chemical, and biological characteristics of north-temperate landscapes (Naiman et al. 1988). Modification of aquatic ecosystems by beaver results in a shifting successional mosaic of aquatic patches embedded within the larger regional landscape (Naiman et al. 1988; Johnston and Naiman 1987; Johnston and Naiman 1990; Pastor et al. 1993; Johnston 1995; Schlosser 1998; Schlosser et al. 1998; Snodgrass 1997; Snodgrass and Meffe 1998). Since beaver ponds are created and maintained by living organisms, they are temporally dynamic as stream reaches are colonized, flooded, and eventually abandoned. This results in a spatially and temporally dynamic mosaic of aquatic patches, including ponds, collapsed ponds, and streams (Naiman et al. 1988). Additional spatial complexity occurs when these aquatic environments are placed in the context of the larger drainage network due to variation in valley shape (Johnston and Naiman 1987) and context within the longitudinal drainage network (Schlosser 1987, 1991, 1995a, b; Schlosser and Angermeier 1995; Osborne and Wiley 1992; Fausch et al. 2002). Considerable variation also occurs in north-temperate aquatic environments associated with seasonal and annual variation in temperature and oxygen regimes (Johnston and Naiman 1987; Schlosser et al. 1998; Tonn and Magnuson 1982; Rahel 1984; Magnuson et al. 1989; Gorman et al. 2014).

We examine how the shifting successional mosaic of local aquatic habitat patches created by beaver influences the number, identity, and relative abundance of fish species and place this in the context of processes operating at the spatial scale of drainage basins. We first examine variation in fish abundance and species rich among the drainages along the Kabetogama Peninsula in Voyageurs National Park in northern Minnesota, USA. We then explore how fish abundance and species richness differs among successional environments associated with beaver activity. Lastly, we examine how composition of fish assemblages varies among drainages associated with non-random distribution of successional environments. Based on these results, we present a hierarchical conceptual model suggesting how geomorphic boundaries and beaver pond succession interact to influence fish assemblage attributes. A more detailed presentation of the data and analysis can be found in Schlosser and Kallemeyn (2000).

9.1 Sampling Fish Assemblages Across a Successional Landscape

Fish were sampled at 23 sites in 12 drainages along the southern edge of the 294 km^2 Kabetogama Peninsula (Fig. 9.1) and included all types of environments associated with beaver pond succession. Distribution of successional sites sampled among

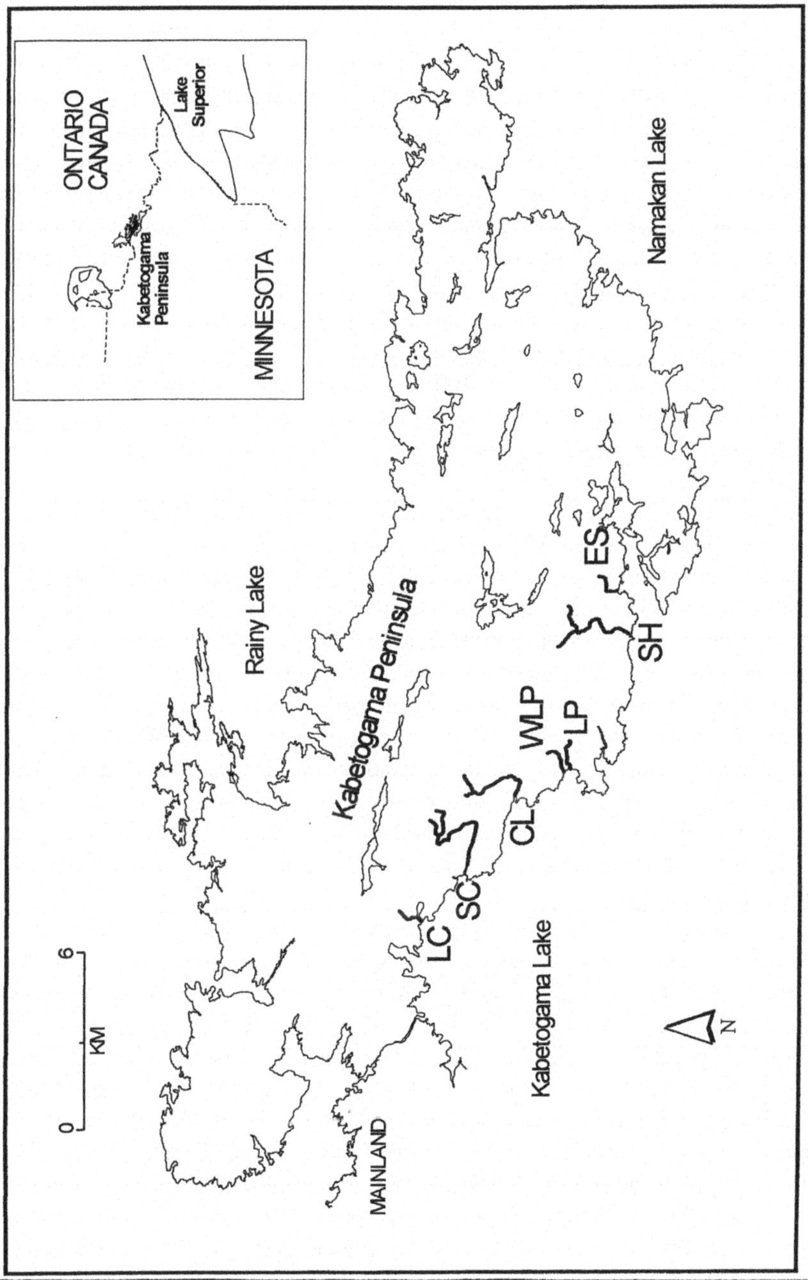

Fig. 9.1 Location of seven drainages sampled along Kabetogama Peninsula—Locator (LC), Sucker Creek (SC), Clyde Creek (CLY), West Lost Ponds (WLP), Lost Ponds (LP), Shoepack (SH), and East Shoepack (ES). From Schlosser IJ, Doeringsfeld MR, Elder JF, Arzayus LF (1998) Niche relationships of clonal and sexual fish in a heterogeneous landscape. Ecology 79 (3):953–968, with permission of John Wiley and Sons

drainages was dictated by their natural distribution on the landscape and could not be randomized among drainages. Each site was sampled once per year between mid-May and late-July with time of sampling randomized within and among drainages to avoid bias due to site by time interactions. Most sites were sampled for 3 years (1993–1995). Twelve un-baited Gee's minnow traps (40 × 19 cm, 2.5 cm opening, and 0.5 cm mesh) were sampled for 3d at each site. Sampling was restricted to minnow traps to allow standardization of sampling effort between sites and because many sites were remote with difficult access. Trapping effort between sites was standardized based on total shoreline length sampled, with approximately 100–150 m of shoreline sampled at each site. Sampling in pond and collapsed pond environments was restricted to littoral regions since previous research indicated the vast majority (>95%) of all fish were captured in these areas and no new species were added when minnow traps were suspended in pelagic zones (Schlosser et al. 1998; He and Lodge 1990; IJ Schlosser, unpublished data). Most fish were identified in the field and a sample of up to 300 fish of each species measured for total length and returned to the water. Based on the total number of fish captured, a catch-per-unit-effort (CPUE: number captured per trap per day) was calculated for each species, site and year.

Twenty-one of the 23 sites could be classified into one of five types of successional environments based on physical-chemical characteristics: upland ponds; lowland ponds; collapsed ponds with partially rebuilt dams; completely collapsed ponds; and streams. The latter three types of environments were easily distinguished based on their unique morphological characteristics. Streams were free-flowing reaches with no remnants of a beaver dam and no pooling of water out of the stream channel (Fig. 9.2a). Completely collapsed ponds were areas where the remnants of a dam was still present with water partially ponded behind the dam, but a least part of the dam was fully breached and water was freely flowing through the breached area (Fig. 9.2d). Collapsed ponds with partially rebuilt dams were areas where the dam had been previously breached but partially rebuilt by beaver so the pool behind the dam was deeper than in a fully collapsed pond and there was no free-flowing stream.

Segregation of upland and lowland ponds required a more objective and quantitative approach. Upland ponds were in the upper reaches of drainages on the Kabetogama Peninsula and were created by short dams in incised valleys (Johnston and Naiman 1987), resulting in narrow but potentially deep ponds (Fig. 9.2b). In contrast, lowland ponds were created by long dams placed across broader and flatter reaches of valley drainages, creating wider and potentially shallower ponds (Fig. 9.2c). Based on these morphological differences one would also expect upland and lowland ponds to exhibit different patterns of oxygen profiles since the relative surface area of water exposed to organically rich pond bottoms would be higher in shallow lowland than deep upland ponds. Six physical-chemical characteristics were used to quantitatively separate ponds into upland and lowland environments, including: (1) distance between the pond and Kabetogama Lake, (2) dam length, (3) mean pond depth, (4) maximum pond depth, (5) maximum dissolved oxygen, and (6) water depth at which the pond became anoxic (<1.0 mg/L O_2). Of the six a

b

Fig. 9.2 Successional environments sampled during the study included: (**a**) streams (Photo courtesy of Thomas D. Gable), (**b**) upland ponds, (**c**) lowland ponds (Photo courtesy of Carol A. Johnston), (**d**) collapsed ponds

Fig. 9.2 (continued)

variables, numbers 1–4 were only measured once, while 5–6 were based on averages over 3 years of sampling. A hierarchical cluster analysis using standardized data, Euclidean distance metric, and single linkage (nearest neighbor) clustering technique (SPSS 1996) clearly distinguished two different clusters representing upland and lowland ponds (Schlosser and Kallemeyn 2000).

Two of the twenty-three sites could not be effectively classified according to successional stage since they were relatively small, recently constructed ponds with standing timber. These two sites were included in the analysis of differences in fish assemblages among drainages but were not included in the analysis of successional environments.

Since data were restricted to 3 years, the primary objective of our analysis was to evaluate spatial variation in fish assemblages. Sites represented the individual replicate in all analyses and species richness and fish abundance for each site were based on mean values calculated from 3 years of sampling. Species richness and total fish abundance across drainages and among successional sites were compared using a one-way ANOVA on $\log_{10}$ transformed data, with the relatively conservative Tukey's multiple comparison follow-up test. Hierarchical cluster analyses, using fish abundance data standardized with z scores, Euclidean distance metric, and single linkage (nearest neighbor) technique, were used to separately cluster drainages, successional environments, and species. We also simultaneously clustered species within drainages and species within successional environments. In the species by drainage and species by successional environment analysis, rows and columns in the data matrices were permuted according to the algorithm in (Gruvaeus and Wainer 1972). All statistical analyses were performed using SYSTAT version 6.0 for Windows (Schlosser and Kallemeyn 2000; SPSS 1996).

9.2 Spatial Variation in Fish Assemblages Across a Beaver-Influenced Successional Landscape

Comparisons of fish abundance and composition among drainages were limited to seven drainages with at least two sample sites per drainage (Fig. 9.3a). Fish abundance differed significantly among the seven drainages, decreasing in the middle of the peninsula, associated with the presence of geological barriers to fish movement at drainage outlets in Clyde Creek and West Lost Ponds (Fig. 9.3a). Among drainages without barriers to fish movement there were still significant differences in fish abundance, with higher fish abundance in East Shoepack than in all other drainages.

Species richness per trap also varied significantly among drainages. Again, species richness per trap was lower in drainages with geological barriers and there was variation among drainages without geological barriers (Fig. 9.3b). Species richness per trap in Sucker Creek tended to be the highest of all drainages but none of the pairwise comparisons of species richness in Sucker Creek were statistically differ-

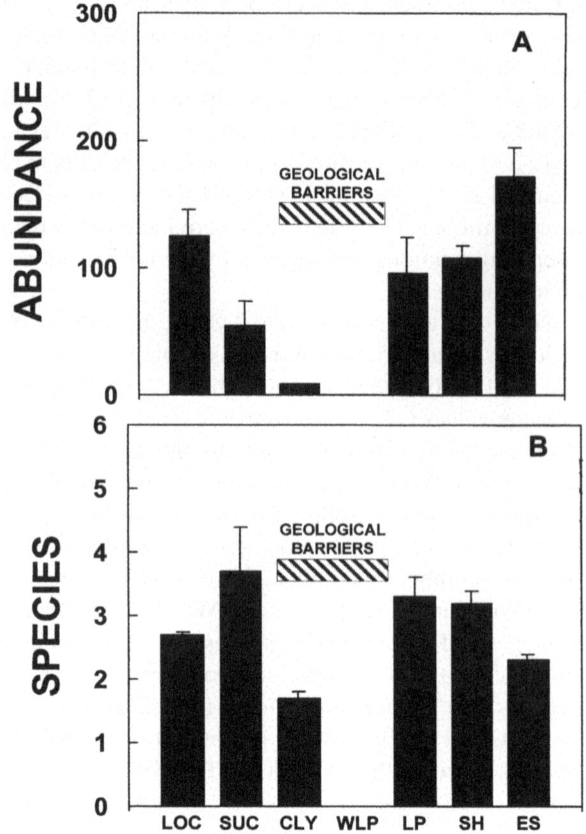

Fig. 9.3 Total fish abundance (**a**) and species richness (**b**) per minnow trap for seven drainages along the Kabetogama Peninsula. Histogram height is the mean and vertical bar is 1 SE. Note that geological barriers occur at the outlets of Clyde Creek and West Lost Ponds. From Schlosser IJ, Kallemeyn LW (2000) Spatial variation in fish assemblages across a beaver-influenced successional landscape. Ecology 81 (5):1371–1382, with permission of John Wiley and Sons

Table 9.1 Total number of fish species captured over a 3-year period (1993–1995) in seven different drainages located along the Kabetogama Peninsula

Drainage	Number of species
Locator	4
Sucker creek	10
Clyde creek	4
West lost ponds	0
Lost ponds	4
Shoepack	4
East shoepack	3

ent from those in the other drainages without geological barriers. The tendency for Sucker Creek to exhibit higher species richness was strongly reaffirmed, however, when total number of species captured in each drainage over the 3 year period were compared (Table 9.1). The number of species captured in Sucker Creek (10 species) was 2–3 times the number of species captured in other drainages (3–4 species). The dramatic difference in species richness when mean values per trap (Fig. 9.3b) and total number of species captured over the 3-year period (Table 9.1) were compared indicates additional species captured in Sucker Creek were relatively rare.

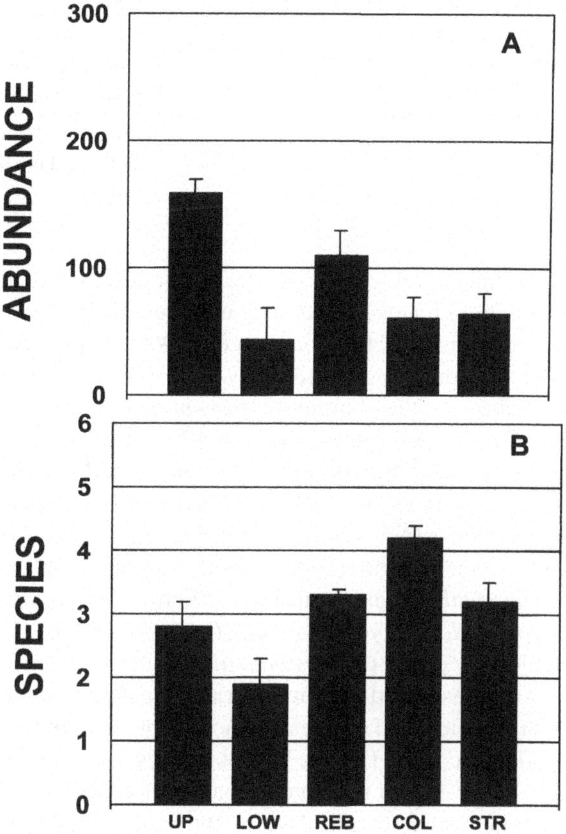

Fig. 9.4 Total fish abundance (**a**) and species richness (**b**) per trap for five different environments associated with beaver pond succession on the Kabetogama Peninsula, including upland ponds (UP), lowland ponds (LOW), collapsed ponds with partially rebuilt dams (REB), collapsed ponds (COL), and streams (STR). Histogram height is the mean and vertical bar is 1 SE. From Schlosser IJ, Kallemeyn LW (2000) Spatial variation in fish assemblages across a beaver-influenced successional landscape. Ecology 81 (5):1371–1382, with permission of John Wiley and Sons

Comparisons of fish abundance and composition among successional environments included all 23 sites in the 12 drainages (Fig. 9.4a). Fish abundance and species richness per trap also varied significantly among successional environments. Fish abundance was nearly four times as high in upland ponds than lowland ponds (Fig. 9.4a), while other environments were not different from each other. Mean species richness per trap was significantly larger and twice as high in collapsed ponds than lowland ponds (Fig. 9.4b), while other environments exhibited intermediate and not significantly different species richness. The tendency for collapsed ponds to exhibit higher species richness than upland, lowland, and partially rebuilt ponds was even greater when the total number of species captured in each environment was compared over the 3-year period (Table 9.2).

Furthermore, although the average number of species per trap was not higher for streams compared to other environments, the total numbers of species collected over the 3 years in streams approached that of collapsed ponds (13 vs. 15 species). As in the comparison among drainages, the large difference in mean species richness per trap (Fig. 9.4b) and total number of species captured in each successional

Table 9.2 Total number of fish species captured over a 3-year period (1993–1995) in five different successional environments located along the Kabetogama Peninsula

Successional environment	Number of species
Upland pond	4
Lowland pond	5
Partially rebuilt pond	4
Collapsed pond	15
Stream	13

environment (Table 9.2) over the 3 year period indicates additional species captured in collapsed ponds and streams were relatively rare.

Comparison of fish species composition in different successional environments indicated four species (redbelly dace (*Phoxinus eos*), brook stickleback (*Culaea inconstans*), fathead minnow (*Pimephales promelas*), and finescale dace (*Phoxinus neogaeus*)) were predominant in all successional environments, with variation in species composition between environments due to the addition of several rare species (brassy minnow (*Hybognathus hankinsoni*), pearl dace (*Semotilus margarita*), golden shiner (*Notemigonus crysoleucas*), blacknose shiner (*Notropis heterolepis*), central mudminnow (*Umbra limi*), burbot (*Lota lota*), white sucker (*Catastomus commersoni*), smallmouth bass (*Micropterus dolomieui*), black crappie (*Pomoxis nigromaculatus*), yellow perch (*Perca flavescens*), Iowa darter (*Etheostoma exile*), and johnny darter (*Etheostoma nigrum*)).

Analyses based on capture rates clustered according to successional environment, species, and species by successional environment (Schlosser and Kallemeyn 2000) indicated 11 of the 12 upland, lowland, and rebuilt pond sites formed a single cluster due to the predominance of redbelly dace, brook stickleback, fathead minnow, and finescale dace in those environments. In contrast, stream and collapsed pond sites in close proximity formed a very unique cluster, with a large number of rare species. Analysis of the size structure of less abundant species in collapsed pond and stream environments further indicated dispersal of early life history stages (age 0) of large lake species (e.g., black crappie, smallmouth bass, yellow perch, and burbot) was critical in determining species richness and species composition in stream and collapsed pond successional environments (Schlosser and Kallemeyn 2000).

Lastly, we examined the role spatial distribution of successional environments played in causing variation in species composition among drainages. Analyses of capture rates for all species standardized with z-scores and clustered according to drainage, species, and drainage by species revealed three patterns: (1) species composition varied dramatically in drainages with and without geological barriers; (2) all drainages without geological barriers had four predominant species (fathead minnow, redbelly dace, finescale dace, and brook stickleback); and (3) variation in species composition between drainages was primarily due to the addition of several less abundant species, indicating the non-random distribution of successional sites (streams and collapsed ponds) was critical in determining variation in species richness and composition among drainages (Schlosser and Kallemeyn 2000).

9.3 A Hierarchical Model of Fish Assemblages in Landscapes Influenced by Beaver

A basic challenge ecologists face is understanding how hierarchically organized environmental variation interacts across spatial scales to influence dispersal and ecological success in successional landscapes (Allen and Starr 1982; Pickett and White 1985; Wiens 1997; Hanski and Simerbloff 1997). Others (Tonn and Magnuson 1982; Tonn 1990) have suggested environmental "filters" acting across the spatial hierarchy from the continental to local scales determine fish assemblage attributes in small north-temperate lakes. Our results extend their model to the uppermost reaches of north-temperate drainages, where geomorphic boundaries act in conjunction with local successional environments created by beaver to strongly influence fish assemblages (Fig. 9.5).

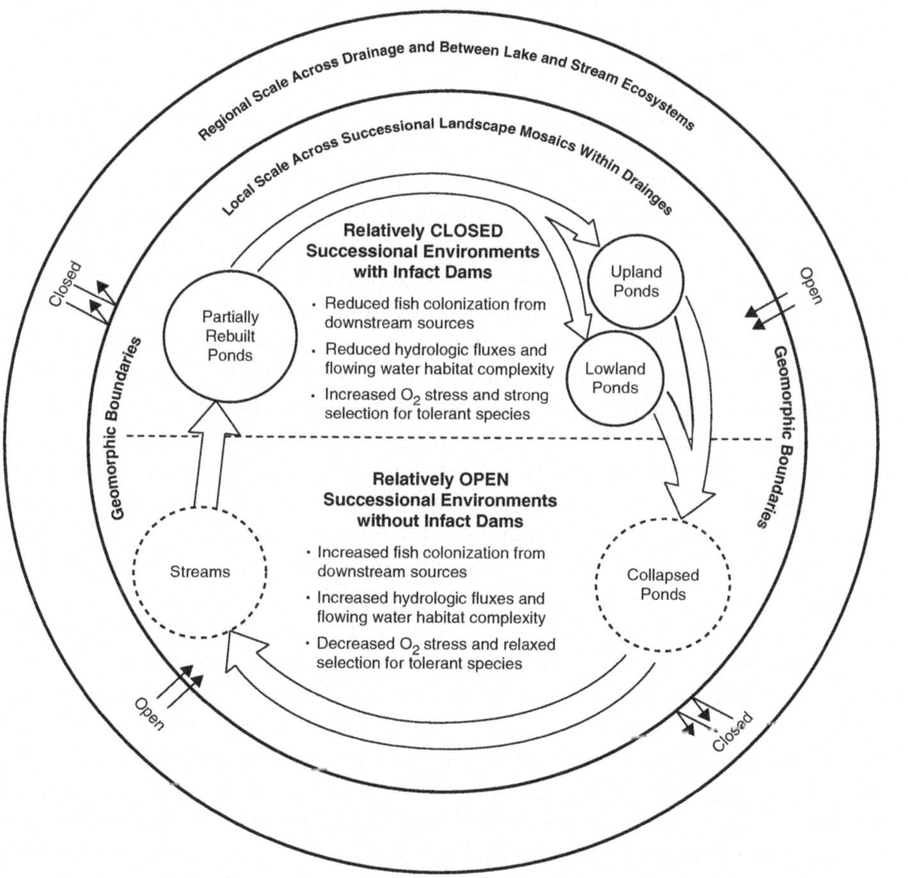

Fig. 9.5 Conceptual framework suggesting how regional geomorphic boundaries and local succession influence fish assemblage attributes in north-temperate landscapes modified by beaver activity. From Schlosser IJ, Kallemeyn LW (2000) Spatial variation in fish assemblages across a beaver-influenced successional landscape. Ecology 81 (5):1371–1382, with permission of John Wiley and Sons

Regional geomorphology is especially important since it controls dispersal corridors between adjacent aquatic environments (Tonn et al. 1990; Tonn 1990; Angermeier and Winston 1998). The Kabetogama Peninsula is the result of continental glaciation approximately 10,000–12,000 years BP, resulting in juxtaposition of large lakes and small peninsular streams, multiple stream drainages of variable size, and potential geological barriers to fish dispersal. These geomorphic attributes strongly influence whether boundaries between adjacent aquatic environments are "open" or "closed" to fish movement, the magnitude of hydrologic fluxes through drainages, and potential composition and age-structure of fish colonizing drainages (Fig. 9.5).

A critical local factor influencing fish abundance and diversity within drainages was habitat succession due to beaver activity (Fig. 9.5). The change from relatively "closed" environments with complete or partially rebuilt dams to relatively "open" environments without intact dams resulted in a fundamental change in assemblage attributes from potentially high fish abundance and low species richness, particularly in deep upland ponds, to low fish abundance and high species richness in collapsed ponds and streams.

The presence or absence of intact beaver dams at the outlet of successional environments influences multiple key factors effecting fish assemblages (Fig. 9.5). Presence of beaver dams reduces the potential for fish colonization from downstream sources, especially in small drainages during periods of low discharge (Schlosser 1995a; Lokteff and Roper 2013). Furthermore, our results and those of others (Magnuson et al. 1998) suggest spatial proximity of downstream colonization sources to successional environments will play a critical role in determining likelihood of successful colonization and subsequent species richness. Presence or absence of intact beaver dams also has dramatic impact on the rate of hydrologic exchange through successional environments and subsequent diversity of physical-chemical habitats (e.g., riffles and pools) available to fish colonizing these sites (Fig. 9.5). Finally, increased rate of hydrologic exchange in collapsed ponds or streams will decrease the potential severity of oxygen stress during both summer (Schlosser 1998; IJ Schlosser, unpublished data) and winter (Tonn and Magnuson 1982; Magnuson et al. 1989; IJ Schlosser, unpublished data), thus reducing the intensity of selection for fish tolerant to anoxia (Fig. 9.5).

Previous studies (Schlosser 1998; Tonn and Magnuson 1982; Rahel 1984; Cooper and Washburn 1946; Gee et al. 1978; Klinger et al. 1982) suggest the predominant species occurring in small ponds are all tolerant of oxygen stress, indicating anoxia is likely to be a particularly strong environmental "filter" determining species composition (Tonn and Magnuson 1982; Rahel 1984; Magnuson et al. 1989; Tonn 1990). Large lake species less tolerant of oxygen stress were only added to fish assemblages in relatively more "open" and highly oxygenated collapsed ponds, streams, and drainages characterized by the presence of these environments (Schlosser and Kallemeyn 2000). The role of winter anoxia in influencing species richness and species composition in beaver-influenced drainages will, however, undoubtedly exhibit variation associated with geographic variation in the severity of winter conditions (Snodgrass and Meffe 1998; Schlosser and Kallemeyn 2000).

Whether or not successional environments associated with intact beaver dams are truly "open" or "closed" will also be strongly influenced by their regional geomorphic and hydrologic context (Snodgrass 1997; Snodgrass and Meffe 1998; Schlosser 1995a; Fullerton et al. 2010; Zeug et al. 2005; Parkos et al. 2011; Lokteff and Roper 2013). Drainage basin size above the ponds, subsurface geology around the ponds, and regional precipitation will strongly influence the magnitude of hydrologic fluxes through drainages and degree to which ponds function as "open" or "closed" environments.

An important remaining question is how successional processes within drainages create differences in fish composition among drainages. A cross-classification of fish assemblage attributes for different successional sites and drainages, along with a comparison of the distribution of successional sites among drainages, suggests a strong relationship between local successional processes and regional attributes of fish assemblages. The most striking examples of this interaction occurred in the Locator, Sucker Creek, and East Shoepack drainages. Locator and East Shoepack are relatively small drainages at the extreme ends of the peninsula with primarily upland ponds. These drainages were characterized by relatively high fish abundance but low species richness. In contrast, Sucker Creek the largest drainage in the center of the peninsula contained only collapsed ponds and streams with low to moderate fish abundance but high species richness. These results indicate processes operating at regional levels between drainage basins (e.g., basin size, magnitude of hydrologic fluxes, successional patterns in terrestrial vegetation, and historic patterns of beaver colonization/abandonment) have strong effects on the distribution of local successional environments and subsequent regional variation in fish assemblages (Schlosser and Angermeier 1995; Creque et al. 2005; Mesquita et al. 2006).

Finally, our results have considerable implications for management of north-temperate landscapes if the goal is to assure both a diverse and productive fish assemblage. The data indicate during beaver pond succession a shift occurs from active pond shorelines with potentially high fish abundance and low species richness to collapsed pond and stream shorelines with low fish abundance but high species richness. Furthermore, this conclusion is conservative since it is based on trap CPUE values sampled over similar shoreline lengths at different successional sites. When beaver invade a regional landscape or humans remove beaver from the landscape they dramatically change the relative amount of shoreline in pond vs. collapsed pond or stream environments more than they change the relative number of different successional environments (Naiman et al. 1988; Broschart et al. 1989). Naiman et al. (Naiman et al. 1988) estimated during re-invasion of the Kabetogama Peninsula by beaver from 1940 to 1986 shoreline length increased 13.6-fold. Consequently, changes in fish abundance/production and species richness associated with increased or decreased beaver activity will be even more accentuated when comparisons are extended from similar shoreline lengths for different environments to actual shoreline lengths for each environment. Based on these data and those of Snodgrass and Meffe (Snodgrass and Meffe 1998) it is almost certain the presence of beaver ponds will substantially increase the overall production of fish on the landscape. Our results also suggest, however, that whether or not beaver activity ultimately increases species richness will be influenced by the subsequent impact of succession on spatial and temporal heterogeneity in the physical-chemical environment. Clearly, if a primary goal of landscape management is to assure the presence of a highly productive and diverse fish assemblage, then the entire mosaic of successional habitats associated with beaver activity needs to be preserved (Hood and Bayley 2008; Kemp et al. 2012; Smith 2012).

9.4 Conclusions

- Ecologists increasingly examine how hierarchically dependent attributes of environmental variation, especially patch structure and boundary dynamics, interact in the context of landscape succession to influence dispersal and ecological success in community assemblages.
- Modification of aquatic ecosystems by beaver results in a shifting successional mosaic of aquatic patches, including ponds, collapsed ponds, and streams embedded within the larger regional landscape.
- The Kabetogama Peninsula is the result of continental glaciation approximately 10,000–12,000 years BP, resulting in juxtaposition of large lakes and small peninsular streams, and multiple stream drainages of variable size.
- These regional geomorphic attributes strongly influence whether boundaries between adjacent aquatic environments are "open" or "closed" to fish movement, the magnitude of hydrologic fluxes through drainages, and potential composition and age-structure of fish colonizing drainages.
- Habitat succession due to beaver activity resulting in a change from relatively "closed" environments with complete or partially rebuilt dams to relatively "open" environments without intact dams is a critical local factor influencing fish abundance and diversity within drainages.
- If a primary goal of landscape management is to assure the presence of both a highly productive and diverse fish assemblage, then the entire mosaic of successional habitats associated with beaver activity needs to be preserved.

References

Allen TFH, Starr TB (1982) Hierarchy. University of Chicago Press, Chicago

Angermeier PL, Winston MR (1998) Local vs. regional influences on local diversityIn stream fish communities of Virginia. Ecology 79(3):911–927

Broschart MR, Johnston CA, Naiman RJ (1989) Prediction of beaver colony density using impounded habitat variables. J Wildl Manag 53(4):929–934

Chesson P, Huntley N (1997) The role of harsh and fluctuating conditions in the dynamics of ecological communities. Am Nat 150(5):519–553

Cooper GP, Washburn GN (1946) Relation of dissolved oxygen to winter mortality of fish in Michigan lakes. Trans Am Fish Soc 76(1):23–33

Creque SM, Rutherford ES, Zorn TG (2005) Use of GIS-derived landscape-scale habitat features to explain spatial patterns of fish density in Michigan rivers. N Am J Fish Manag 25(4):1411–1425
Endler JA (1991) Genetic heterogeneity and ecology. In: Berry RJ et al (eds) Proceedings of the 33rd Symposium of the British Ecological Society. Blackwell Scientific Publications, Oxford, pp 315–334
Fausch KD, Torgerson CE, Baxter CC, Li HW (2002) Landscapes to riverscapes: bridging the gap between research and conservation of stream fishes. Bioscience 52(6):483–498
Fullerton AH, Burnett KM, Steel EA, Flitcroft RL, Pess GR, Feist BE, Torgerson CE, Miller DJ, Sanderson BL (2010) Hydrological connectivity of riverine fish: measurement challenges and research opportunities. Freshw Biol 55(11):2215–2237
Gee JH, Tallman RF, Smart HJ (1978) Reactions of some great plains fish to progressive hypoxia. Can J Zool 56(9):1962–1966
Gorman O, Kallemeyn L, Maki R (2014) Biogeogrpahic patterns of inland lake fish communities at Isle Royale, Voyageurs, and Sleeping Bearn Dunes national park units. Natural Resource Technical Report NPS/GLKN/NRTR-2014/893. National Park Service, Fort Collins
Gruvaeus G, Wainer H (1972) Two additions to hierarchical cluster analysis. Br J Math Stat Psychol 25(7):200–206
Hanski I, Simerbloff D (1997) The metapopulation approach, its history, conceptual domain, and application. In: Hanski I, Gilpin M (eds) Metapopulation biology: ecology, genetics and evolution. Academic, San Diego, pp 5–26
He X, Lodge DM (1990) Using minnow traps to estimate fish population size: the importance of spatial distribution and relative species abundance. Hydrobiologia 190(1):9–14
Hood GA, Bayley SE (2008) Beaver (Castor canadensis) mitigate the effects of climate on the area of open water in boreal wetlands in western Canada. Biol Conserv 141(2):556–567
Johnston CA (1995) Effects of animals on landscape pattern. In: Hansson L, Fahrig L, Merriam G (eds) Mosaic landscapes and ecological processes, Landscape ecology study series, vol 2. Chapman & Hall, New York, pp 57–80
Johnston CA, Naiman RJ (1987) Boundary dynamics at the aquatic-terrestrial interface: the influence of beaver and geomorphology. Landsc Ecol 1(1):47–57
Johnston CA, Naiman RJ (1990) Aquatic patch creation in relation to beaver population trends. Ecology 71(4):1617–1621
Kemp PS, Worthington TA, Langford TE, Tree ARJ, Gaywood MJ (2012) Qualitative and quantitative effects of reintroduced beavers on stream fish. Fish Fish 13(2):158–181
Klinger SA, Magnuson JJ, Gallep GW (1982) Survival mechanisms of the central mudminnow (Umbra limi), fathead minnow (Pimephales promelas) and brook stickleback (Culea inconstans) for low oxygen in winter. Environ Biol Fish 7(2):113–120
Lokteff RL, Roper BB (2013) Do beaver dams impede the movement of trout? Trans Am Fish Soc 142(4):1114–1125
Magnuson JJ, Pazkowski CA, Rahel FJ, Tonn WM (1989) Fish ecology in severe environments. In: Sharitz RR, Gibbons JW (ed) Freshwater wetlands and wildlife conference. DOE Symposium Series No. 61, USDOE Office of Scientific and Technical Information, Oak Ridge, pp 487–515
Magnuson JJ, Tonn WM, Banerjee A, Toivonen JT, Sanchez O, Rask M (1998) Isolation versus extinction in the assembly of fishes in small northern lakes. Ecology 79(8):2941–2956
Mesquita N, Coelho MM, Filomena MF (2006) Spatial variation in fish assemblages across small Mediterranean drainages: effects of habitat and landscape context. Environ Biol Fish 77(2):105–120
Naiman RJ, Johnston CA, Kelly JC (1988) Alteration of north American streams by beaver. Bioscience 38(11):753–762
Naveh Z, Lieberman AS (1984) Landscape ecology: theory and application. Springer, New York
Osborne LL, Wiley MJ (1992) Influence of tributary position on the structure of of warmwater fish communities. Can J Fish Aquat Sci 49(4):672–681
Parkos JJ, Ruetz CR, Trexler JC (2011) Disturbance regime and limits on benefits of refuge use for fishes in a fluctuating hydroscape. Oikos 120(10):1519–1530

Pastor J, Bonde J, Johnston CA, Naiman RJ (1993) A Markovian analysis of the spatiallydependent dynamics of beaver ponds. In: Gardner RH (ed) Predicting spatial effects in ecological systems. American Mathematical Society, Providence, pp 5–27
Pickett STA, White PS (eds) (1985) The ecology of natural disturbance and patch dynamics. Academic, Orlando
Rahel FJ (1984) Factors structuring fish assemblages along a bog-lake successional gradient. Ecology 65(4):1276–1289
Rahel FJ (2002) Homogenization of freshwater faunas. Annu Rev Ecol Syst 33:291–315
Schlosser IJ (1987) A conceptual framework for fish communities in small warmwater streams. In: Matthews WJ, Heins DC (eds) The ecology and evolution of north American stream fish communities. University of Oklahoma Press, Norman, pp 17–24
Schlosser IJ (1991) Stream fish ecology: a landscape perspective. Bioscience 41(10):704–712
Schlosser IJ (1995a) Dispersal, boundary processes, and trophic level interactions in streams adjacent to beaver ponds. Ecology 76(3):908–925
Schlosser IJ (1995b) Critical landscape attributes influence fish population dynamics in headwater streams. Hydrobiologia 303:71–81
Schlosser IJ (1998) Fish recruitment, dispersal, and trophic interactions in a heterogeneous lotic environment. Oecologia 113(2):260–268
Schlosser IJ, Angermeier PL (1995) Spatial variation in demographic processes of lotic fishes: conceptual models, empirical evidence, and implications for conservation. Am Fish Soc Symp 17(1995):392–401
Schlosser IJ, Kallemeyn LW (2000) Spatial variation in fish assemblages across a beaver-influenced successional landscape. Ecology 81(5):1371–1382
Schlosser IJ, Doeringsfeld MR, Elder JF, Arzyus LF (1998) Niche relationships of clonal and sexual fish in a heterogeneous landscape. Ecology 79(3):953–968
Smith JM (2012) Beaver dams maintain native fish biodiversity via altered habitat heterogeneity in a coastal stream network: evaluating gear, quantifying fish assemblages, and testing ecological hypotheses. Dissertation University of Massachusetts, Amherst
Snodgrass JW (1997) Temporal and spatial dynamics of beaver-created patches as influenced by management practices in a south-eastern north American landscape. J Appl Ecol 34(4):1043–1056
Snodgrass JW, Meffe GK (1998) Influence of beavers on stream fish assemblages: effects of pond age and watershed position. Ecology 79(3):928–942
SPSS (1996) SYSTAT 6.0 for windows: statistics. SYSTAT, Chicago
Tonn WM (1990) Climate change and fish communities: a conceptual framework. Trans Am Fish Soc 119(2):337–352
Tonn WM, Magnuson JJ (1982) Patterns in the species composition and richness of fish assemblages in northern Wisconsin lakes. Ecology 63(4):1149–1166
Tonn WM, Magnuson JJ, Rask M, Toivonen J (1990) Intercontinental comparison of small-lake fish assemblages: the balance between local and regional processes. Am Nat 136(3):345–375
Wiens JA (1997) Metapopulation dynamics and landscape ecology. In: Hanski I, Gilpin M (eds) Metapopulation biology: ecology, genetics, and evolution. Academic, San Diego, pp 43–62
Zeug SC, Winemiller KO, Tarim S (2005) Response of Brazos River oxbow fish assemblages to patterns of hydrological connectivity and environmental variability. Trans Am Fish Soc 134(5):1389–1399

Chapter 10
Beavers as Engineers of Wildlife Habitat

Steve K. Windels

Abstract Beavers are ecological engineers but also keystone species, whose presence in the boreal forest is critical for creating and modifying habitat for a myriad of other species. Beaver impoundments are used extensively by moose and white-tailed deer for feeding, cooling, predator avoidance, and relief from biting insects. Gray wolves focus use on beaver impoundments when preying on beavers, moose, and deer. Wolves also use beaver meadows for rendezvous sites and abandoned lodges for den sites. Several other species of semi-aquatic mammals and mesocarnivores have been reported using abandoned lodges and dams for den sites. Beaver impoundments produce a diverse array of environmental features that provide habitat for many different bird guilds or functional groups. Trumpeter swans and other waterfowl are attracted to open water habitats created by beaver impoundments. Great blue herons and osprey nest almost exclusively on dead trees killed by beaver-flooded forests. Woodpeckers and other cavity nesters also benefit from dead trees in flooded areas. Beaver meadows are attractive to many passerine species that prefer shrubby or grassy habitats. Open water in ponds also provide excellent habitat for amphibians and turtles. In combination, the effect of beaver activity on terrestrial vertebrate diversity in Voyageurs National Park is impressive. Inventory, monitoring, and research efforts at Voyageurs National Park from 1973–2016 documented ≥124 species of terrestrial vertebrates using portions of beaver-affected wetlands for at least part of their life history, representing at least 61% of mammals, 30% of birds, 100% of amphibians, and 20% of reptiles species extant in the park. More directed inventories at beaver impoundments would likely increase these estimates. Future assemblages of terrestrial vertebrates in this boreal ecosystem will change as beaver abundance and their resultant influence also changes across the landscape.

S.K. Windels (✉)
National Park Service, Voyageurs National Park,
360 Hwy 11 E, International Falls, MN 56649, USA
e-mail: steve_windels@nps.gov

C.A. Johnston, *Beavers: Boreal Ecosystem Engineers*,
DOI 10.1007/978-3-319-61533-2_10

10.1 Introduction

Other chapters in this book have laid the foundation for why beavers are so transformative in boreal ecosystems. Beavers construct dams across small streams that create wetland complexes and dramatically alter biogeochemical processes (Pollock et al. 1995; Naiman et al. 1986). They cut trees for foraging or construction materials that modify woody plant communities. They selectively feed on aquatic plants that can alter the trajectory of wetland plant succession. They dig canals in ponds that alter shorelines and change the terrestrial–aquatic interface. And they transport coarse woody debris from uplands to wetlands, creating habitat for aquatic invertebrates and other species. Therefore, beavers are not only ecosystem engineers, they are "keystone species" (Naiman et al. 1986). Named for the single stone in the center of an archway that prevents the structure from collapsing, a keystone species' removal likewise can result in a cascade of changes for other species in the system (Paine 1969).

Ecological changes wrought by beaver activity occur across a spectrum of spatial scales (Rosell et al. 2005). For example, beavers remove single trees at fine spatial scales for foraging but their works increase in spatial hierarchy to individual dams or lodges, then individual ponds and pond complexes, and finally to landscapes of 1000s of square kilometers pockmarked with beaver impoundments of varying size, shape, age, and biogeophysical properties (Naiman et al. 1986). It follows then that the creation of habitat for other wildlife species is also scale-dependent. Beavers cutting a small patch of hardwood trees may create or modify habitat for the tiny salamander, for example, but not for wide-ranging mammalian carnivores. Further, the juxtaposition or heterogeneity of components at individual spatial scales will also determine which species are affected and how (Tews et al. 2004). The cumulative effects of beaver activities on other species are reflected in measures of biodiversity, which are also scale-dependent. Generally, the presence of beavers results in an increase in species diversity across a wide range of taxa, especially at larger scales (Rosell et al. 2005). But as mentioned earlier, heterogeneity matters. Indeed, the impact of beaver activity on wildlife diversity might be less in landscapes where freshwater wetlands occur independent of beaver activity, or conversely, where beaver impoundments dominate the landscape and non-beaver-affected wetlands are rare (Wright et al. 2004).

The term "habitat" is often used to generally mean a particular type of vegetation (e.g., forest habitat) or substrate (e.g., rocky habitat), when in fact it has a more specific definition in ecology: an area with the right components (e.g., food, water, cover) and environmental conditions (e.g., temperature, snowfall, absence of competitors) that an individual species needs to survive and reproduce (Morrison et al. 2012). Thus "habitat" is specific to each species. This distinction is important when talking about what kinds of wildlife habitats are created or modified by beaver activity.

Several excellent reviews exist of beaver effects on other species (Pollock et al. 1995; Rosell et al. 2005) and there is no need to present a similar exhaustive review here. Rather, my goals in this chapter were to (1) briefly highlight how beaver activities can modify boreal ecosystems in ways that affect other mammals, birds, and reptiles through changes in their habitats, and (2) present specific examples of these effects from work conducted in Voyageurs National Park. It should be noted some of the information presented here is derived from unpublished data collected by Park staff during the period 1973–2016.

10.2 Mammals

There are seven extant orders of mammals in Minnesota, with only the members of the opossum order (Didelphimorphia) having no interactions with beavers in Voyageurs National Park (Hazard 1982). Mammals that interact with beavers or beaver impoundments include the small-bodied members of the shrew, bat, rabbit, and rodent orders, as well as the generally larger-bodied members of orders containing carnivores and ungulates. Some of the affected mammals are highly mobile, such as bats, and others such as larger carnivores and ungulates maintain large home ranges. In both instances, beaver activity can create heterogeneous landscapes that can meet multiple life history needs of individual species. Through a combination of observations by Park staff from 1973–2016, unpublished Voyageurs National Park data, and specific mammal surveys in beaver impoundments for small mammals and bats (Jannett et al. 2007; Schaberl and Route 2013; West 2015), 35 species of extant mammals have been documented using beaver impoundments as part of their life history, representing 61% of the mammal diversity in Voyageurs National Park (Table 10.1).

10.2.1 Moose and White-Tailed Deer

Moose (*Alces alces*) are common throughout the boreal forest, whereas white-tailed deer (*Odocoileus virginianus*) reach their northern range limit in the southern portions of the boreal forest. Moose in Voyageurs National Park are primarily found in the Kabetogama Peninsula, with numbers ranging from 40 to 51 from 2009–2016 (0.13–0.17 moose/km^2; Windels and Olson 2016). Deer are more abundant and evenly distributed throughout the park, with densities between 2–5 deer/km^2 most years (Gogan et al. 1997; Windels and Olson 2016; Gable et al. 2017). Both species regularly use beaver-created wetlands for a variety of life history needs such as foraging, escape from predators or biting insects, and cooling on hot days. Moose are particularly known for their affinity for beaver ponds and other wetlands (they are even called "swamp donkeys" in some parts of their range). Recent research that included GPS-collared moose in the Park demonstrated that moose require heterogeneous landscapes where they can easily move between cover types to access foraging habitat, thermal refugia, or escape cover at hourly or daily time scales

Table 10.1 List of extant mammal species in Voyageurs National Park that have been observed using beaver impoundments as part of their life histories

Order	Common name	Scientific name
Insectivora	Arctic shrew	*Sorex arcticus*
	Masked shrew	*Sorex cinereus*
	Northern water shrew	*Sorex palustris*
	Short-tailed shrew	*Blarina brevicauda*
Chiroptera	Big brown bat	*Eptesicus fuscus*
	Silver-haired bat	*Lasionycteris noctivagans*
	Eastern red bat	*Lasiurus borealis*
	Hoary bat	*Lasiurus cinereus*
	Little brown bat	*Myotis lucifugus*
	Tri-colored bat	*Perimyotis subflavus*
	Northern long-eared bat	*Myotis septentrionalis*
Lagomorpha	Snowshoe hare	*Lepus americanus*
Rodentia	Meadow vole	*Microtus pennsylvanicus*
	Red-backed vole	*Myodes gapperi*
	Deer mouse	*Peromyscus maniculatus*
	Southern bog lemming	*Synaptomys cooperi*
	Least chipmunk	*Tamias minimus*
	Meadow jumping mouse	*Zapus hudsonius*
	Muskrat	*Ondatra zibethicus*
	American beaver	*Castor canadensis*
Carnivora	Coyote*	*Canis latrans*
	Gray wolf	*Canis lupus*
	Gray fox*	*Urocyon cinereoargenteus*
	Red fox	*Vulpes vulpes*
	Canada lynx*	*Lynx canadensis*
	Bobcat*	*Lynx rufus*
	Striped skunk*	*Mephitis mephitis*
	River otter	*Lontra canadensis*
	American marten	*Martes americana*
	Fisher	*Pekania pennanti*
	Mink	*Mustela vison*
	Raccoon	*Procyon lotor*
	Black bear	*Ursus americanus*
Artiodactyla	Moose	*Alces alces*
	White-tailed deer	*Odocoileus virginianus*

This list does not include mammals that use beavers' impoundments incidentally. Several species that are present in VNP that likely use beaver impoundments but which direct confirmation of their use of beaver ponds is lacking are denoted with *. Data were derived from observations from National Park Service staff, unpublished Park data, and published sources (Jannett et al. 2007; Gogan et al. 1997; Windels and Olson 2016)

(Street et al. 2016). Voyageurs National Park presents similar opportunities to study how the distribution and abundance of beaver-created wetlands within moose home ranges affects aspects of moose life history and behavior.

Moose and deer are considered browsers, with teeth and digestive systems well adapted to subsisting on poorly digested woody twigs in winter or more palatable leaves and herbaceous plants in summer. They are less adapted to grazing, and therefore seldom graze on emergent vegetation such as sedges and grasses often found along margins of beaver ponds or throughout beaver meadows (Morris 2014). Moose and deer will consume submergent and floating-leaf aquatic plants found in beaver ponds (Morris 2014). The long legs and snout of a moose make them especially adapted to underwater feeding. They have even been observed diving to a depth of 5.5 m and staying submerged for 50 s during feeding bouts (Peterson 1955). Based on data collected from 20 random locations in each of 18 beavers ponds on the Kabetogama Peninsula in 2012, mean depth in active beaver ponds was 1.13 m (SD = 0.40). The mean maximum depth—which was almost always closest to the middle of the dam—was 1.72 m (SD = 0.55). These results suggest that moose can readily forage throughout all portions of ponds, either by standing or diving.

Not all wetlands are the same when it comes to producing aquatic foods for moose, which can be categorized as submergent plants like pondweeds (Potamogeton spp.) or floating-leaf plants like pond lilies (Nuphar spp.). Research on the Kabetogama Peninsula showed that beaver ponds were 2–4× greater in richness and biomass of moose forage than lacustrine (lake) wetlands (Morris 2014). When considering just beaver ponds, pond age and size also influenced production of moose forage. Young (6–14 years old) and medium-aged ponds (21–38 years old) were greater in richness of aquatic plants than old ponds (>50 years old) (Morris 2014). Larger ponds (>1 ha) also had higher species richness values compared to the smaller ponds (<1 ha) for both the submersed and floating species. Biomass of submergent and floating-leaf plants were nearly twice as abundant in large ponds than small ponds.

Beaver ponds also provide refuge from predators or biting insects (Morris 2014; Franzmann and Schwartz 2007). Moose are excellent swimmers and can generally escape from wolves (*Canis lupus*) when they are in deep enough water. However, a moose swimming in a beaver pond on Isle Royale was attacked by wolves, suggesting they are not immune to predation when in water (Jordan et al. 2010). Biting insects such as mosquitoes (Culicidae), black flies (Simuliidae), horse flies (Tabanidae), and moose flies (Muscidae) are annoying pests for moose and deer (Morris 2014). Moose and deer have been observed using beaver ponds to submerge deep enough to limit insect bites. Moose flies, which cause open sores on the hocks of moose, are persistent enough that they have been reported hovering in swarms over submerged moose waiting for them to expose their rumps to re-alight (P.A. Jordan, personal communication).

Moose, a cold-adapted species, are sensitive to warm temperatures, especially during the ice-free season (Renecker and Hudson 1989). Beaver impoundments affect the thermal landscape for moose by reducing tree and shrub cover that

Fig. 10.1 Cross-section of a black globe thermometer showing temperature data logger inserted into a 15-cm diameter copper toilet bowl float painted matte black. Black globe thermometers measure "operative temperature." Loggers were hung 0.75 m aboveground 15 cm from trees to mimic operative temperature experienced by a bedding moose (Olson et al. 2014)

normally would absorb or reflect significant amounts of solar radiation (Olson et al. 2014). Colleagues and I quantified the thermal landscape for moose on the Kabetogama Peninsula using a system of "black globe thermometers" (Fig. 10.1) which combine ambient temperature, relative humidity, and solar radiation to measure operative temperature, an approximation of the temperature that a moose would actually feel on its skin (Olson et al. 2014; Olson 2013). Not surprisingly, open habitats such as beaver ponds and meadows were as much as 8 °C hotter than closed forests during mid-day in summer (Fig. 10.2). Conversely, these open habitats also cooled off more than closed forests at night, creating potential thermal refuges that moose exploited during warm periods in summer (McCann et al. 2016). When projected at the landscape scale, it becomes apparent that beaver activity greatly alters the thermal landscape for moose (or other wildlife) through variation in the size, distribution, juxtaposition, and temporal availability of thermal patches (Fig. 10.3). The net effect of this heterogeneity on the ability of moose to respond to extreme temperatures is not yet fully understood.

Beaver activities also affect thermal dynamics for moose by providing deep pools where moose can immerse themselves to cool off (Franzmann and Schwartz 2007), or moistening soils in lowland areas which moose can use to thermally regu-

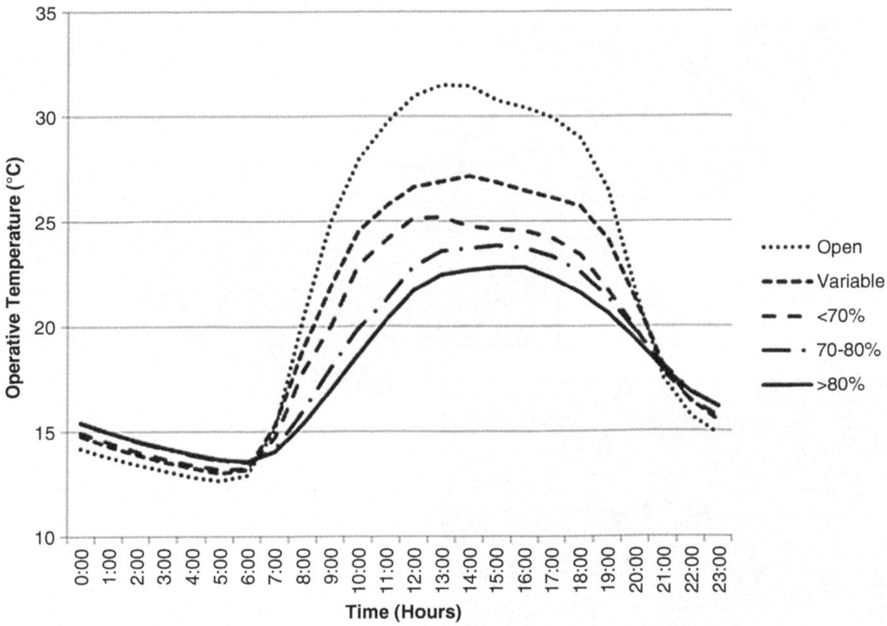

Fig 10.2 Mean operative temperatures across varying amounts of canopy cover in summer over a 24-h period during 1 June–31 August 2010 in Voyageurs National Park, MN, USA. "Open" areas are predominantly beaver ponds and meadows where all tree cover has been removed from flooding or cutting (Olson et al. 2014)

late through bedding (McCann et al. 2016). Colleagues and I measured summer water temperatures in a number of different types of beaver ponds to understand the thermal properties of beaver ponds in relation to moose thermoregulation (Fig. 10.4). Size (<1 ha or >1 ha) or age (young, medium, old—see above) of beaver ponds did not significantly affect temperature regimes, but water temperatures did closely track ambient air temperatures (as measured at a nearby meteorological station in Voyageurs National Park), which explained 64% of daily maximum pond temperatures and 44% of daily minimum pond temperatures (Fig. 10.5). Water has a much higher specific heat than air and therefore changes in water temperature in ponds are much slower and of less amplitude than of the ambient air around them. Maximum daily water temperatures were similar to maximum daily air temperatures on most days and minimum daily water temperatures were warmer than minimum daily air temperatures. In other words, based on temperature alone, there would appear to be little benefit of a moose to use beaver ponds for cooling. However, because of the high specific heat of water, water is much better at absorbing heat from a moose than air at a similar temperature, and therefore even "warm" water temperatures in beaver ponds may provide effective conditions for thermoregulation by moose. Conversely, the unsubmerged portions of a moose in open beaver ponds can be exposed to direct sunlight, which can result in high thermal inputs. Thus, there are

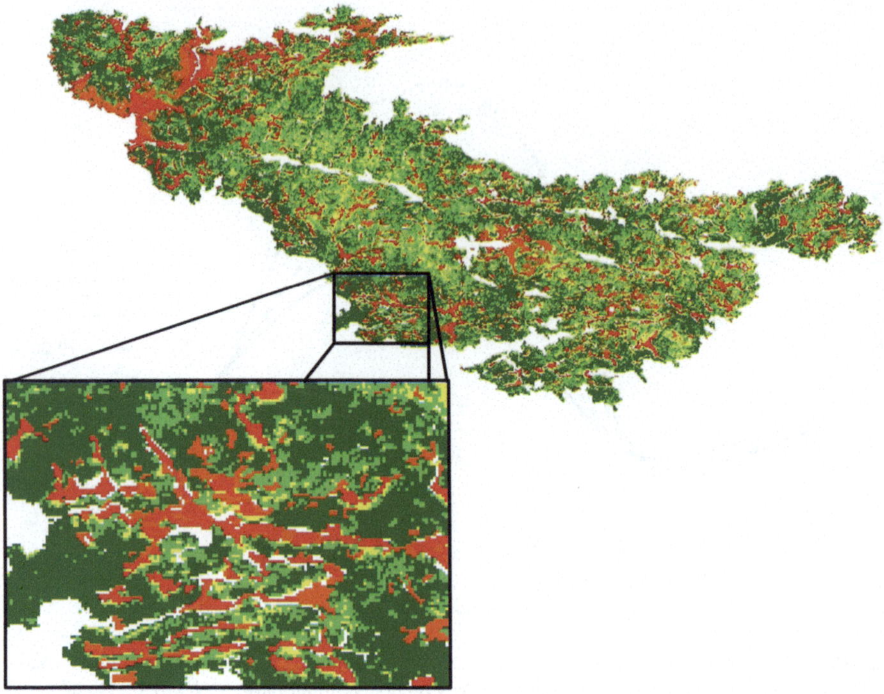

Fig. 10.3 Modeled operative temperature of the Kabetogama Peninsula, Voyageurs National Park, Minnesota, during the hottest 3 h of the day (12-3p) in July. *Red pixels* (30-m) are warmest areas, usually beaver ponds or meadows, and *dark green pixels* are coolest areas, generally closed canopy conifer forests. *Inset* shows a close-up of a particularly thermally heterogeneous part of the park. The juxtaposition of thermal habitats is created by beaver activity and is especially important for thermally sensitive species like moose that need to balance thermal regulation with foraging and other needs (Elmore et al. 2017)

numerous interactions that can take place in the aquatic interface that make understanding the thermoregulatory dynamics of moose and other large-bodied ungulates in beaver ponds challenging. More research is needed to better understand what conditions (time of day, season, difference in air and water temperature, behavior, etc.) moose would be expected to use water bodies such as beaver ponds as thermal habitat.

There is some evidence that moose prefer to give birth and rear calves in areas next to water bodies (McGraw et al. 2011, 2014), either to provide escape cover from predators or to provide easy access to drinking water during lactation. Preliminary data analysis of 14 GPS-collared cow moose in Voyageurs National Park from 2010–2015 does not suggest that moose are selecting calving areas in proximity to beaver ponds (S. Windels, unpublished data). However, given the number of ponds on the landscape, most calving areas were within several hundred meters of a pond, and therefore may still provide cow moose with young calves access to this habitat feature when needed.

Fig. 10.4 Temperature data logger array deployed in 18 beaver ponds and 6 lake bays in Voyageurs National Park in 2011. Temperature loggers were enclosed in a beaver-resistant case and suspended 0.5 m below the water surface from a floating buoy attached to an anchor (*inset*). Arrays were deployed in the deepest part of the beaver pond, typically next to the center of the dam. See Morris (2014) for more details about pond and lake sites sampled (Photo by S. Windels)

Wetlands such as beaver ponds can play an important role in the life cycle of some important diseases and parasites of wildlife. Giant liver flukes (*Fascioloides magna*) are parasites that primarily infect members of the deer family (Lankester and Samuel 2007). The definitive hosts of liver flukes are white-tailed deer but they also require intermediate gastropod hosts to complete their life cycle; host snails are almost exclusively aquatic snails that are commonly found in beaver ponds and other wetlands (Fig. 10.6). Deer are seldom affected by fluke infections but they pass fluke eggs through their feces, which can infect snails. Infected snails are incidentally ingested by moose foraging on aquatic plants and cause infections that can result in significant mortality and morbidity (Lankester and Samuel 2007). The interplay between deer, aquatic snails, and wetlands (e.g., beaver ponds) are therefore important to the transmission of liver flukes from deer to moose. Research conducted in Voyageurs National Park demonstrated that rates of liver fluke infection in deer were positively correlated with the amount of pond habitat in a deer's home range, specifically ponds dominated by floating-leaf aquatic plants typical of beaver impoundments (Vanderwaal et al. 2015). Further, wetlands with standing water (e.g., beaver ponds) had a greater effect on infection rate than other types of wetlands. More research is needed to better understand how the type and distribution of beaver-influenced wetlands can affect transmission of diseases between wildlife species using these wetlands.

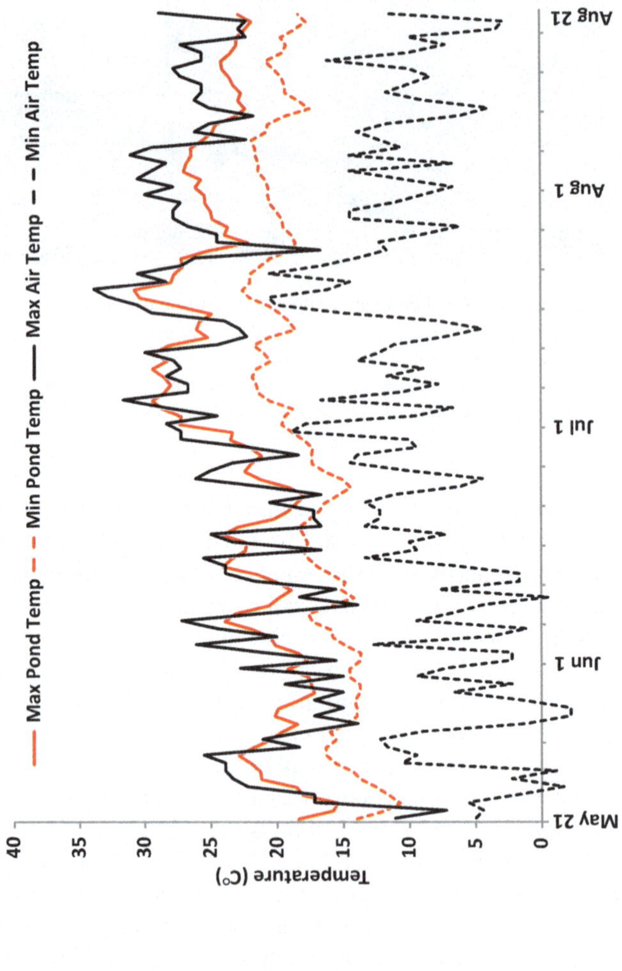

Fig. 10.5 Daily maximum (*solid black*) and minimum (*dashed black*) ambient air temperatures in International Falls, MN and mean daily maximum (*solid red*) and minimum (*dashed red*) water temperatures for 18 beaver ponds at Voyageurs National Park during May–August 2011

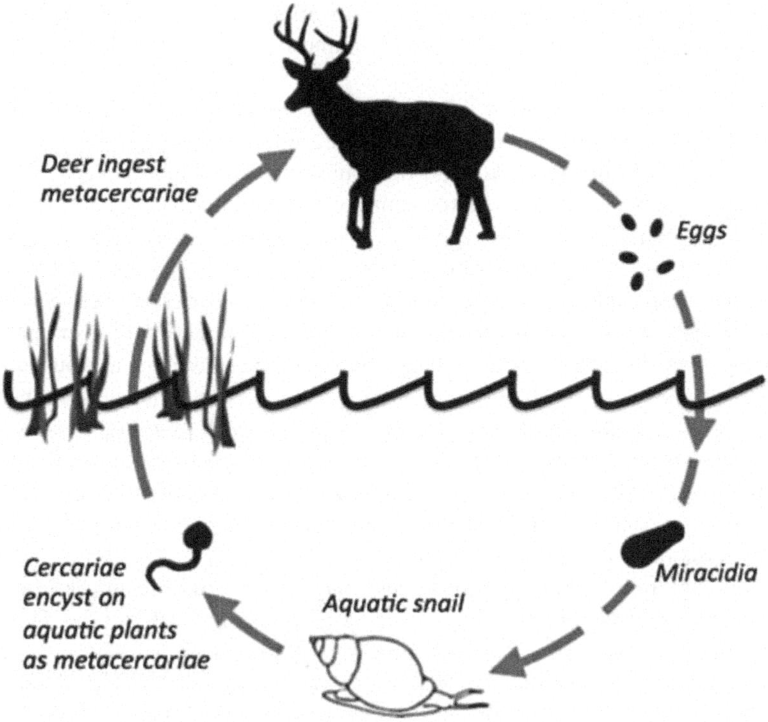

Fig. 10.6 Life cycle of the giant liver fluke (*Fascioloides magna*), showing the relationship between white-tailed deer and aquatic snails in wetlands such as beaver ponds. Moose foraging on aquatic vegetation in beaver ponds accidentally ingest infected snails or metacercariae of flukes, and become infected themselves. White-tailed deer rarely suffer any adverse effects of infection but fluke infections often cause morbidity or mortality in moose. From VanderWaal KL, Windels SK, Olson BT, Vannatta JT, Moen R (2015) Landscape influence on spatial patterns of meningeal worm and liver fluke infection in white-tailed deer. Parasitology 142(5):706–718, with permission from Cambridge University Press

10.2.2 Gray Wolves

Gray wolves are the top mammalian carnivore in Voyageurs National Park. The extremely dense beaver population in Voyageurs National Park, at least relative to many other places in the United States, presents a unique opportunity to better understand the dynamic between wolves and beavers. Similar to other places in wolf range, the annual diet of wolves in the Park consists primarily of deer and moose. But given the abundance of beavers available on the landscape, beavers comprise a substantial portion of wolf diets in the ice-free season (Gogan et al. 2004; Gable et al. 2017). Not surprisingly, wolves use beaver impoundments to hunt for beaver prey. One common hunting behavior entails lying below active dams to ambush beavers moving between ponds or conducting dam maintenance (Gable et al. 2016). Wolves also will kill beavers that are moving along the narrow canals beavers dig around the perimeter of ponds to improve water access to forage, or attack them while they move up to 90 m inland to cut trees for food or building materials.

Other prey species of wolves that utilize beaver ponds may be at increased risk of predation during times when wolves are focused on foraging in beaver habitats. Adult caribou (*Rangifer tarandus*) mortality in summer in Alberta was a function of wolves selecting beaver habitats in the summer presumably to hunt beavers (Latham et al. 2013). Spatial overlap between wolves and caribou increased as a result, and consequently incidental predation of adult caribou by wolves increased. Observations at Voyageurs National Park suggest similar dynamics are occurring between wolves, beavers, deer, and moose in the park, where incidental predation of deer and moose increases when they are in active beaver habitats where wolves are hunting. One hypothesis to be tested is whether predation risk for deer and moose is a function of distance of the ungulate prey to active beaver ponds, with predation risk declining as ungulates are further away from pond sites. Current investigations of this topic in the Park by Park staff and university collaborators have the potential to reveal exciting new insights into the role the beavers and beaver-created habitats play in predator–prey dynamics of a range of species.

Wolves maintain large home ranges and they spend a significant portion of their time (and energy) traveling throughout their territories in search of food. During the winter months, beavers are mostly immune to wolf predation as long as they remain safely hidden inside their lodges. Thus most wolf hunting activity in winter is not directed at beavers, but rather at vulnerable ungulate prey. Data from GPS-collars deployed on wolves in Voyageurs National Park suggests that wolves frequently use beaver ponds as travel corridors, especially in winter (Voyageurs National Park, unpublished data). This behavior is similar to wolf use of frozen lake surfaces, where wind and sunlight can reduce the amount of snow cover relative to forested habitats and result in more energy-efficient travel. Beaver ponds are also typically along stream courses and thus are simply expanded versions of linear features that wolves seem to prefer for travel (Mech 1970; Dickie et al. 2017). Scent-marking in wolves is an effective means of communication among wolves, particularly in delineating territory boundaries (Mech 1970). "Scent posts" are typically rocks, stumps, dirt mounds, or other objects that present a conspicuous visual cue to other wolves. Tracks and urine in the snow are frequently observed on beaver lodges in the Park in winter, and most scent marking is focused toward the top of the "cone" of the lodge (Fig. 10.7). Beavers do occasionally leave the safety of their lodges in winter in search of food. Therefore, it is also possible that wolves visit individual lodges with the dual intent of using lodges as scent posts and searching for signs of vulnerable beaver prey.

Beaver impoundments also play an important role in the reproductive ecology of wolves. Wolves select den sites in February or March, or 4–6 weeks before pups are born. Many dens are in soil embankments or the base of rock cliffs but wolves will also modify existing dens of other mammals, including beaver lodges (Mech 1970). In 2015, Voyageurs National Park staff documented wolves using an abandoned lodge in a beaver meadow near Cranberry Bay, Rainy Lake for a den site (Fig. 10.8;

Fig 10.7 Wolf tracks are visible around an active beaver lodge in Voyageurs National Park. Wolves frequently use beaver lodges and other conspicuous objects to scent-mark their territories. Beavers are secure within the confines of their lodges in winter, but become highly vulnerable to predation when they are forced to leave in search of forage if their food cache is depleted (Photo by S. Windels)

T. Gable, personal observation). About 8–10 weeks after pups are born, wolves will move them from the den site to temporary areas called "rendezvous sites," where the pups remain alone while the adult pack members go hunting (Mech 1970). In the boreal forest region, rendezvous sites are typically in bogs or grassy areas in wetland complexes. In Voyageurs National Park, wolves show a high affinity for using inactive beaver impoundments for rendezvous sites (Voyageurs National Park, unpublished data). Abandoned ponds or adjacent beaver meadows are often lush with grasses and sedges that wolves may prefer for hiding pups (Fig. 10.9).

10.2.3 Other Mammals

Though perhaps less charismatic than moose, deer, or wolves, other mammals also use the array of habitat features available within beaver pond complexes (Rosell et al. 2005). Semi-aquatic mammals are those that spend a large portion of their lives in aquatic habitats such as those created by beaver impoundments, which

Fig. 10.8 Voyageurs National Park staff observed a pack of wolves in the Cranberry Bay area of the Kabetogama Peninsula using an abandoned beaver lodge as a den site in spring 2015. The den entrance can be seen at the front of the lodge. [*Inset*] There were four pups with the pack at the time of the observation (Photo by T. Gable)

create stable water levels, abundant food resources, and suitable den sites that many semi-aquatic mammals, including beavers, require. Muskrats (*Ondatra zibethicus*) are similar to beavers in their water requirements and will readily colonize beaver ponds near source wetlands, especially older ponds with a more developed emergent plant zone. River otters (*Lontra canadensis*) and mink (*Neovison vison*) are both common mammalian carnivores on the Kabetogama Peninsula that require water for most of their life history needs and readily use beaver ponds (Tumlison et al. 1982). A field study in the 1980s documented that otters readily moved between the large lakes and inland beaver ponds on the Kabetogama Peninsula during summer, but only a small proportion of otters chose to overwinter in beaver ponds (Route and Peterson 1988). The omnivorous raccoon (*Procyon lotor*), a relative newcomer to the Kabetogama Peninsula sometime in the mid-twentieth century, is also a regular visitor to beaver ponds and other wetlands where it hunts amphibians, reptiles, and aquatic invertebrates. Other authors have suggested that the expansion of the wetland-loving raccoon in parts of the United States has coincided with the recovery and expansion of beavers (Bailey and Stephens 1951).

Many species of smaller-bodied mammals also readily utilize the diversity of habitats provided by beaver impoundments. For bats, beaver complexes are attractive because (1) they create large openings near water sources where bats prefer to forage for flying insects, (2) small wetlands produce tremendous numbers and diversity of flying insect prey, and (3) exfoliating bark on flood-killed trees may

Fig. 10.9 Wolves in Voyageurs National Park show a high affinity for using abandoned beaver ponds or beaver meadows for rendezvous sites—areas where wolves can hide their pups in thick grass while they are away hunting. Wolves frequently relocate their pups over the course of the summer. Even if only used for a few weeks at a time, rendezvous sites often accumulate bones of deer, moose, and beaver prey that are brought back to the pack for food (Photo by T. Gable)

provide excellent roosting habitat for some species (Rosell et al. 2005). Seven species of bats [big brown bat (*Eptesicus fuscus*), little brown bat (*Myotis lucifugus*), tri-colored bat (*Perimyotis subflavus*), northern long-eared bat (*Myotis septentrionalis*), hoary bat (*Lasiurus cinereus*), eastern red bat (*Lasiurus borealis*), and silver-haired bat (*Lasionycteris noctivagans*)] inhabit the Kabetogama Peninsula (Schaberl and Route 2013; West 2015).

Many species of mice and shrews prefer the grassy or shrubby habitats which are abundant in beaver complexes. An inventory of small mammals was conducted in Voyageurs National Park from 1996–2001, including sampling of a single beaver meadow in 1997–1998 (Voyageurs National Park, unpublished data). Ten species of small mammals were trapped over the 2 years in the beaver meadow, but six of the species represented 96% of total captures ($N = 222$ total individuals). Meadow voles (*Microtus pennsylvanicus*) composed 34% of captures, followed by 18% arctic shrews (*Sorex arcticus*), 14% short-tailed shrews (*Blarina brevicauda*), 11% masked shrews (*Sorex cinereus*), 10% red-backed voles (*Myodes gapperi*), and 9% deer mice (*Peromyscus maniculatus*) (Table 10.1). Less than three individuals were captured of southern bog lemming (*Synaptomys cooperi*), water shrew (*Sorex palustris*), meadow jumping mouse (*Zapus hudsonicus*), and least chipmunk (*Tamias minimus*).

10.3 Birds

Voyageurs National Park is situated along the transition between the northern hardwood and boreal forest zones that stretches from Maine to Minnesota. Bird diversity in this zone is higher than anywhere north of Mexico, and the park itself harbors some of the highest bird diversity in the United States with >100 species of breeding birds and >100 more non-breeding visitors documented (Voyageurs Region National Park Association 2001). At a finer scale, the high landscape heterogeneity within Voyageurs National Park, much of which can be attributed to beaver activity, contributes to the high bird diversity documented here. The Voyageurs National Park landscape, and the beaver ponds and meadows within, provides a diversity of structural features and vegetation communities for many bird species. This includes wetland obligates like trumpeter swans (*Cygnus buccinators*) or great blue herons (*Ardea herodias*), or more generalist species like chestnut-sided warblers (*Dendroica pensylvanica*) that like the "edges" of beaver-modified habitats. In addition to the littoral areas of the ponds themselves (Littoral Zone), most active beaver complexes also include a periphery of water-tolerant shrubs (Shrub Zone), and patches of cattails and other emergents (Emergent Zone; Fig. 10.10). Although abandoned ponds are typically a mosaic of these three zones, grass and sedge-dominated "beaver meadows" are also a common feature. Compiled from long-term forest breeding bird monitoring plots (Egan et al. 2015) and observations of Park staff from 1973–2016, 77 species of birds have been observed in beaver ponds and meadows (Table 10.2), representing 30% of the total number of breeding and visiting birds documented in the Park. The Shrub Zone was the most common vegetation type used (36 species total), dominated by species of jays, flycatchers, thrushes, wood warblers, and sparrows. The Emergent Zone was used primarily by waterbirds (e.g., bitterns and yellowlegs), raptors, flycatchers, and blackbirds (21 species total). Waterfowl (14 species) and the belted kingfisher (*Ceryle alcyon*) were the most common birds in the Littoral Zone. Dead Trees were primarily used by cavity nesters such as woodpeckers or nuthatches, and also by great blue herons and ospreys (*Pandion haliaetus*) that build nests on top of dead trees (12 species total). Additional information about individual taxonomic groups of birds that have been the subject of long-term monitoring or focused research in Voyageurs National Park are presented below.

10.3.1 Passerines and Woodpeckers

Beaver works produce abundant invertebrates on which many species of passerines feed. Beaver works provide a wide array of habitat features important for these species, including foraging and perching sites, nesting substrates, and escape cover (Fig. 10.11). Standing live, dead, or dying timber in flooded forests create excellent nesting and foraging habitat for woodpeckers and other cavity nesters (see Great

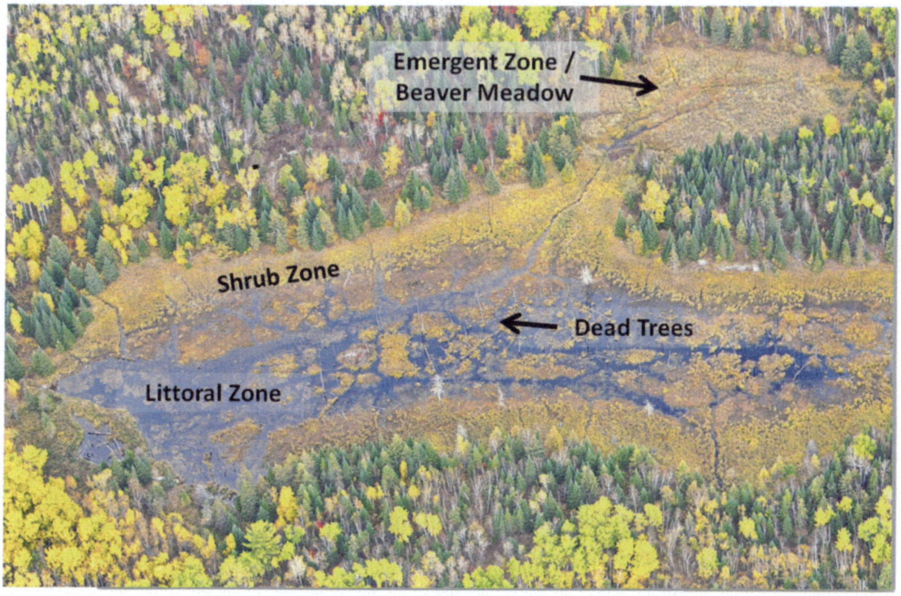

Fig. 10.10 Beaver impoundments provide a wide variety of vegetation types or other habitat features that can attract a diverse assemblage of birds. In addition to the littoral areas of the ponds themselves (Littoral Zone), most active beaver complexes also include a periphery of water-tolerant shrubs (Shrub Zone), and patches of cattails and other emergents (Emergent Zone). Areas that were previously dewatered often revert to the classic beaver meadow form dominated by grasses and sedges. See Table 10.2 for a list of birds documented using each of these habitat zones or features in beaver impoundments in Voyageurs National Park

Blue Heron section below for more details) (Rosell et al. 2005). Voyageurs National Park staff have documented 49 species of passerines and woodpeckers using various portions of the beaver pond complex (Table 10.2). A nonsignificant increase of 7% in abundance of wetland birds was observed on survey routes from 1995–2015 in Voyageurs (Egan et al. 2015). However, most survey routes only contained a few beaver-influenced wetlands within the detection radius so the direct tie of beaver activity to wetland bird abundance in the Park is unknown at this point.

10.3.2 *Swans, Geese, and Ducks*

Beaver activity creates an abundance of shallow, open water habitat preferred by nesting and migrating waterfowl (McCall et al. 1996). Trumpeter swans, once near the brink of extinction, have rebounded in northern Minnesota over the last 30 years, perhaps in part due to the presence of beaver-created wetlands. Swans feed primarily on submergent vegetation, a common staple of beaver ponds, and substrates and structural materials for nest building, such as old beaver lodges and emergent

Table 10.2 List of extant bird species confirmed to use beaver impoundments as part of their life histories

Family	Common name	Scientific name	Shrub zone	Emergent zone	Littoral zone	Dead trees
Anatidae	Canada goose	*Branta canadensis*			●	
	Trumpeter swan	*Cygnus buccinator*			●	
	Wood duck	*Aix sponsa*			●	
	American black duck	*Anas rubripes*			●	
	Mallard	*Anas platyrhynchos*			●	
	Blue-winged teal	*Anas discors*			●	
	Green-winged teal	*Anas crecca*			●	
	Ring-necked duck	*Aythya collaris*			●	
	Bufflehead	*Bucephala albeola*			●	
	Common goldeneye	*Bucephala clangula*			●	
	Hooded merganser	*Lophodytes cucullatus*			●	
Phasianidae	Ruffed grouse	*Bonasa umbellus*	●			
Podicipedidae	Pied-billed grebe	*Podilymbus podiceps*			●	
Phalacrocoracidae	Double-crested cormorant	*Phalacrocorax auritus*			●	
Pelecanidae	American white pelican	*Pelecanus erythrorhynchos*			●	
Ardeidae	American bittern	*Botaurus lentiginosus*		●		
	Great blue heron	*Ardea herodias*		●		●
Pandionidae	Osprey	*Pandion haliaetus*				●
Accipitridae	Northern harrier	*Circus cyaneus*		●		
	Broad-winged hawk	*Buteo platypterus*		●		
Falconidae	American kestrel	*Falco sparverius*		●		
	Merlin	*Falco columbarius*		●		
Rallidae	Sora	*Porzana carolina*		●		
Gruidae	Sandhill crane	*Grus canadensis*		●		

(continued)

Table 10.2 (continued)

Family	Common name	Scientific name	Shrub zone	Emergent zone	Littoral zone	Dead trees
Scolopacidae	Spotted sandpiper	*Actitis macularia*		●		
	Greater yellowlegs	*Tringa melanoleuca*		●		
	Wilson's snipe	*Gallinago delicata*		●		
Caprimulgidae	Common nighthawk	*Chordeiles minor*		●		
Trochilidae	Ruby-throated hummingbird	*Archilochus colubris*	●	●		
Alcedinidae	Belted kingfisher	*Ceryle alcyon*			●	
Picidae	Yellow-bellied sapsucker	*Sphyrapicus varius*				●
	Downy woodpecker	*Picoides pubescens*				●
	Hairy woodpecker	*Picoides villosus*				●
	Northern flicker	*Colaptes auratus*				●
	Pileated woodpecker	*Dryocopus pileatus*				●
Tyrannidae	Olive-sided flycatcher	*Contopus cooperi*	●			
	Yellow-bellied flycatcher	*Empidonax flaviventris*	●			
	Alder flycatcher	*Empidonax alnorum*	●	●		
	Eastern phoebe	*Sayornis phoebe*	●	●		
	Eastern kingbird	*Tyrannus tyrannus*	●	●		
Corvidae	Gray jay	*Perisoreus canadensis*	●			
	Blue jay	*Cyanocitta cristata*	●			
	American crow	*Corvus brachyrhynchos*	●			
	Common raven	*Corvus corax*	●			
Hirundinidae	Tree swallow	*Tachycineta bicolor*				●
Paridae	Black-capped chickadee	*Poecile atricapilla*	●			
Sittidae	Red-breasted nuthatch	*Sitta canadensis*				●

(continued)

Table 10.2 (continued)

Family	Common name	Scientific name	Shrub zone	Emergent zone	Littoral zone	Dead trees
Troglodytidae	Sedge wren	*Cistothorus platensis*		●		
Regulidae	Ruby-crowned kinglet	*Regulus calendula*	●			
Turdidae	Veery	*Catharus fuscescens*	●			
	Swainson's thrush	*Catharus ustulatus*	●			
	American robin	*Turdus migratorius*	●			●
Mimidae	Gray catbird	*Dumetella carolinensis*	●			
Parulidae	Ovenbird	*Seiurus aurocapillus*	●			
	Northern waterthrush	*Seiurus noveboracensis*	●	●		
	Black-and-white warbler	*Mniotilta varia*	●			
	Nashville warbler	*Vermivora ruficapilla*	●			
	Common yellowthroat	*Geothlypis trichas*	●			
	American redstart	*Setophaga ruticilla*	●			
	Northern parula	*Parula americana*	●			
	Magnolia warbler	*Dendroica magnolia*	●			
	Blackburnian warbler	*Dendroica fusca*	●			
	Yellow warbler	*Dendroica petechia*	●			
	Chestnut-sided warbler	*Dendroica pensylvanica*	●			
	Yellow-rumped warbler	*Dendroica coronata*	●			
	Black-throated green warbler	*Dendroica virens*	●			
	Canada warbler	*Wilsonia canadensis*	●			

(continued)

Table 10.2 (continued)

Family	Common name	Scientific name	Shrub zone	Emergent zone	Littoral zone	Dead trees
Emberizidae	American tree sparrow	*Spizella arborea*	Migration only			
	Chipping sparrow	*Spizella passerina*	●			
	Clay-colored sparrow	*Spizella pallida*		●		
	Song sparrow	*Melospiza melodia*	●			
	Swamp sparrow	*Melospiza georgiana*	●			
	White-throated sparrow	*Zonotrichia albicollis*	●			
	White-crowned sparrow	*Zonotrichia leucophrys*	Migration only			
	Dark-eyed junco	*Junco hyemalis*	●			
Cardinalidae	Scarlet tanager	*Piranga olivacea*	●			
Icteridae	Red-winged blackbird	*Agelaius phoeniceus*		●		●
	Common grackle	*Quiscalus quiscula*		●		●
	Brown-headed cowbird	*Molothrus ater*	●			
		Total species by zone	36	21	15	12

Specific vegetation zones/habitat features of beaver impoundments used by each species are also denoted—see Fig. 10.10 for details. Data were derived from observations from National Park Service staff and unpublished Park data

vegetation, are abundant. Though there has never been a systematic trumpeter swan monitoring program at Voyageurs National Park, many observations have been recorded since 1974 by Park staff during the course of aerial monitoring for other species of interest such as bald eagles, osprey, and great blue herons. Few, if any, swans were observed in the Park before the Minnesota Department of Natural Resources initiated a statewide reintroduction program in 1986. The first pair of successfully breeding trumpeter swans in the Park were documented in 2002 in a beaver pond on the west end of the Kabetogama Peninsula, where they produced five cygnets (Fig. 10.12). From 2002–2011, at least three more successful breeding attempts have been documented in beaver ponds in the Park, representing all known breeding attempts in the park during that time. Other waterfowl species are also commonly observed in beaver ponds in Voyageurs National Park. Canada geese (*Branta canadensis*) frequently use beaver ponds for breeding areas and beaver

Fig. 10.11 Beaver ponds and associated vegetation communities provide habitat for a variety of bird species, including the eastern kingbird (*Tyrannus tyrannus*) which feeds on flying insects common in wetland areas (Photo by T. Gable)

lodges as nest platforms. Eight other species of ducks and mergansers have been observed in beaver complexes in the Park (Table 10.2).

10.3.3 Great Blue Herons and Other Waterbirds

Great blue herons are regular summer residents in Minnesota. They are colonial nesters, and their rookeries (or heronries) are almost exclusively built in mature live or dead trees with sturdy, exposed limbs on which nests can be constructed. Great blue herons are dependent not only upon mature trees for rookeries, but also upon the juxtaposition of those rookeries with quality aquatic foraging areas, where they feed on small fish, amphibians, reptiles, small mammals, and aquatic invertebrates (Eckert 1981). Potential heron rookeries are created, then, whenever beavers flood areas with standing timber. Lowland forests dominated by black ash (*Fraxinus nigra*) are the most common forest type flooded by beaver activity in Voyageurs National Park and these are readily used by nesting herons (Fig. 10.13). Starting in 1973, Voyageurs National Park biologists began documenting great blue heron rookeries by aircraft. The locations of rookeries were recorded, along with the total number of active (i.e., with eggs or incubating adults) and inactive nests observed. All 31 of the heron rookeries documented and monitored in the Park were in beaver

Fig. 10.12 Trumpeter swans (*Cygnus buccinators*) frequently use abandoned beaver lodges as nest platforms. Surrounded by water, these platforms provide protection from mammalian predators and clear visibility of approaching avian predators. All four documented cases of breeding trumpeter swans in the Park from 1973–2016 occurred in beaver ponds (Photo by Voyageurs National Park staff)

impoundments, and 23 of these had at least 1 active nest during the period 1973–2013.

How long do dead trees persist in beaver flooded forests? Using a combination of ground-based observations of heron rookeries and time series of aerial photos, I was able to estimate maximum duration that dead trees remain standing in six different ponds in Voyageurs National Park. All of these ponds had numerous (i.e., >30) dead trees still standing after 25 years. In some older ponds, as many as three individual dead trees could still be seen standing as much as 43–66 years after pond creation.

Great blue herons often re-nest in the same area year after year, and old nests may be enlarged and reused (Eckert 1981). Of the 23 active rookeries documented in beaver ponds, mean number of active nests colony^{-1} year^{-1} was 15.1 and the mean rookery lifespan (conservatively estimated as the number of years from the first detection with active nests to the last year detected with active nests) was 9.6 years. By contrast, three active rookeries monitored outside of Voyageurs National Park on pine-covered islands of Namakan and Rainy Lakes had mean annual nest counts of 54.6 and a mean lifespan of 6.8 years.

Fig. 10.13 Newly created beaver dams often result in high tree mortality, such as in this lowland black ash forest in Voyageurs National Park. Standing dead trees provide excellent nesting structures for the colonial nesting great blue heron (*Ardia herodias*), which are clearly visible atop many of the trees in this photo. All 31 great blue heron rookeries that have been documented in the Park from 1973–2016 were located in beaver-created impoundments (Photo by L. Grim)

Other waterbirds will readily use wetland environments created by beavers. Those documented in Voyageurs National Park include American bitterns (*Botaurus lentiginosus*) and sandhill cranes (*Grus canadensis*) (Table 10.2).

10.3.4 Osprey and Other Birds of Prey

Starting in 1973, Voyageurs National Park biologists documented all known occurrences of osprey nests in the Park and monitored for occupancy and productivity annually until 2012. Most osprey nests recorded during this period (142 of 172; 83%) were found in beaver impoundments. The majority of nests were in beaver ponds above open water, but some nests also occurred in flooded swamps or in beaver meadows. Nearly all nest trees were dead (98%), and for those nests where the nest tree was identifiable, 42% of nests were in spruce (*Picea* spp.), 29% in ash (*Fraxinus* spp.), and 18% in white pine (*Pinus strobus*) (Fig. 10.14). By comparison, osprey that nested in non-beaver habitats (e.g., lake shorelines) chose dead white

Fig. 10.14 An osprey (*Pandion haliaetus*) nest sits atop a dead white pine next to an active beaver pond in Voyageurs National Park. More than 80% of all osprey nests documented in the Park from 1973–2012 were found within beaver impoundments (Photo by L. Grim)

pines as their nest tree 61% of the time, followed by 13% dead spruce and 4% dead ash. Productivity (fledged chicks year^{-1}) was similar between nests in beaver ponds (mean [±SE]) = 0.95 ± 0.06) and non-beaver habitats (0.84 ± 0.13), but nest trees in beaver sites tended to be extant for longer periods (i.e., conservatively estimated as the time between nest discovery and nest destruction; mean [±SE] = 5.2 ± 0.4 years) than trees in non-beaver sites (3.8 ± 0.7 years).

Other raptors also use beaver ponds and meadows for breeding, foraging, or other life history needs. Broad-winged hawks (*Buteo platypterus*), northern harriers (*Circus cyaneus*), merlins (*Falco columbarius*), and American kestrels (*Falco sparverius*) are commonly observed foraging in the open areas above beaver ponds and meadows (Table 10.2). Nine species of owls, including at least six breeding species, have been documented in Voyageurs National Park, though we have no specific records of them in beaver impoundments. Similar to woodpeckers, several of these species prefer nesting in cavities in dead trees.

Table 10.3 List of extant amphibian and reptile species in Voyageurs National Park confirmed to use beaver impoundments as part of their life history

Class	Family	Common name	Scientific name
Amphibia	Bufonidae	American toad	*Bufo americanus*
	Hylidae	Gray treefrog	*Hyla versicolor*
		Spring peeper	*Pseudacris crucifer*
		Boreal chorus frog	*Pseudacris maculata*
	Ranidae	Green frog	*Rana clamitans*
		Northern leopard frog	*Rana pipiens*
		Mink frog	*Rana septentrionalis*
		Wood frog	*Rana sylvatica*
	Ambystomatidae	Blue-spotted salamander	*Ambystoma laterale*
	Salamandridae	Eastern newt	*Notophthalmus viridescens*
Reptilia	Chelydridae	Snapping turtle	*Chelydra serpentina*
	Emydidae	Western painted turtle	*Chrysemys picta*

Data were derived from observations from National Park Service staff, unpublished Park data, and published sources (Sadinski and Roth 2009a, b)

10.4 Amphibians and Reptiles

Amphibians and turtles in the boreal region require water for reproduction and/or other components of their life history (Moriarty and Hall 2014). Beaver ponds and seasonally flooded wetlands are necessary for these species, and thus reinforces the notion of beavers as a keystone species. Intensive monitoring was conducted from 2002–2005 to assess the population status of amphibians and reptiles in Voyageurs National Park, with extant beaver ponds ($N = 15$) as one of the wetland types surveyed (Sadinski and Roth 2009a, b). All ten species of amphibians present in the park were found in beaver ponds: boreal chorus frog (*Pseudacris maculata*), American toad (*Bufo americanus*), gray treefrog (*Hyla versicolor*), northern leopard frog (*Rana pipiens*), spring peeper (*Pseudacris crucifer*), wood frog (*Rana sylvatica*), mink frog (*Rana septentrionalis*), central newt (*Notophthalmus viridescens*), and blue-spotted salamander (*Ambystoma laterale*) (Table 10.3). The authors stated that non-lacustrine wetlands, including beaver ponds, provided the majority of amphibian habitat in the Park. Amphibian physiology and development is closely linked to water temperatures, and factors impacting the thermal characteristics of beaver ponds (e.g., pond age, amount of tree cover) may also impact amphibians in ponds (see Moose and Deer section) (Rosell et al. 2005).

Common snapping turtles (*Chelydra serpentina*) and western painted turtles (*Chrysemys picta*), the two species of turtles in Voyageurs, were also present in sampled beaver ponds (Sadinski and Roth 2009a, b) (Table 1). Snapping turtles have been observed using beaver dams as basking areas (S. Windels, personal observation). Likewise, painted turtles are often visible basking on beaver-felled trees and logs in beaver ponds (S. Windels, personal observation).

10.5 Beaver Lodges and Dams as Wildlife Habitat

Beavers also create habitat for other species when dams, lodges, or bank dens are abandoned. In addition to wolves (see Sect. 10.2.2 above), female black bears (*Ursus americanus*) have been documented using abandoned beaver lodges and dams for winter dens in Voyageurs National Park (D. Garshelis, personal communication). Muskrats have been observed inhabiting abandoned lodges in Voyageurs National Park but co-occupation of an active beaver lodge has also been observed (S. Windels, unpublished data). A fisher (*Martes pennanti*) was observed using a dewatered beaver food cache (i.e., when water levels receded in late winter) on Rainy Lake as a rest site (S. Windels, personal observation). It was not clear if the fisher was using the lodge as well, or if the beavers themselves were still occupying the lodge. Many species of birds will use abandoned lodges as nesting or loafing sites (see Bird section above).

The ectoparasitic beaver nest beetle (*Platypsyllus castoris*) is also a frequent denizen of beaver lodges, and presents itself as an interesting invertebrate species to discuss in relation to the concept of beaver lodges as habitat. Blind and wingless, nest beetles are beaver obligates but often drop off in the lodge interior, perhaps during grooming by beavers (Peck 2006). More than anywhere else, lodges may act as dispersal hubs where nest beetles can move between related beavers within a family unit (colony) or between unrelated individuals when lodges are abandoned and later reoccupied by unrelated individuals. Lodges are also the most likely location where nest beetles have been transmitted to river otters, the only other mammal reported with *P. castoris* (Peck 2006).

10.6 Conclusions

- Beavers are not only ecosystem engineers, they are also "keystone species" whose activities have a significant impact on the diversity and abundance of a host of other wildlife species in Voyageurs National Park.
- Beaver activities and the resultant impacts occur at several spatial scales. Consequently, effects on habitat for other wildlife species also occur across multiple spatial scales.
- Inventory, monitoring, and research efforts at Voyageurs National Park from 1973–2016 documented ≥124 species of terrestrial vertebrates (i.e., non-fish species) using beaver impoundments for some part of their life history; more intensive survey efforts aimed specifically at beaver impoundments would likely identify more species. Still, the list represents 61% of extant mammals species documented in the park (35 of 57 mammals), 30% of birds (77 of 254 species), 100% of amphibians (10 of 10 species), and 20% of reptiles (2 of 5 species).
- Moose and white-tailed deer use beaver impoundments for foraging, escape cover for themselves and their young, and relief from biting insects.

- Moose are dark, large-bodied mammals that are sensitive to warm temperatures. Beaver activities on the Kabetogama Peninsula have dramatically shaped the thermal landscape for moose.
- Beavers constitute a large component of gray wolf diets in Voyageurs National Park. Wolves focus predation efforts on dams and feeding trails where beavers are most likely to go when they leave the water. Moose and deer that utilize beaver impoundments for other life history needs may be at higher risk of predation from wolves that are hunting beavers.
- Wolves travel more efficiently along linear features. Beaver activities widen existing stream networks, providing commonly used travel networks for wolves year-round.
- Wolves frequently use beaver meadows as rendezvous sites in summer, i.e., places where pups are "stashed" while older pack members are out hunting. Wolves will also use abandoned beaver lodges as den sites.
- Other semi-aquatic mammals like muskrats, river otters, and mink readily use beaver impoundments, which often provide sufficient water conditions to satisfy the life history needs of these species.
- Many species of rodents and shrews live in beaver ponds and meadows. All of the bat species documented in Voyageurs National Park utilize open areas around beaver ponds for foraging.
- Beaver impoundments create a diversity of habitat features that attract a wide host of bird species that use them for some aspect of their life history, including waterfowl that feed and rest in the littoral zone to woodpeckers that nest and feed on flood-killed trees.
- There are three common and charismatic bird species of Voyageurs National Park that almost exclusively use beaver impoundments for nesting sites. All documented nest sites for swans (4 of 4) were in open beaver ponds. Osprey (142 of 172 nests; 83%) and great blue heron (31 of 31 rookeries; 100%) breeding areas were primarily in treed beaver ponds. Moreover, the longevity of both individual nest trees and nesting colonies was greater in beaver impoundments than for other sites such as dead white pines along rocky shorelines of lakes.
- Beaver impoundments create a diversity of habitat features that attract a wide host of other bird species, including waterfowl that feed and rest in the littoral zone to woodpeckers that nest and feed on flood-killed trees.
- Beaver ponds create excellent habitat for amphibians and reptiles. All ten species of amphibians and both species of turtles in Voyageurs National Park rely on beaver ponds for habitat.
- Beaver lodges and dams provide excellent denning or resting habitat for a number of species, particularly wolves, bears, muskrats, and otters. They also provide unique points of transfer for beaver nest beetles, an ectoparasite endemic to beavers, between individual beavers or beaver colonies.

Acknowledgements I wish to thank my colleagues who have directly contributed to the research `scribed here (alphabetically): Thomas Gable, Brian McLaren, Ron Moen, David Morris, Bryce ‵n, William Severud, Kim VanderWaal, and Jerry Warmbold. I especially acknowledge Leland

"Lee" Grim for his dedicated service to Voyageurs National Park from 1973 to present, during which time he collected much of the data on birds presented in this chapter.

References

Bailey R, Stephens RF (1951) Effects of beavers on fish and wildlife. West Va Conserv 15(6):11–16

Dickie M, Serrouya R, McNay RS, Boutin S (2017) Faster and farther: wolf movement on linear features and implications for hunting behaviour. J Appl Ecol 54(1):253–263

Eckert AW (1981) The wading birds of North America. Doubleday, New York

Egan AT, Gostomski T, Ferrington LC Jr (2015) Landbird population trends for seven Great Lakes Network national parks. National Park Service, Fort Collins

Elmore RD, Carroll JM, Tanner EP, Hovick TJ, Grisham BA, Fuhlendorf SD, Windels SK (2017) Implications of the thermal environment for terrestrial wildlife management. Wildlife Soc Bull 41:000–000 (early access online)

Franzmann AW, Schwartz CC (2007) Ecology and management of the North American moose, 2nd edn. University Press of Colorado, Boulder

Gable TD, Windels SK, Bruggink JG, Homkes AT (2016) Where and how wolves (*Canis lupus*) kill beavers (*Castor canadensis*). Plos One 11(12):e0165537

Gable TD, Windels SK, Olson BT (2017) Estimates of white-tailed deer density in Voyageurs National Park: 1989-2016. National Park Service, Fort Collins

Gable TD, Windels SK, Bruggink JG (2017) The problems with pooling poop: confronting sampling method biases in wolf (*Canis lupus*) diet studies. Can J Zool 95:000-000 (early access online)

Gogan PJP, Kozie KD, Olexa EM, Duncan NS (1997) Ecological status of moose and white-tailed deer at Voyageurs National Park, Minnesota. Alces 33:187–202

Gogan PJP, Route WT, Olexa EM, Thomas N, Kuehn D, Podruzny KM (2004) Gray wolves in and adjacent to Voyageurs National Park, Minnesota: research and synthesis 1987–1991. National Park Service, Denver

Hazard EB (1982) The mammals of Minnesota. University of Minnesota Press, Minneapolis

Jannett FJ, Broschart MR, Grim LH, Schaberl JP (2007) Northerly range extensions of mammalian species in Minnesota. Am Midl Nat 158(1):168–176

Jordan PA, Peterson RO, LeDoux KA (2010) Swimming wolves, *Canis lupus*, attack a swimming moose, *Alces alces*. Can Field-Nat 124(1):54–56

Lankester MW, Samuel WM (2007) Pests, parasites and diseases. In: Ecology and management of the North American moose, 2nd edn. Smithsonian Institution Press, Washington, pp 479–517

Latham ADM, Latham MC, Knopff KH, Hebblewhite M, Boutin S (2013) Wolves, white-tailed deer, and beaver: implications of seasonal prey switching for woodland caribou declines. Ecography 36(12):1276–1290

McCall TC, Hodgman TP, Diefenbach DR, Owen RB (1996) Beaver populations and their relation to wetland habitat and breeding waterfowl in Maine. Wetlands 16(2):163–172

McCann NP, Moen RA, Windels SK, Harris TR (2016) Bed sites as thermal refuges for a cold-adapted ungulate in summer. Wildl Biol 22(5):228–237

McGraw AM, Moen R, Schrage M (2011) Characteristics of post-parturition areas of moose in Northeast Minnesota. Alces 47:113–124

McGraw AM, Terry J, Moen R (2014) Pre-parturition movement patterns and birth site characteristics of moose in Northeast Minnesota. Alces 50:93–103

Mech LD (1970) The wolf: the ecology and behavior of an endangered species. Natural History Press, Garden City

Moriarty JJ, Hall CD (2014) Amphibians and reptiles in Minnesota. University of Minnesota Press, Minneapolis

Morris DM (2014) Aquatic habitat use by North American moose (*Alces alces*) and associated richness and biomass of submersed and floating-leaved aquatic vegetation in north-central Minnesota. Lakehead University, Thunder Bay

Morrison ML, Marcot B, Mannan W (2012) Wildlife-habitat relationships: concepts and applications. Island Press, Washington

Naiman RJ, Melillo JM, Hobbie JE (1986) Ecosystem alteration of boreal forest streams by beaver (*Castor canadensis*). Ecology 67(5):1254–1269

Olson BT (2013) Characteristics and modeling of the thermal landscape for moose at Voyageurs National Park. Bemidji State University, Bemidji

Olson B, Windels SK, Fulton M, Moen R (2014) Fine-scale temperature patterns in the Southern boreal forest: implications for the cold-adapted moose. Alces 50:105–120

Paine RT (1969) A note on trophic complexity and community stability. Am Nat 103(929):91–93

Peck SB (2006) Distribution and biology of the ectoparasitic beaver beetle Platypsyllus castoris Ritsema in North America (Coleoptera: Leiodidae: Platypsyllinae). Insecta Mundi 20(1-2):85–94

Peterson RL (1955) North American moose. University of Toronto Press, Toronto

Pollock MM, Naiman RJ, Erickson HE, Johnston CA, Pastor J, Pinay G (1995) Beaver as engineers: influences on biotic and abiotic characteristics of drainage basins. In: Jones CG, Lawton JH (eds) Linking species and ecosystems. Chapman & Hall, New York, pp 117–126

Renecker LA, Hudson RJ (1989) Seasonal activity budgets of moose in aspen-dominated boreal forests. J Wildl Manag 53(2):296–302

Rosell F, Bozsér O, Collen P, Parker H (2005) Ecological impact of beavers *Castor fiber* and *Castor canadensis* and their ability to modify ecosystems. Mammal Rev 35(3):248–276

Route WT, Peterson RO (1988) Distribution and abundance of river otter in Voyageurs National Park, Minnesota. USDI National Park Service, Midwest Regional Office, Omaha

Sadinski W, Roth M (2009a) Surveys of amphibians, abnormalities, pathogens, triazines, breeding-site characteristics, and reptiles in five areas managed by the National Park Service and the U.S. Fish and Wildlife Service in the Upper Midwest, 2002-2007. National Park Service, Fort Collins

Sadinski W, Roth M (2009b) Surveys of amphibians, abnormalities, pathogens, triazines, breeding-site characteristics, and reptiles in five areas managed by the National Park Service and the U.S. Fish and Wildlife Service in the upper Midwest, 2002-2007: data supplement. National Park Service, Fort Collins

Schaberl JP, Route WT (2013) A cursory survey of bats in Voyageurs National Park, Minnesota. National Park Service, Fort Collins

Street GM, Fieberg J, Rodgers AR, Carstensen M, Moen R, Moore SA, Windels SK, Forester JD (2016) Habitat functional response mitigates reduced foraging opportunity: implications for animal fitness and space use. Landsc Ecol 31(9):1939–1953

Tews J, Brose U, Grimm V, Tielbörger K, Wichmann MC, Schwager M, Jeltsch F (2004) Animal species diversity driven by habitat heterogeneity/diversity: the importance of keystone structures. J Biogeogr 31(1):79–92

Tumlison R, Karnes M, King AW (1982) The river otter in Arkansas: II. Indications of a beaver-facilitated commensal relationship. J Ark Acad Sci 36:73–75

Vanderwaal KL, Windels SK, Olson BT, Vannatta JT, Moen R (2015) Landscape influence on spatial patterns of meningeal worm and liver fluke infection in white-tailed deer. Parasitology 142(5):706–718

Voyageurs Region National Park Association (2001) Birds of Voyageurs National Park: a guide to the Minnesota-Ontario border country. University of Minnesota Press, Minneapolis

West I (2015) Summary of acoustic bat population monitoring data. Great Lakes Forested Parks Network. WEST, Inc., Laramie

Windels SK, Olson BT (2016) Moose population survey at Voyageurs National Park: 2016. National Park Service, Fort Collins

ight JP, Gurney WSC, Jones CG (2004) Patch dynamics in a landscape modified by ecosystem gineers. Oikos 105(2):336–348

Index

C.A. Johnston, *Beavers: Boreal Ecosystem Engineers*,
DOI 10.1007/978-3-319-61533-2

The manufacturer's authorised representative in the EU is Springer
'ure Customer Service Centre GmbH, Europaplatz 3, 69115 Heidelberg,
ny. If you have any concerns regarding our products, please
roductSafety@springernature.com

d by CPI Group (UK) Ltd, Croydon, CR0 4YY
25/09/2026
02222836-0009